Potency of the Common

Challenges of Life

Essays on philosophical and cultural anthropology

Edited by
Gert Melville and Carlos Ruta

Volume 3

Potency of the Common

Intercultural Perspectives about Community and Individuality

Edited by
Gert Melville and Carlos Ruta

Editorial Manager
Laura S. Carugati

DE GRUYTER
OLDENBOURG

ISBN 978-3-11-045735-3
e-ISBN (PDF) 978-3-11-045979-1
e-ISBN (EPUB) 978-3-11-045746-9

Library of Congress Cataloging-in-Publication Data
A CIP catalog record for this book has been applied for at the Library of Congress.

Bibliographic information published by the Deutsche Nationalbibliothek
The Deutsche Nationalbibliothek lists this publication in the Deutsche Nationalbibliografie; detailed bibliographic data are available in the Internet at http://dnb.dnb.de.

Typesetting: Michael Peschke, Berlin
Printing: CPI books GmbH, Leck
♾ Printed on acid free paper
Printed in Germany

www.degruyter.com

Gert Melville and Carlos Ruta

Preface

This volume harkens back to a symposium that took place in Buenos Aires from 10 to 13 November in 2015 under the academic supervision of Carlos Ruta, President of the National University of San Martín, and of Gert Melville, Director of the FOVOG at the University of Dresden, Germany. Following upon its predecessor conferences dedicated to the topic of "Life Configurations" and "Thinking the Body as Basis, Provocation, and Burden of Life", this symposium was the third in a series of international meetings with the title "Challenges of Life". Throughout these meetings, specialists with different scientific horizons have been invited from all over the world to reflect upon problems and topics considered essential for a manifold understanding of the human being, of its place in the world, and of the fundamental challenges that life presents in various forms.

The central question of the last conference was: to what extent do the "common" and the "community" – its institutionalized form – present a challenge in the life of the individual?

The single person needs the common and hence must live in community, which can, however, only be achieved under the influence of vast impositions upon his or her own individual autonomy, which in turn stipulate strenuous acts of continual socialization. For the purpose of verifying this assessment, both theoretical reflections as well as the adoption of a broad spectrum of empirical fields of investigation of various epochs and civilizations are needed.

Communities arise when their members „view themselves as part of a collective" (Ferdinand Tönnies) and when the „configuration of social action ... depends upon the subjectively experienced (affective or traditional) shared identity of the parties concerned" (Max Weber), whereby social action can be thoroughly directed by rationally founded rules. This requires an emotionally founded identification of the individual with the goals and values of the whole, which can also be expected of the other members as partners, as well as an inner willingness of the individual to submit him or herself to shared rules.

Communities may consist of a family, clan, tribe as well as occupational (work teams, guilds, unions), military (brigade groups), political (parties, confederations, princely courts), cultural or religious (e.g. monasteries) organizations, but also alliances in the event of emergency and danger („unions born of necessity") or of isolation (early ship's crews, emigrant trails, etc.).

Communities are thus based upon social formations of significantly differing ranges (spanning up to apparent „national community" or „class community" in accordance with pertinent ideologies). A direct face-to-face communication is not an absolute prerequisite for the cohesiveness of a community, which can also be produced by means of distance-spanning media.

Communities can cover entire spheres of an individual's life (e.g. tribal communities in indigenous cultures; monastic communities) or only parts thereof (e.g. in political and occupational communities). Moreover, they have a diverse normative density as well as a varied binding force upon their members. In this regard one may compare, for instance, a monastic convent with the community of a school class or the familial clan of an indigenous ethnic group with the team community of an occupational assembly in modern societies.

Communities offer their members resources, security, and relief as well as making demands. They submit the interaction of their members to rules, which are either hierarchically imposed or negotiated by all or rather set down by custom. This leads to the relief of individual action in the community but also to enormous burden if the rules must be perennially set down anew or run counter to individual needs. Communities offer social and symbolic capital in the form of resources, which are put at the disposal of each member in interaction with the community as a whole as well as with other members. They lend members identity, rank and reputation, protection, life orientation, and the meaningful opportunity to implement life goals and values on a more permanent basis than an independent individual could do. They generally seek a transcendent anchor, which presents itself in symbolical fashion and can lead to the development of a transpersonal identity. This transpersonal identity can either stand against individuals or unify them. But communities – if one understands them as "autopoietic, self-referential systems" (NIKLAS LUHMANN) – also pursue strongly pronounced self-interests, which they vehemently represent in relation to other communities and their members in the form of a clear demarcation. As a result, they create ‚free spaces' for the assured maintenance of their offering of resources.

As a *zoon politikon* (ARISTOTLE), or rather an *animal sociale* (THOMAS OF AQUINUS), humankind is only equipped with limited and diversely oriented competencies, respectively. The common enables the aggregation and the synergistic supplement of these individual competencies. It also takes on institutional forms (ARNOLD GEHLEN; KARL-SIEGBERT REHBERG) in order to ensure stability.

Communities place demands on members, which have consequences for their individual autonomy. Of their members, communities demand an inward and outward readiness to comply (obedience), cooperation, mutual solidarity, and integration, which – in part or in full (e.g. monastery) – can mean the commitment of individual autonomy (of the freedom of choice, conduct, and expression). Communities are a „form of social affirmation" (FERDINAND TÖNNIES), that is, affiliation requires individual voluntary action, which can, however, oscillate in a range from the natural development of childhood up to an explicit act of decision. Deficiencies of behavior are expected of individuals in communal settings: it is practically impossible for individuals to accomplish any conduct within a community or rather to coordinate all of their objectives in a certain community. Conflicts of roles and values thus occur, which can lead to compromise or revocation of readiness to comply (rebellion) with corresponding sanctions and disciplinary measures.

Consequently, common life must be learned and rehearsed. Adequately performing the conventions, norms, and goals of a community necessitates a socialization, which consists in the ‚learning of communal life.' Socialization can occur in institutionalized fashion by means of a „novitiate", that is, a more or less systematic preparatory phase. In addition, socialization may require an inner conversion, a transformation of the ego to the end of willingness to fulfill the communal goals or a finalizing „rite de passage" (ARNOLD VAN GENNEP). The same applies in the case of charismatically lead communities, in which a state of being spontaneously "swept up" may suffice for entrance. Yet socialization does not end with these sorts of acts of entrance. Rather, it prevails by means of the entire life in the form of adaption to individually-changing conditions (corresponding to stages of age and education) or to the ever-changing requirements of a community.

In particular, the following systemic aspects or guiding questions were at the basis of the conference and the papers given there, which are presented in elaborated form in the current volume:

- How does the relationship of services and of impositions work out in communal life?
- How high is the degree of self-abandonment of individual autonomy in a community?
- To what extent do communities span all spheres of individual life? To what extent must they share their competencies with others and thereby enable a greater autonomy of the individual?
- What repertoire (inward or outward) of conformity to the community or of rebellion against the community, respectively, does the individual possess?
- To what extent can an individual evade the life of a community? To what extent are there 'cultures' of individuals?
- To what extent does the identity of an individual define itself according to affiliation with a community?
- To what extent does the binding of an individual to a community allow otherness or diversity, which can lead to the fracturing of an identity or to its alteration, respectively?

Content

Cultural Identities

Philosophical and Sociological Basics

Carlos Ruta

The Hermeneutical Constitution of "the Common"

> We had asked the voice imitator [...] to present us something entirely different [...] to imitate quite different people [...] In fact, the voice imitator did imite voices of quite different people—all more or less well known [...] We were allowed to express our own wishes, which the voice imitator fulfilled most readily. When, however, at the very end, we suggested that he imitate his own voice, he said he could not do that.
>
> Thomas Bernhard. *The Voice Imitator.*[1]

> Like all lovers, they spoke much of themselves, as if they might thereby understand the world which made them possible.
>
> John Williams. *Stoner.*

How to avoid outlining, in a preliminary fashion, some of the lines in the horizon for the search and expectations that sustain the questioning and the development of the arguments that give some density to this essay? To thread the string of questions that weaves its course.

In a diverse and previous region from the one defined by the category of community, in any of its instances, the question of "the common" is raised. This territory of shared contexts that founds any chance of a human life that aims for a more complete sense of its deployment and maturity. However, where is the possibility of "the common" as a stage prior to any configuration of community founded? Is it in the density of content or in the lack that life itself involves since its inception? What features does this determination spread to the configuration of the tasks imposed in the creation of a common space? How much, how and in what direction is it necessary to infer a hermeneutic imprint of such configuration? If this is possible, where does its art play that hermeneutical mood that configures the common? Which determinations implies the hermeneutical experience from or towards the community of life that supports it? If language is its own space, and if language shows its own determination in a phenomenology of dialogue, which of its own features point out the imprint of the hermeneutical configuration of the constitution of the common? If that hermeneutic experience is an event of words in the specular dialectics of conversation, then, how should the event of common be analyzed from the event of a hermeneutic experience?

---

1 Thomas BERNHARD, *The Voice Imitator*, translated by Kenneth J. Northcott, p. 2.

1 From the community to the common: the complexity of its polysemy.

We often face multiform distrust regarding any discourses which emphasize, put forward or base their grammar on the semantic field articulated by the different interpretants of the signifier "community".[2] The rejection of (or discomfort with) the use of the interpretative key of "community" is not foreign, perhaps, to the polysemy implied in its concept. If we add to that the suspicions which are in many cases created by those appropriations which in the eyes of many interpreters represent dangerously conservative or reactionary tendencies, then the complexity of its conceptual content and the multiplicity of its uses make it arduous and on many occasions insubstantial to debate its sense and meaning. In any case, it should not be forgotten that despite these facts, the notion of community has remained the object of multiple epistemological orientations, especially in disciplines such as sociology, anthropology and political philosophy. On the other hand, within the context of social life we can observe the constant recreation of relationships, of bonds intended to elude the market and the State, labeled under the category (as experience) of community.[3] In other words, it is confirmed that the need for a community always revives, though with new apparel, even, for its own aporias.

It is worth recalling at this point that the urgency involved in confronting the unintentional effects of modernization (even with a view to ruling over and controlling them) constitutes, by all means, an adequate context for the growth of a need for a community as shelter from uncertainty and risk. However, in many instances this phenomenon emerges as a reactive feeling against the threats of the global horizon and it is burdened with all of its risks and aporias. What lies behind this contemporary development is also related to intellectual reactions, consequences and positions which came into being as a result of the historic epilogue of the compromise between capitalism and democracy which had characterized the social democratic era of the 20th century. Thus, the new "sacrifice situations" would evidence the crisis experienced by the social contract which served as the foundation of such historical cycle. There was an exponential increase in new pathological phenomena which uncovered the truth of an abstract solidarity which led to the severance of any community bonds and to the decadence of active citizenship. The mass media, the money and the power had given rise to a growing rationality and colonization of the life- worlds, thus undermining the

2 Francesco FISTETTI, *Comunità*.

3 Francesco FISTETTI, *Multiculturalismo*.

symbolic infrastructures of the civil society and the structures of the critical-reflexive consciousness. The reactions have already drawn attention and can be seen throughout the entire planetary geography. Facing threats and fears, to lock oneself up in a communal sentiment would in many cases raise as a reactive protective strategy. Thus, the emergence of the desire for a community also meant hazards, lies and blocked roads that led to the obsession for identity, to pervert strategies of a scapegoat or the reclusive logic of friend-enemy.

Facing this complexity of historic, social, semantic, cultural bias it is worth exploring other theoretical itineraries that help to focus this field of issues with further precision. From the temporal tensions sustained by any present, emerges, in this case, the possibility to face those community desires reconsidering not only, or not further, the underlying concept of community but its own entire background. The conditions, of diverse order, that sustain its eventuality. That would previously demand questioning the correct approach for the issue about every or any "us". Or maybe, preliminary, a question that highlights diverse instances, essential dimensions, and the self dynamics of the emergency of the bonds that sustain and enable any "us". In any case, this also leads after all to rethink that "common" that funds and sustains the possibility of a shared life and therefore any alternative that may be close to, in its asymptote, an experience of the community.

The current situation makes evident a pressing and dramatic feature that the challenge of living is also part of the challenge of living "in common".[4] The magnitude of this task, claims, in no small way, to explore an acquired design about how to live "the common" or something in common. And maybe, just initially, to consider the tacit demands to the pass of its foundation. Such pragmatic sketch involves precisely weighing what is implicit at the time when all the aspects of its eventuality forge a self "figure" as a vital articulation of the possible ways of life. In this case, of a life in common.[5] We know that in this case the typology of "figure" becomes a hermeneutic category of the idiosyncrasies that life generates and develops as that essential self mobility that enables it to guarantee and improve creatively its mere persistence above its ephemeral character. The questioning demands then, at some point, to stop to gain further clarity regarding the way in which "something" becomes in "common" grounds from where it is possible to sustain the attempt to live together. However, the previous and cardinal stage would require an analysis of the phenomena in which the possibility of the common sustains; or maybe even further its unavoidable feature. From there, the task could become, to consider "the common" at the stage or at the moment

4 Richard SENNET, *Thogether. The Rituals, Pleasures and Politics of Cooperation.*
5 Carlos RUTA, "Contributions for Considering the Possibility of All Life Configurations", 3–24.

of its configuration. With further precision, in our case, what could be named, anticipating precisely one of its radical ways and paths, its "hermeneutic configuration", even when the sense of this determination still appears unclear, blurred or even erroneous.

2 The foundation of the common

2.1 How should we think then the creation, the constitution or the advent of the common or of something in common? In any of these categories, that indicate also diverse conceptual and praxis perspectives, there is also the imposition of a moment for the articulation of layers and dynamics that, due to its uneven congruence, seem to be heterogeneous, and appear in need of a minimum sediment of consistency that reaches a "figure" that is appropriate for the life that is there at stake. The configuration of something common that enables a shared life. It is a questioning about how to consider that basis of participation that sustains and determines the mode of being-with and that therefore seem to be the only viable manner to imagine a joint life.

Obviously, the presence of something "common" in the interaction does not prevent questioning about the modes of its emergence. We know that, regarding this, all naming already implies a particular bias in the way of it being thought or experimented. However, the "advent of the common", beyond the multiple forms in which it can be named and considered, presupposes and lies on the immediate and almost superficial fact of the existence of the duality. Of the second person. Of the other. And in that, the link that defines them and co-implies from its origin. All denominations that, beyond its evidence, hide within the different layers of meanings and senses.

2.2 The work by Hermann Cohen has here a position, both unique for its early effort to conceptualize the problem, as insinuating for the perspectives it shows on its own traditions and those that it portrays beyond its own epochal horizon.

Even in the 1904 *Ethik des reinen Willens*[6] does it show glimpses of the direction of the parable that his thought would describe and from where would derive new categories of thought about the sense of the second person as the determining connotation of an "us". Thus, it requires attention both the conceptual consolidation that is created as well as the unique movement of the arguments. One of the conceptual knots that underlies in the plot tension of the text, and that it

6 Hermann COHEN, *Ethik des reinen Willens.*

expands to later texts, precisely focuses its concern in identifying the specific character of the second person.

In that text of 1904 the analysis of this category, that summarizes incipiently the issue over duality and plurality, is significantly framed in the section that studies the "Self consciousness of the pure will". The chapter gets to this concept through a search of the self specificity of the "will" (Wille) opposed to the "Thought" (Denken) and "Desire" (Begehren). What pure will refers to, also to support its specific feature, is the Subject of that will. The own "interior" that this concept prescribes is translated by Cohen through a concept of "conscious" that also requires precision regarding its specific content. The conscious engaged in the pure will is clearly deployed in the acting (Handlung) and has at the Subject itself and its self content. Therefore, the subject of this objectivation as content of the conscious when acting, frames that "issue" that makes a clear difference in respect of the conscious of any other object and must be described for Cohen, under the title of "self-consciousness" (Selbstbewusstsein). In spite of these accuracies, the issue over the content that now defines the key concept of "self consciousness" is still open, where is also at stake, the same issue of the "unity of the self". In the preamble to a brief description of the history of the problem, Cohen underlines that since early times personal pronouns were used for the self-conscious as key words for that idealism and its variations. In plural in Plato, the us; the first person singular in Plotinus. Beyond the cut of the historic sequence he makes, it is important to state that the focus of the argument direction. Cohen takes here the distinction developed by Fichte between the Me and the Not me and asks for its meaning for an ethical theory (Sittenlehre). In it the No Me can only be used as an original concept (Ursprungsbegriff) of the Me. What is then at stake will be a judgment about the origin. And if the key question of ethics has to be expressed under the formula, what is self conscious? Or, as it was hinted above, what is I? Then it seems essential to define as well, how to determine fairly the origin of that me? Obviously for Cohen the correlation (Korrelation) are the "things" it does not reach or where it does not seem satisfactory. Therefore, the Not-Me can here only refer to the concept of "mankind", only from where the Me can have its origin. Also, this concept of mankind has to refer to a plurality (Mehrheit) of men; even more all can be deducted from this plurality. However, it is still worth asking: Where does that plurality of men come from? Where does the second man ...der Nebenmensch..., that who is at our side, come from? The indication of terminology indicated here by Hermann Cohen opens a new chapter of semantic precision paths that seek to determine with further clarity what is implicit in the concept of the second person as well as the sense of that same question. On one side, it is hinted here the issue of the unity of that plurality confronted with the consolidation of the idea itself of man and, therefore, at the same time, the issue regarding the unity of the human genre. In this way from man to the concept of man, where Philo from Alexandria fulfills for Cohen an exceptional role, even if it is formulated for the first time the idea of a homogeneous man (...den einheitlichen Menschen als Idee...), there still remains to reply to the question about from where does the Nebenmensch come from, the second person, that who is by my side and, as we now know, shares with us the same genre (human). Even if the reply leads to suppose the experience as the place itself of provenience, it is necessary to clarify – for Cohen – that the Nebenmensch is not only just one of the many other men that are by my side. What it refers to is the alterity itself, the Other, the concept of Other. And that Other is also not just one more but he is thought of in a precise correlation or rather in a relation of continuity of the I. The Other, the alter ego, is the origin of Me. Precisely for this reason it is to be confirmed the possibility to be reconstructed by ethics and therefore it cannot be only referred to experience. If now we go back to the ethical problem of the self-consciousness viewing in it the same problem of the pure will, if we remember that that will is made in acts and those acts can only refer to two subject: then the self-consci|ousness cannot be limited to the isolated individual. Rather it has to refer to the conscious of the other, about the other. In the unity made of the other with one what really produces is this self-consciousness, as self-consciousness of the pure will. And in that sense, the self-conscious is the key reason for the ethics of the pure will. For that same reason its basis is found in the pure thinking.

It is at this moment where Cohen starts with the attempt to give place in the argument process, to his own religious tradition. A task that he exercises in a serious criticism to his own concepts but that, at the same time, looks to open to the questioning and provocation that those traditions establish as a challenge to the ethics of the pure will. Also in monotheism reappears the issue of the other, but in a diverse sense, maybe complementing it. For the Prophets, specially, the other initially takes the appearance of the foreigner, with all the difficulties this provoked. However, through a conceptual and historic complex course of time, in a posthumous text of Cohen itself he will expand on, it discovers the Nebenmensch, associated now to the entire humanity, through love as an expression of the high methodic value of religion. Through this path, it will move from love to the otherness, to transform in love to the other. Beyond the inconveniences inherent to these semantic passages, the essential for us – as Cohen also indicates – is to perceive here that the unique and real problem is the Selbst itself, the self itself. It is the open issue of the question for self-conscious of pure will. A question that aims to shed light on the need of the other, as other, to reach and precise the owned self.

In the text *Der Begriff der Religion im System der Philosophie* of 1914, Cohen, when expanding on the reflection that gives room to religion in its intellectual universe, thinking as well the meaning of the originary moments of religion deepens in the scope, content and profile of the concept of man.[7] If his research provides, though that path, the discovery of the process in which it is conceived precisely the expansion or plenitude of the concept of man, that forces and allows for the introduction of a series of statements and lematic increments in which that genesis is precisely considered and the projection of its sense. Behind that maturity of argument lays maybe the conscious of the "lack of ethics as its concept of man".[8] Precisely, in the boundaries of richness that implies the concept of humanity as a universal category to establish the foundations of man, it emerges as well the concern that that lacking exposes.[9] As Cohen himself asks: "What is man and what does he lack if he only is humanity?"[10] If it emerges from the experience from the Nebenmensch , in the path that leads from the plurality to the totality, the universal idea of humanity, now a new genesis is registered from there. It is that in the Nebenmensch, the concept of humanity offers or launches, as well the category of Mitmenschen. A term that has precedents in the 1786 Adelung dictionary, that evidences the existence of the term by 1777 already.[11] Its meaning refers there to *any person that shares with me the condition of being human*. However, Cohen rethinks and re-elaborates its semantics precisely from the moment in space that he himself grants to the religious experience generated in the history of Judaism. Thus he would sustain that, even when we are under a "epoch turn" determined by the shadow of the commandment of love to the next of kin (Nächstenliebe), neither the concept of the Mitmenschen nor the one of love emerge from the obvious movement of the human consciousness.[12] Here the term Mitmenschen relates to the reflection that ties the "next of kin" inserted in the commandment of love with the expansion and deepening of a new concept of man. This is possible in the recognition of a self lack (der Erkenntnis des Mangels)[13] coherent at the same time with the concept of human being, and therefore, root of the lack itself that is relevant for ethics. Besides, it might be convenient not to forget that, for Cohen, in the process of this conceptual increase there is a

---

7 Hermann COHEN, *Der Begriff der Religion im System der Philosophie*.
8 Ibidem, N. 38, 52.
9 Ibidem, N. 39, 52.
10 Ibidem, N. 39, 53.
11 Johann Christoph ADELUNG, *Versuch eines vollständigen grammatisch-kritischen Wörterbuches der hochdeutschen Mundart*.
12 Hermann COHEN, *Der Begriff der Religion im System der Philosophie*. N. 87, 75.
13 Ibidem, N. 50, 57.

key mediation of the recognition of the "poor":[14] ..."I cannot think of myself as man, unless I have taken my concept of man from the human concept of poor" .[15] A thought that takes his plenitude beyond a mere theoretic sphere, to project it in the experience of suffering origin as well as of compassion (Mitleid). It can be observed an argument movement that bonds its power with the matured reflection about the lack that inhabits and surrounds the theoretic and practical experience of human life.

This theoretical deploy will find a new state of maturity in the posthumous work of Cohen, *Die Religion der Vernunft aus den Quellen des Judentums*, of 1919.[16] To the expense of the conceptual and argument wideness developed in it, what is relevant at the moment to strengthen or complete is some reflections already mentioned. The central core that might require highlighting might be the sense itself implicit in the emergency of the concept of Mitmenschen. Its genetic bond with the historic experience provides, for Cohen, an inescapable density. If the individual emerges, under the denomination of Nebenmenschen, as the element of a series, is in that it appears embedded the problem of the Mitmenschen, what is there at stake is finally the possibility of a statement about "the full problem of the human being". For Cohen, to think of man as Mitmensch means to think the man as complete, ...der ganze Mensch... .[17] Then, under the term and concept of Mitmensch underlies the possibility of the origin itself of the concept of man.[18] Here is operating a social conception of human being in which the instrument of its motivation the experience of poverty.[19] This way the compassion emerges (Mitleidenschaft) as the heuristic concept to discover the human being to achieve a new concept of human being that essentially includes the practical correlation between man and man from where the human being is born as a "Me".[20] The feeling of humanity in which we create (erfinden) "man, Mitmenschen and men in general."[21]

2.3 If Cohen's concepts re elaborate under multiple categories the problematic issue of the second person and, in that, the issue of the other and is genetic bond with the me, what is at stake then essentially is the sense of that link that defines finally a replenished meaning of human being. We can speculate that the inter-

14 Ibidem, N. 89, 77.
15 Ibidem, N. 87, 76.
16 Hermann COHEN, *Die Religion der Vernunft aus der Quellen des Judentums.*
17 Ibidem, 132.
18 Ibidem, 159.
19 Ibidem, 104.
20 Ibidem, 109.
21 Ibidem, 167.

twining of this reflection sustains or seeks a sense for that essential bond that can also be considered as the meaning of "being with" or "be with" that defines and profiles, from the others, the modes of being, living and acting of men in its complete selfness. We know that this question reappears uniquely in the representative work of Heidegger *Sein und Zeit*.[22] The Analytic that redefines its boundaries also has behind it the task of elucidating the conditions that enable a philosophical answer to the question of "what is man?"[23] Obviously the statement unravels here is opposite from an original horizon that transforms its fundamental design and therefore its sense. Therefore, even though one could imagine partial resemblances of the semantic implication beating in the lexical field of the Mitmenschen, now the circle of meanings is established in a new vocabulary that refers to the mutation of the horizon on which it is believed. Its density portrays, at the same time, an unavoidable shadow to discuss the configuration of all possible space of "common" meaning.

Heidegger's view seems to evoke in perspective a somber image of the human condition, determined by his own ontological interpretation of the world. This aspect of opacity relates primarily to the phenomenon of being-in-the-world but expands, from there, its shadow to the net that forms all the structures of the Dasein from its originary way of being. The final apex in which the radium meet, where that view reflects the image of Dasein is profiled, seems to be characterized by the impersonal determination that defines and names the subject of the everydayness as "Man", the One, the they-self.[24] But its conceptual and phenomenological basis resides in the way of being dominant in which this is found in respect of the world.[25] This way of being (and being in in which it is founded) is constituted in the fact that the Dasein, in its everydayness, is immediate and primarily absorbed by its world. The text precisely tries to draw a path that enables to unveil upon the eyes this new phenomenical field of the everydayness of Dasein and, in this way, reply to the question about "who is the Dasein that appears in the everydayness". The phenomenon that replies to this question evidences those structures of the Dasein that co-originate with the being-in-the-world. These establish at the same time the new vocabulary that thinks the specific character of the thread of links that conform the "being-among-one-another" of Men, their common room, the mutual belonging and, above all, its own being itself: Mitsein, Mitdasein. But then that somber profile that affects the human condition expands to the nexus that essentially relate human beings between them and in

22 Heidegger, Martin, *Sein und Zeit*.
23 Ibidem, § 9, 41–45.
24 Ibidem § 27, 126–130.
25 Ibidem, 113.

whose plot emerges the "being oneself". This way, all considerations projected on the Dasein seem to include essentially the two sides of the opposites that articulate it. Therefore, even when the Dasein, when referring to itself, always states "this is me", that happens also when it is not. That is that, even when the Dasein discovers in its lost itself (Selbstverlorenheit), even then, it confesses always a determinate manner of being itself. This aspect is shown, at the same time as conclusive to re think the phenomena underlying any possible configuration of commonness.

The question that enquires about the "who" of the everydayness, about the "subject" of the everydayness, it already derives the search toward the interpretation of a type of entity entirely different than the useful and the things that, at the same time for its kind of being Dasein, is in itself in the world. However, the clarification that the analytics has made of the fact of of being in the world shows that there is no immediate nor it will ever be given a mere subject without world. But neither will it happen in an immediate manner a me isolated without the others.[26] Even though the others are always in the "being in the world" coexisting (mit da sind), for Heidegger, the ontological structure of what is thus given should not be considered obvious. Thus it should be stated that, then, the task is precisely to clarify from a phenomenical perspective the nature of that coexistence (Mitdasein) in the immediate everydayness. It does correspond above all, to once again admit that the encounter with the Dasein of the others has to differentiate this latter from the entities that are at hand or that are just there. This reflexion elaborates a thought of the Dasein that embraces uniquely and fundamentally its determination from its characteristic of the encounter with the others. That expands, in consequence, in a re characterization both of the world of the Dasein as of the being in itself intra-world. Both in their intricate co membership should be reconsidered from the "existing with" (Mit auch sein) that determines it. Therefore, the world of the Dasein is a shared world (Mitwelt) and its being in itself in the world is to "exist with" the other Dasein (Mitdasein). These self statements of its analysis suppose that for Heidegger, the Dasein is essentially "existing or being with" in an existence ontological sense, that is that it corresponds to Dasein form itself. The Mitsein determines then the Dasein existentially in as much as even the being alone of the Dasein is a Mitsein, a "being-with" in the world.[27] The lack and absence are modes of Mitsein. Here we see that for Heidegger the "being with others" is existentially constituent of the "being in the world" and therefore should be interpreted from the key phenomena of it, i.e.: the care. In this case its own and existential mode becomes the "request", the Fürsorge. That will deter-

26 Ibidem, §25, 116.
27 Ibidem, §26, 120.

mine the different existential modes of living together, the Miteinandersein. But always on the basis that the Dasein, especially while it “is” such, has a way of being the living-together, of the Miteinandersein; and therefore not only, even, by the real presence of the others on it. This framework enables to hint, at least, two decisive instances to rethink those portray questionings to the phenomena of the configuration of the common. That is, the sense of the determination of the human being as existence Existenz;[28] and also the consideration of the way in which the impersonal character of the subject of the everydayness sketches the immediate interpretation of the world and of being in the world. This will question the urgency to reconsider the senses in which the world and the others constitute, for the Dasein, the very reference of meaningfulness.

2.4 The reflexions previously explored, already provide hints about the perspective of a questioning that tries to establish if that common, introduced from an existential co-membership, does not require at the same time to be analysed from a broader framework, and to track the tacit reverberance for the configuration of the common and the efforts for radicalization of those questions specific to phenomenology. That is, to rethink it on the horizon conformed by the interrelations between the subject and its world, whereas the institution of something common is always sustained on the net of relations that pertain to the world and other subjects. In phenomenological terminology, it means (or supposes) the questioning about the universal correlation between the entity and its subjective modes. Here it is key to think or question about the subject of the correlation. A subject that cannot be thought of either as a transcendental conscious outside the world nor as an entity of the world amongst others. Consequently, the question that emerges primarily is the questioning regarding the sense of being of the subject, as subject to that correlation.

Heidegger’s consideration, previously questioned, somehow moves towards this same scape line. And they try to examine, from the only angle that he considers valid, an ontology of “be-with”, or “being-with” that falls in existential terms on both extremes of his composition. This way, we have seen that the everydayness of the Dasein cannot be understood without revealing the existential intertwining that ties the itself with the others, “sein” and “mit”. But that inferred that common territory of a world shared under whose rules any phenomena in its own phenomenality would be subject. In spite of its apparent solidness of this path attempted by Heidegger, we know it will be subject to a substantial criticism whose key core aims to recover the view of a key phenomena for the Dasein. In that sense, the work by Michel Henry takes a special prerogative.

28 Ibidem, 117.

If it refers to the analysis of the original structures that constitute the subject, for this new line of thought, the first determination that is definitely recorded about this subject is living, in the double sense of being in life and experimenting life. This way, the phenomenology of correlation can only be considered as a phenomenology of life.[29] Precisely, Michel Henry focuses all his philosophical reflexion in the elaboration of this proposal. However, his effort attempts precisely to radicalise the predominance of the phenomenology over the ontology by positioning in the first stage of critical analysis the issue of appearing.[30] Now, the criticism to Heidegger will focus in the privileged or unique character that he grants to appearing in the world. Henry's reproval aims to the core itself of the phenomenology as it questions the implicit senses not questioned of the intention. For him, to prevent an endless regression of intention to intention, would another revelation have to be considered to view the intention? The question will end reverting on the definitions that reveal the self appearance of life in its essential specificity. And the self revelation of life opposites bit by bit to the self appearance in the world. If the world reveals in the "beyond itself" in such a way that what is revealed is external and different, on the contrary, the first decisive aspect of the revelation of life is that it does not make any estrangement, it never differs from itself, it does not reveal any other thing but itself. Because life is "self revelation".

All the philosophy of life that Michel Henry develops is built against what he himself called "an ontological monism" that sees in the view in the distance the fundamental condition of appearing, in the intention of the law of the phenomenology. The mode specific to the phenomenology of life, away from the exterior of the visibility of the object of the world, is the pure "impression-sensation of the flesh.[31] Thus, the task of a radical phenomenology will be to disassociate these two modes of phenomenalization, and free the phenomenality corresponding to the "impression" of the intention to highlight its self status totally genuine and constitutional. For Henry, the appearance of the impression is always at the same time the self-revelation of life: the impression persists in life, as it is life itself that appears in it. Precisely in that intersection of regimes of appearing pertaining to life and the world, is where there opens a contradictory gap that demands the reinterpretation of Michel Henry's statement. Otherwise there is a risk of a conception of life concealed in the boundaries of immanence, without further possibility of evidencing the effective bonds with the world, ignoring the determinations that the "body", as its essence, imposes.

29 Michel Henry, *De la phénoménologie. Tome I. Phénoménologie de la vie.*
30 Michel Henry, *L'essence de la manifestation.*
31 Michel Henry, *Incarnation. Une philosophie de la chair*, p. 74.

In this path, Henry's philosophy confronts its own limit. As Barbaras will indicate, even when Henry's original proposal to treat the body from the *flesh*, that is life, is accepted there continues to be perceived a knot of questionings that are constrained in it.[32] The issue is precisely

> to know what is the sense of life that testifies the phenomena of *flesh*, to test this *flesh*, that it is body, the initial determination of life, Affection and Immanence. In as much it makes in a body of *flesh* that, under this title, is essentially capable of *moving*, should not life be understood in a totally diverse manner than as that originary element that remains indifferent to the kingdom of the exterior?

And these same limitations, inherent to Henry's perspective, reverberate in the way he himself believes the foundation of common life.

However, what Barbaras considers an inconsistent logic, remains open to the key question that allow to understand the entire movement of life, i.e. the issue about the intertwining between a subjective power and a movement that it is an entrance in the exteriority. Henry's position leaves exposes the need to re establish the same status of the body, of the body as pure body i.e. as an accessible object to other in the center of the world. Because the movement inherent to the live being, in its self "per se", appears torn apart due to the anticipation that tears it away from itself and prevents it to be itself it it is not going outside of it. The subject of movement does not apprehend itself but being dispossessed of itself. As intention of movement, the "me" is the same passage to the exterior, of the manner in which going outside of it is that it becomes to be itself.

In this sense, Barbaras' proposal will state that essence of life insofar as it encompasses the possibility of consciousness, should be regarded as Desire[33], which is not to be understood as an empirical feeling but, rather, as that which constitutes the ultimate condition of possibility of living and experimenting with or trying that. The attempt here is to prove that only on the basis of such characterization of the subject's living as Desire is it possible to account for the correlation insofar as such correlation is not evidently an encounter between two pre-given beings; instead, such correlation originally connects one with the other, which is what truly gives rise to its possibility. Desire befits the correlation inasmuch as it names its effective mode and to such extent reveals the true sense of intentionality, which is thus the condition of possibility of appearance as appearance of a transcendent.

Now, in which sense is the subject of living a subject of desire? Here Desire does not refer to something which reaches an already constituted subject. This

32 Renaud BARBARAS, *La vie Lacunaire.*

33 Renaud BARBARAS, *Introduction à une phénoménologie de la vie.*

is not about a possibility among others. What characterizes Desire, as understood herein, is the involvement of the subject themself and if desire is desire of a subject, it is so in a renewed sense. In order to understand it, it should be contrasted with the concept of need. Need refers to a definite, circumscribed lack, to an objective gap (lack) the persistence of which may compromise the subject's autonomy. When referring to a definite reality, need must be satisfied. In that sense, it constitutes a transitory state of imbalance: it is the sign of a definite lack and an appeal for the fulfillment thereof. Need forwards an already constituted subject, the pursuit of which need it enables. It is a reiteration instance and, as such, it allows the subject to remain what he is. Nothing new can come to the subject thereby. This is not the case with Desire which, in contrast, cannot be fulfilled. The desired object fulfills it but at the same time fuels it further, which shows that the subject of desire is not a constituted subject. The desire cannot be fulfilled because the subject itself is the lack, is what is missing. The difference between need and desire lies in the fact that the subject-desire relationship does not fall within the having category, but within that of being. What is at stake here is the subject's own being. It is its own mode of fulfillment. Because of desire, the subject becomes what it is. In that sense, desire is always desire for itself, as what is always in its perspective is the subject themself. Unlike need, desire is not subordinated to itself as a simple intermediary but it is, rather, the very condition of its emergence, the place in which it is constituted: desire is not, like need, a return to the self by means of the other but a quest for the self in the other. This is what gives rise to the dissatisfaction: the self is announced in the other but as that whose advent will never occur. This is why when it comes to Desire; the other is identically what it manifests to the self of the subject and what denies it. The insatiable nature of desire as such precisely results from the fact that the other does not present the self of the subject but as an absence. This pertains to the constitutive irreducibility of the self to the other. Everything happens as if, through desire, the subject were in quest of its self-fulfillment, its completeness, in a world which, as such, is at the same time the denial. As a quest for the self in the other, desire is never-ending as the other is at the same time the place of a loss of the self. To desire is to believe that I have found myself in the other in which I cannot but lack, be missing, be absent. This is why it can be said that desire is the proof, the experimentation of the lack of the self in the other or, in other words, "the other's" experience of its own lack. Desire is not, thus, a quest for completeness but the experience of lack: it is not a quest lacking the self but, rather, the access to the self as lack. This is why the other is the place of a revelation (of the self as lack) and, at the same time, the place of disappointment. Desire is the experience of its own lack and such experience is not possible in any other place but that where the advent of such lack may occur and where such lack is revealed, that where I am not but where I can find myself, namely, in the other. If need equates to appropriation and therefore, somehow, to the destruction of the otherness, in contrast, desire, which is never fulfilled, is the experience of the impossibility of the fulfilled, and in that it is experience of the other as that for which the subject can be. This is why possession, the denial of the other, is impossible: this would necessarily entail denying the self. Even though it can be said that desire implies a certain appropriation of the other, there dwells, however, an impossibility of denial; it should be better said that desire always implies a recognition and preservation of its otherness. Desire is precisely the experience of the other as such. This test (épreuve) of the other cannot be simply passive; it necessarily entails "doing." As a condition of advent of the subject, it presupposes an active relation to the other and, to some extent, an accomplishment movement in the other, or an active quest for the self in the other. In line with the inquiry into the subject's sense of being, desire can be defined as "lack of the subject," which should be understood in two ways: as the lack experienced by the subject but in such manner that in that experience it is the subject themself what is lacking. To say that desire is lack of the subject is to acknowledge that the subject exists in such manner that its experience, experience which, in a way, is what the subject is, is the experience of its own lack. The subject exists as experienced lack. But as experience of an objective lack, it relates to itself (and it experiments with itself) on the basis of the mode of its own lack, of its own absence. This is not about understanding this in the sense of a life experience and, therefore, of the immanence of something as an internal perception. This is why the absence of the self, referred to here, is not the content of the experience but its mode. It should be said that the experience consists in such lack. It is because of this lack and within this lack that the subject experiments, i.e., it is in that way or there that ipseity is fulfilled. It is worth noting that lack in itself is experience; there is no experience if not as lack. Absence is, therefore, the subject's own way of existing as such and, as a result, the condition for experimentation. That is why desire, thus understood, reveals a deeper level (from which the distinction between lack of the subject in its objective side and lack of the subject in its subjective side derives). Here, desire is not just another affection but the condition of possibility of every affection. It is the very shape of receptivity, an indefinite openness insofar as it is open but determined for nothing. It is, thus, openness as being. This pure openness in which and by which everything can emerge is the absolute acceptance by which and in which there may be an otherness. That is why desire is not that which is experienced; on the contrary, to experience is to desire, i.e, to lack, and the empirically experienced desire is the reflection of shape at the content level. In this sense, the subject does not advent as the subject of an experience unless it exists in accordance with the nature of that lack. That is why he is able to receive and experience. The lack of the self is the only sense of being that is functional to that pure acceptance that it is sometimes called "subjectivity." The absence of the self and the acceptance of the other mutually call for each other and, in the end, are the two sides of a same mode of being. There is no appearance other than that which takes place against a background of lack, which is always a lack of the self, i.e., Desire.

What has been said so far regarding the subject's sense of being as Desire means, as we have suggested, that the subject advents in desire. That is to say, it is constituted in itself by this desire. That means that the subject of the correlation can only be thought of through the correlation itself. The advent of the subject occurs in that correlation. What matters here is to rethink the correlation in itself and its implications. The main idea has been to show that Desire –which is just another name for intentionality, or rather, its originary mode– makes it possible to think the correlation rigorously. This same correlation understood precisely as a correlation between a transcendental being and its subjective ways of such given being constitutes that transitivity, as openness to an otherness different from itself that is a part of the being of consciousness. Now, to think life as desire is to acknowledge that essential transitivity of consciousness, giving it a true foundation. What is crucial for us, due to the implications contained therein, is to understand that the correlation is not just a mere encounter between two given realities but constitutive of each of its terms, the relationship prevailing over the terms. And it is worth noting that each of the terms is not itself but for the other. Deeper than consciousness and reality is that which relates one to the other, the co-belonging element that is no other thing than phenomenality itself. Although they may look different, here, there is no doubt that the nature of the dangers is always the same: the risk of being so out of focus that we may not be able to think the relationship in itself. The challenge, essential for our own horizon, is to understand the way of being of a bond that while growing tighter also keeps the distance from that which it connects, that unifies without compromising difference, like a uniform fabric in which a tear is implicit. In other words, the correlation establishes proximity in such a way that the distance is preserved therein and, therefore, to think the correlation is, precisely, to assign a meaning to the proximity. To think the correlation as Desire aims precisely at visualizing the constitutive conjunction of coincidence and distance. Desire intends to be defined here as the originary installation of Proximity. The sense of this Proximity can only be captured taking into account that the condition of the implied relationship is the co-belonging in respect of a place or a common element. The originary proximity is precisely that being-in-common that is deeper than every other relationship because it is constitutive of every reality in relationship: a being-together rigorously entails the common as being. As a result of this first Proximity, things are close to each other and, thus, there is no separation other than in communication, in the sense that distance is a type of Proximity. This Proximity is that which establishes Desire. It opens an original approach, the very dimension of proximity. To desire is to give rise to the advent of proximity. It establishes the community of the subject and that which occurs around him. It is the dimension of a being-in-common, of being together. Proximity is to be desired as such. That ontological community of commitment is a unity in multiplicity. There is a fundamental commitment of Desire that, as aspiration or as directed to the other, brings together and constitutes the set. Understood in this way, as that which establishes Proximity, Desire establishes the community of the subject and what advents near him. It is the dimension of a being-in-common, of being together. Desire commits its whole self and thereby designs the open Totality in which realities may appear: it establishes the originary proximity that is in charge of its relationships. It is Desire that removes the self from the ontological night, that brings the self close to it and displays the dimension of our reunion, thus allowing every encounter to take place in its midst.

In addition to the specificity of all life, of its determination as Desire, it is worth asking about the specificity which characterizes human beings. Here the attribute assigned by Barbaras takes up a characterization found in an expression coined by Herder: man as an incomplete being.[34] In this sense, man's specificity originates in a negativity. The anthropological difference is described on the basis of a denial of natural determinations. Humanity would thus arise from weakness, from a differentiation by retreat or loss in respect of nature or the ability of instinct. However, this constitutive incompleteness comes with a pulsion which is beyond the scope of natural needs. This excess of gratuity, like a burst of energy, is a force that is revealed as the counterpart of a weakness. This pulsion, this movement which is always reborn and which exceeds the order of needs is what Barbaras refers to as Desire. From the point of view of the ontological incompleteness that characterizes man, the homo faber should be defined as homo desiderans. While man shares with all living beings the incomplete essence of life itself, he is the proof of his own lack. In this sense, man keeps his distance from the other living beings due to his proximity to life itself in its own incomplete essence. This feature emphasizes his dissatisfaction. Therefore, desire is but desire for nothing. And this is why man can receive everything. Man's relentless dissatisfaction is pure acceptance: the depth of the world opened is measured by the power and the indetermination of man's aspiration. It is the power of acceptance itself, the form of receptivity, the activity inherent to passivity. Man's con-

---

34 Renaud BARBARAS, *La vie Lacunaire.*

dition is proof of his incompleteness and is realized but as a movement: it opens the depth of the world, it becomes intentional.

As it can be seen, as from these considerations of that which sustains the life in common, it should not only be considered since its roots in life itself as self experience, but also it is discovered registered in various layers depending on the degree of concreteness in it s consideration. We then discover that way the layer of the last foundation both in Henry's mutual phenomenological interiority as in Barbaras' phenomenology of desire and lacking. But it also corresponds to think of this pragmatic extension that recovers the real deployment of the everydayness. On the layer of the foundation of the possibility of the commonality, it is built as self expansion, in the layer of the human praxis that consolidates the joint constitution of the commonality and therefore the constitution or configuration of a life with others. Here are linked the questionings that address to thinking of the hermeneutic nature of the determinations that weight over that configuration task. A perspective that comes from the self and inescapable imprint of Desein itself.

3 Hermeneutical configuration

In a recession from 1929 to the work by K. Löwith *Das Individuum in der Rolle des Mitmenschen,*[35] young Gadamer already suggested, on the basis of the analysis of such review, the decisive coordinates of the problem about the fundamental phenomenon of *Miteinanderseins.*[36] About the being-together, the being-in-common, the living-with-each-other. The characterization made by Gadamer in his paper already showed the features that situate this phenomenon in a field specifically delimited by a hermeneutical horizon: what emphasizes this "contribution to the anthropological foundations of ethical issues" in Löwith›s phenomenology is, first of all, the necessary execution of a phenomenology of the *Mitsein* to understand the constitution of the *Selbst* (of the self as a person and individual). A determination that carries, in turn, a categorical reference to its essential overlap in ethical responsibility. And this phenomenon can neither be analyzed nor understood outside its founding linguistic determination. These features arising from a phenomenological analysis therefore base their description not on the common thread of the conceptual tradition of ethics, but on the deployment of the ambiguities of everyday human life itself and which, from

35 Karl Löwith, *Das Individuum in der Rolle des Mitmenschen.*
36 Hans-Georg Gadamer, *Gesammelte Werke 4. Neuere Philosophie II*, 234 sq.

them, reveal the root of all their conditionings. For Gadamer, Löwitz's phenomenological statements already involve decisive consequences to understand the meaning and possibility of any objective understanding (sachlichen Verstehens). But also, subtly appearing in the background, as a condition to any possibility, is a new awareness of human finitude. Precisely, the reflection on the constitutive meaning of "You" for a concept of the human individual, of the being of the person, shows, in the web of relationships that interweave it, a constituent unit set by its radical and inescapable finitude.

Barbaras' own observations, bounded within a phenomenological perspective, form a possible horizon (among others) to understand the meaning and the deep roots (phenomenological) of that finitude. That constitutive lack of human existence, its vital determination, its incomplete uniqueness —as to the determination of the defining trace of all subjectivity— concomitantly reveal the imprint chiseled by the defining matrix that embraces, precisely, the configuration of "the common". Openness to the world and to the others in the pulsion which arises, and which Barbaras calls Desire, is the essential pattern of human life that designs from itself all other configuration parameters of the common while it specifies its essential overlapping in the phenomenon of life. This not only means that any development of the common happens under the terms of a hermeneutical configuration, but it also determines the specific imprint of such configuration, the attributes that mark from the very inception its unique possibility. Even if they may seem obvious, these considerations certainly require a more cautious analysis. It is clear that in them lies our current concern regarding the common nature of our existences. However, as clearly stated by Jean-Luc Nancy, it seems that today we are letting go this simple and essential condition of being, just as we are also letting go all its foundation attempts and all the totems erected as a guarantee of a common being or, at least, as a guarantee of a common existence.[37] This precision casts an unfolding of the "the common", already encompassed, however, in the aforementioned phenomenological descriptions. Because "the common" may indicate both the element of a primary and irreducible equality as well as the shared in the sense of "taking part," playing a role, a phenomenon similar to the participation in a game, in a conversation, in a drama, in family life. That happens only in the context of plural relationships. That knot of bonds tied and, in turn, twisted within a defining de-bondage. Therefore, Nancy essentially believes that what constitutes the common is sharing the finitude. In this way, the community names the fact of a continuous sharing which does not distribute anything given, but merges with the condition of being-exposed (Ek-sistenz), and which thus excludes any constituted community. Rather, it refers to a con-

37 Jean-Luc Nancy, *La Communauté désavouée.*

stituent or constitution community, in which "the common" –as such– becomes the movement and event from which it is born. What is communicated there is not a common substance but the fact itself of being in a relationship. That communication does not transmit anything but the fact that there is transmission, passage, co-participation. This unfolding may allow, in turn, thinking or placing in its proper horizon the double territory of background and pragmatics that is transcribed into the question of "the common."

That always open communicative constitution of participation on which "the common" is based expresses, in turn, the "thrownness" condition of human existence. That is, the breaking of the ideal of full self-possession and self-awareness, with which an additional layer in the depth of finitude is revealed. Based on this trace (inscribed in Heidegger's drafts) the hermeneutics of the early twentieth century, recovering Kierkegaard's critique of Hegel, clears the path to the primacy of the phenomenon of "the other". This imposed, in counterpart, the awareness of the linguisticality of our orientation in the world. Learning to speak is our first orientation in the world. That linguisticality of our being-in-the-world actually articulates all the space of experience.

Now, to speak is always to speak to someone.[38] In this sense, the speech does not belong to the sphere of I, but to the sphere of we. In this line, if it is possible to accept that the true self of language lies in the conversation (as a space of we) and, therefore, in the exercise of mutual understanding, then the constituent phenomenon of "the common" is configured in a hermeneutical matrix. It is precisely in the experience of the closeness of dialogue where we may observe the birth of a first "commonality" (Gemeinsamkeit) between the partners which interweaves, in the language, the experience of "the common." However, the hermeneutical prescription of this phenomenon of "the common" in its own configuration is suggestive only if one retraces its various levels of meaning which range from, among others, the descriptive dimension to its pragmatic projection.

If language's effective mode is dialogue, if the decisive chapter of its study –for a hermeneutical theory intended to understand the configuration of the "the common"– refers to the analysis of the language of everydayness, even to the intellection of the link between language and the life-world experience (as explained by Husserl), then we should summarily focus on it (dialogue) to understand something of its own figure as the matrix of the hermeneutical phenomenon.

Perhaps we may more acutely observe the hermeneutical bias (and its implications) of that "commonality" interweaved by dialogue precisely at the moment of the "non- understanding," of the shutting of the mutual understanding between

38 Hans-Georg GADAMER, *Wahrheit und Methode*, 449.

the partners. As Schleiermacher explains from the beginning of his "General Hermeneutics" of 1809–10: "Hermeneutics rests on the fact of the non-understanding of discourse."[39] It is the moment when dialogue is diluted in the depletion of what is produced by its meaning: mutual understanding. Schleiermacher explains that the back side of this phenomenon as the beginning of any art of Interpretation (Auslegungskunst) is also the will to understand the discourse of the other.[40] The link between hermeneutics Theory as *Praxis der Auslegung* and everyday speech was, for Schleiermacher, crucial in order to understand the former on the basis of the common principles governing both. Human beings' everyday speech, its discursive instance, encounters the same understanding difficulties and is governed by the same pragmatic drive inherent to an "operation" that is always an "art of Interpretation" (Auslegungskunst), i.e., a hermeneutical exercise. Here, a hermeneutics of "the common" becomes a hermeneutics of dialogue (in the double sense of a hermeneutics of all, of dialogue itself, and of a hermeneutics exercised in the practice of that dialogue).

The tasks of such Theory are also undoubtedly inscribed as Theory of the Praxis of the understanding. A philosophical reflection of an enormous scope that extends its speculative effort to the entire dimension of the challenges of dialogue subscribed in the challenge of living involved by the configuration of "the common". Although it is not possible here to shred the elements, the articulations, the conditions of dialogue to be analyzed and discussed by such Hermeneutics, we should discuss what seems to be the decisive knot of its study. Entering the sphere of dialogue, in its own hermeneutical circle, is under no circumstances an experience without resistances, objective incapacities, defining denials. However, being able to enter into dialogue may be, nevertheless, a privileged experience of the true and full humanity of man.

But this cannot disregard the shadow side that constitutes, holds and rests on any human experience or relegate it to an illegitimate area. It is therefore crucial to find the stress field outside of which it is unwise or naive to display any theoretical essay on the subject. A Hermeneutics of dialogue only makes sense as an endeavour to understand human finitude. Only in this way may some light be shed on the event which makes dialogue happen, thereby enabling the dialogue partners to talk about things without the fog always inscribed in language. And, in any case, to better understand the reasons why this is impossible.

---

39 Friedrich SCHLEIERMACHER, *Kritische Gesamtausgabe. Vorlesungen. Vorlesungen zur Hermeneutik und Kritik*, *Vol.* 4, 73.

40 Wilhelm DILTHEY, *Gesammelte Schriften. Leben Schleiermachers. Schleiermachers System als Philosophie und Theologie*, vol. 2, 691 sq.

It is here also where it is worth not leaving aside Heidegger's analysis in its most somber aspect. The Impersonal characteristic that determines the subject of the everydayness does not limit its somber to the phenomena of language. Language too, it is essential economy of reciprocity becomes an instrument that articulates and holds in depth that alienation resulting of the absorption of the subject in the world and in others. It is a speaking configures in the matrix of the "being said", of the impersonal speak, of the dialogue that weakens the "self" of the subject in the strangeness of the others and their world.

The analysis will here provide a task that reveals the premeditated concealments that simulate transparency in the tricks of the language. There, the veils of power and interests continuously show their faces. Therefore, when it seems crucial, for those seeking to ensure a dignified common life for all, to reunite in cooperative efforts, the capacity to dialogue is not a secondary issue. What is at stake there is the decisive praxis of any society, or at least of any gathering of those who seek to cooperate in the common struggle for equality.[41] Certainly, understanding the possibilities and difficulties of dialogue from the territory of human finitude opens theoretical ways to understand and facilitate such decisive social praxis: the praxis of solidarity among those who only possess the weapons granted by the full power of their language and their bodies.[42]

Maybe the theoretical horizon, that disposes a hint towards the possibility of thinking dialogue as an instance of the configuration of the common, should recover and reconsider some of the decisive knots that articulate Heidegger's purpose as it is already shown in the Sein und Zeit. Undoubtedly, there the determining vanishing point will always be the intrinsic finitude of Dasein and, therefore, of its own understanding. As all understanding of the being is shown finite from a phenomenological perspective and it that it reveals also its last sense. Even obviously for the same Dasein. But here it corresponds maybe to highlight its own singularity; that which leads to define it as form its existence. Dasein can only understand itself from its existence, from a possibility of itself; to be itself or not.[43] Therefore, the Dasein does not access itself but as a possibility and once there, existing is always and each time a projection. To take the risk of freedom, to take the same existence based in truth as liberty. Its sense, and therefore the sense of its finitude, can only be fulfilled as openness. In it is at stake the mood of its history and that link with language that originates in the net of interpellations that sustain it.[44] Undoubtedly for Heidegger, the impersonal character that

41 François DUBET, ¿Por qué preferimos la desigualdad? *(aunque digamos lo contrario).*

42 Richard SENNET, *Together. The Rituals, Pleasures and Politics of Cooperation.*

43 Martin HEIDEGGER, *Sein und Zeit*, § 4, 12.

44 Martin HEIDEGGER, *Brief über den Humanismus.*

characterizes the human being as a self originary phenomenon imposes restrictions or decisive determinations. It moves it away from itself and imposes, at the same time, the immediate sketch of the interpretation of the world, it articulates the remission context of all meaningfullness.[45] Then how can it open, for itself, its own way of being, and its own authenticity? Here it relates to the possibility of discovery and openness (...entdeckt und sich nahebringt...) of the world and of itself exercised as a breaking of the simulations with which the Dasein closes in front of itself.[46] Then, the final coda or the subsequent link in the chain requires reconsidering whether one should return here to the very point from where man misleads authenticity from its everydayness and gets subsumed in the impersonal primacy of his understanding. To return to its originary constitutive Mitsein. Return there where others are instances of alienation and rethinking if precisely by reason of its finitude it is not there where it should also search for the possibility of openness and the discovery of itself and the world. To own both in its truth (that extract in whose direction we have also found, under different perspectives Cohen, Löwith, Barbaras). An existential task that is only possible, it is worth remembering, because man as Dasein only understands always from its own self as coexistence (Mitsein), the originary understanding coexistence.[47] Then dialogue has a possible sense that lies on, at the same time, the edge of the finitude that sustains, questions and threatens. Finally, the hint of expectations that the dialogue awakens, even the guarantee of its consistency is also sustained in the risk that makes it possible. As if there was not any more certainty (nor art) than the thrownness of the existence.[48]

Bibliography

Adelung, Johann Christoph, *Versuch eines vollständigen grammatisch-kritischen Wörterbuch der hochdeutschen Mundart*, 5 vols., Leipzig, Breitkopf, 1774–1786.

Barbaras, Renaud, *Introduction à une phénoménologie de la vie*. Paris, Vrin, 2008.

Barbaras, Renaud, *La vie Lacunaire*. Paris, Vrin, 2011.

Bernhard, Thomas, *The Voice Imitator*, translated by Kenneth J. Northcott, Chicago, University of Chicago Press, 1997.

Cohen, Hermann, *Der Begriff der Religion im System der Philosophie*. Hildesheim/ Zürich/New York, Georg Olms Verlag, 1996 (Nachdruck der 1. Auflage 1915).

45 Ibidem § 27, 129 sq.

46 Ibidem §27, 129 sq.

47 Ibidem §26, 124: "Das Sichkennen gründet in dem ursprünglich verstehenden Mitsein."

48 Hans-Georg GADAMER, *Wahrheit und Methode*, 371: "Eine Methode, fragen zu lernen, das Fragwürdige sehen zu lernen, gibt es nicht."

Cohen, Hermann, *Die Religion der Vernunft aus der Quellen des Judentums*. Leipzig, Gustav Fock, 1919.
Cohen, Hermann, *Ethik des reinen Willens*. Zürich/New York/Hildesheim, Georg Olms Verlag, 2012 (Vierter Nachdruck der 2. Auflage).
Dilthey, Wilhelm, *Gesammelte Schriften. Leben Schleiermachers. Schleiermachers System als Philosophie und Theologie*, vol. 2. Göttingen, Vandenhoek & Rupprecht, 1985.
Dubet, François, ¿Por qué preferimos la desigualdad? (aunque digamos lo contrario). Buenos Aires, Siglo Veintiuno Editores, 2015.
Fistetti, Francesco, *Comunità*. Bologna, Il Mulino, 2003.
Fistetti, Francesco, *Multiculturalismo*. Novara, Utet, 2008.
Gadamer, Hans-Georg, *Gesammelte Werke, vol 4. Neuere Philosophie II*. Tübingen, Mohr Siebeck, 1987.
Gadamer, Hans-Georg, *Gesammelte Werke*, vol. 1: *Hermeneutik I. Wahrheit und Methode. Grundzüge einer philosophischen Hermeneutik*. Tübingen, Mohr, [6]1990.
Heidegger, Martin, *Brief über den Humanismus*. Bern/München, Francke Verlag, [3]1975.
Heidegger, Martin, *Sein und Zeit*, Tübingen, Max Niemeyer Verlag, [11]1967.
Henry, Michel, *Incarnation. Une philosophie de la chair*. Paris, Seuil, 2000.
Henry, Michel, *L'essence de la manifestation*. Paris, Épiméthée PUF, [2]1990
Henry, Michel, *De la phénoménologie. Tome I. Phénoménologie de la vie*. Paris, Épiméthée PUF, 2003.
Löwith, Karl, *Das Individuum in der Rolle des Mitmenschen*. Freiburg/München, Verlag Karl Alber, 2013.
Nancy, Jean-Luc, *La Communauté désavouée*. Paris, Galilée 2014.
Ruta, Carlos, "Contributions for Considering the Possibility of All Life Configurations", in *Life Configurations*, edited by Gert Melville and Carlos Ruta (Challenges of Life. Essays on Philosophical and Cultural Anthropology, 1). Berlin/Boston, De Gruyter,2014, 3–24.
Schleiermacher, Friedrich D. E, *Kritische Gesamtausgabe. Vorlesungen zur Hermeneutik und Kritik*. Vol. 4, Berlin/Boston, De Gruyter. 2012.
Sennet, Richard, *Together. The Rituals, Pleasures and Politics of Cooperation*. New Haven & London, Yale University Press, 2012.

Pablo de Marinis

The Multiple Uses of 'Community' in Sociological Theory

Historical Type, Ideal Type, Political Utopia, Socio-technological Device and Ontological Foundation of 'Society'

1 Introduction

The vocabulary of community has always permeated the political paroles, religious demonstrations, slogans hoisted by all kinds of groupings in which comes into play anything that is somehow related to 'identity' and 'culture'. Thus, community is not just a word, but it also becomes a sensation. In this point Bauman is right, at least when he says that community produces 'a good feeling'.[1]

Sociology, as a discipline and as cultural institution, has always tried to catch up with the crossroads and political-cultural challenges of its time, and it has often succeeded on doing it. Since its founding times, this discipline has included community as a key concept, notion or idea, with different levels involved in each one of them, ranging from the specificity, with scientific pretensions expected of a concept to the volatility and vagueness that ideas often possess. The present paper will put the focus selectively on some concepts, notions or ideas of community that are present in sociological discourses produced in very different historical periods and cultural contexts. Its purpose is to offer some interpretative keys, which could allow a comparative analysis between different sociological perspectives around community.

The debates on theoretical logic and metatheory, which among others have been carried out by well known sociologists such as Jeffrey Alexander and Georg Ritzer,[2] have certainly enlightened our work. We think that the multidimensionality of sociological reflection can become meaningful in a historical-conceptual

---

1 Bauman has argued that "we have the feeling that community is always something good". According to this author (who has even devoted an entire book to the problem) community is not just a word that has meaning, but it also produces a "good feeling". Zygmunt BAUMAN, *Comunidad*, 7 (my translation). In a similar vein, Williams says that "unlike all other terms of social organization (state, nation, society, etc.) it (the community, PdM) seems never to be used unfavorably, and never to be given any positive opposing or distinguishing term". Raymond WILLIAMS, "Community", 76.

2 Jeffrey ALEXANDER, *Theoretical Logic in Sociology*; George RITZER, "Metatheorizing in Sociology" and ID., "The Legitimation and Institutionalization of Metatheorizing in Sociology".

and comparative study, like the one we have begun some years ago in our research group at the University of Buenos Aires, and which we wish to summarize here.[3] In this context, a number of analytical dimensions of community have emerged, which we have called 'registers', 'uses' or 'meanings' of community, and which we had applied for the interrogation of various authors and works, therefore continuously redefining our original analytical scheme.

In recent years, the debates certainly more resonant about the problem of the community have been carried out in the field of philosophy, by famous names such as Jean Luc Nancy, Giorgio Agamben, Roberto Esposito, Antonio Negri, etc. But sociology also has a long and rich tradition of problematizations about it, which is often ignored or forgotten by those authors, or even in the broader contemporary cultural debate about the community. Modestly, this work has the aim to recover it. But it will not be able to do it in an exhaustive form in the limited space available for this article. Therefore, the resulting exercise must necessarily be selective.

The argument will be deployed in two successive steps. First, in the next section of the article, a sort of rather general interpretative keys will be presented. They are related, on one hand, with the above mentioned 'registers', 'uses' or 'meanings' of community, and on the other hand, with the problem of the 'cultural semantics' of community.[4] After this, in the third – somewhat more extended - part of the work, these different uses of community will be addressed one by one, providing some textual evidences, and also introducing - through all of them - this analytical angle of 'cultural semantics'. Finally, in the conclusions, some brief statements will be brought into account, with the only intention to stimulate a debate on community which, we suspect, will never come to end.

---

3 The 'Grupo de Estudios sobre Teoría Sociológica y Comunidad', which I have lead since 2007 at the Instituto de Investigaciones Gino Germani (Facultad de Ciencias Sociales, Universidad de Buenos Aires, Argentina), as its name suggests, has been devoted to the study of the concept of community in classical and contemporary sociological theory. See the different (finished and ongoing) research projects in http://webiigg.sociales.uba.ar/iigg/miembrosDetalle.php?id=11&opcion_area=2&opcion_rol=&page=2&ipp=10. As an example of the publications generated in that context, see all articles included in the monographical issue of March 2010 of the journal *Papeles del CEIC*, in http://www.ehu.eus/ojs/index.php/papelesCEIC/issue/view/1177, and in the book of DE MARINIS, *Comunidad: estudios de teoría sociológica*. Some of these texts will be again specifically quoted below.

4 We anticipate now that five registers and two semantics will be addressed here.

2 Five dimensions of community in sociological theory; the problem of the 'cultural semantics' of community

In this section of the work, the five communitarian dimensions that we have found in our research process are presented. According to them, community can be a) historical type; b) ideal type; c) political utopia; d) technological device; e) ontological foundation of sociality. In each of them, sociology presents different purposes, each one articulating itself differently with other disciplines and social practices. In a similar way, it relates in diverse manners with another concept, often considered as its opposite in an indivisible couple: society. In some of the analyzed authors, it will be possible to find all dimensions at a time, in others less than five of them, but always more than one.

On the other hand, departing from a kind of 'political geography of knowledge',[5] according to which cultural contexts stimulate the emergence of a certain 'climate of ideas' and not another one, we think it is possible to identify different 'cultural semantics' of community, in which the concept acquires distinctive profiles. There are lots of other semantics, of course, and this has a special importance in Latin America, where we have always been strongly influenced by all of them at once, without parochialisms, without encapsulations and without ignoring any of them.[6]

In this work, only two of these cultural semantics will deserve special attention: the German and the Anglo-Saxon. We believe that the analysis will prove that the mere mention of a certain word in a certain context does not guarantee proximity or similarity among the meanings it carries with it in another context. This must be especially taken into account when trying to translate a concept for another one, or while analyzing 'receptions' or 'simultaneous problematizations'.[7]

Thus, besides all of the common features they may share, a 'Gemeinschaft' is not precisely a 'community', although both words could be translated into

---

5 I borrow the expression from Dick PELS, "Three Spaces of Social Theory", although I do not follow all his valuable theoretical suggestions.

6 The last two projects carried out by the research team that I lead, both still in progress, have been devoted to the study of Latin American sociological semantics about the community, but they have not yet generated significant publications.

7 From an approach of 'simultaneous problematizations', the research team I lead intends to launch in 2016 a new project in which sociological theories of the 'North' and the 'South' will be addressed simultaneously, putting the focus on the concepts of 'mass/es', 'crowd/s' and 'multitude/s'.

Spanish without great difficulties as 'comunidad', into Italian as 'comunità', into French as 'communauté', etc. With no intention of falling into cultural determinisms of any kind, we will see how different authors place themselves in relation to different cultural semantics that would be inherent to them, according to their own and to their partners' institutional present position in the cultural debate, and to the context/horizon of their intellectual influences of the past. Thus, we can see how these writers represent or recreate these cultural semantics more or less faithfully, or make efforts to escape from their influences, or approach to another semantics.[8]

2 a) Community as historical precedent of modern society (Sociology as explanation of the emergence of modernity)

The studies in the academic field of the history of sociology tend to underline -with good reasons- that the main purpose of the different generations of this disciplines' classics was to achieve a complete account of the emerging capitalist and modern social order. Even contemporary authors did not abandon this concern of explaining a long process, whose origins had been several decades or even centuries before the time at which they wrote. In this sense, some classical sociologists place community as a starting point of the process of modernization, and therefore define 'society' as the present, modern, capitalist condition.

This historical register of community has been highlighted by renowned historians of sociology like Robert Nisbet, who insistently and, with a dangerous one-sidedness, gave the whole sociological enterprise a conservative, romantic and nostalgic tone. In his view, the sociological classics were not only in search of an aseptic description of a historical process, for they presented an entirely somber and negative judgment of its consequences as well.[9] But a more careful scrutiny of the work of many authors belonging to the sociological tradition does not show such homogeneity, since the valuations ascribed to past and present conditions may vary largely from one case to another.

---

8 A pioneer reflection on the meaning and scope of these semantics can be appreciated in the brief but very important work of Raymond ARON, *La sociología alemana contemporânea*, aimed to show to a French audience the 'news' he had found during his research stay in Germany in the early 30s. Aron shows that he had great clarity on the issues of translation and interpretation between cultures. Thus he provides his French readers with an important key to understanding the true scope of the concept of community in German sociology, saying that there the word community, "unused in the political language of France, has for German ears the same resonance that 'justice and equality' for the French" (31; my translation).

9 Robert NISBET, *The Sociological Tradition.*

The conceptual resources through which this sequential and historical scheme was developed show a high diversity of references (historicism, biological evolutionism, cybernetic theories, etc.), but in all of them the process is described as a kind of 'evolution' or 'progress' from simple to complex forms, from mechanical groups to organic articulations, from undifferentiated aggregations to functional differentiation. Community, of course, is often located in the first of the paired elements, and is showed as our past, as what we ceased to be.

2 b) Community as ideal type opposed to society (Sociology as the science of social relationships)

Sociology always pretended to scientifically explain the attributes of the emergent social relationships, increasingly predominant in modern societies. In a conventional sense, doing science is nothing other than constructing concepts, typologies and models. Unlike the previous use of community, which had an eminently historical character, generality, abstraction and a relative ahistoricity are the main features of this other kind of sociological experience.

The second dimension of community in sociology is presented as ideal-type of inter-individual relationships, of collective forms of aggregation. In this way, communitarian relationships are not only seen as entities proper of the pre-modern past, but also constitutive of the modern present times.

The ideal-type of community usually implies attributes of stability, warmth, affection, localization, ancestry, tradition, status, adscript roles and co-presence. On the other hand, societal relationships tend to be depicted as artificial, evanescent, cold, distant, territorially expandable, contractual, impersonal, and with typically acquired roles.

2 c) Community as utopian solution to the pathologies of the present (Sociology as promoter of political praxis)

In tension with the above mentioned opposition between the (often nostalgic) remembrance of the past and the (allegedly aseptic) observation-description of the present, a third dimension of community emerges, and it points towards the future. Here, community is depicted not only as the entity which can solve the evils that modern rationalization has brought about (alienation, anomie, depersonalization, loss of meaning, etc.), but also as a utopian project of a future that can transcend this present, or, at least, that may suppress its most painful consequences.

We are now no longer confronted with an historical sociology of long-time processes, or with an abstract and relatively ahistorical systematic sociology. We are dealing instead with a discipline that is strongly normatively 'charged', oriented in showing the ways in which social practice (and above all political action) could and should follow, in order to overcome the present state of moral malaise and mediocrity. In this case, sociology appears as a discipline that does not respect anymore the distinction between 'statements on facts' and 'value judgments', or it simply destroys it.

2 d) Community as technological device for the reconstitution of the social bond (Sociology as social engineering)

Besides its enduring scientific ambitions, sociology had from its inception a very practical stance, clearly oriented towards social intervention. In this orientation, sociology has promoted different kinds of 'communitarian devices' in order to achieve the reconstitution of the bonds of social solidarity which modernization processes have broken or deteriorated. These devices are more humble, localized and particularized, and less utopian and general than the ones mentioned in 2 c). They tend to move populations' energies in a positive sense, affirmative of the bonds in danger of dissolution or corrosion.

2 e) Community as moral foundation of life in common (Sociology as social philosophy)

Despite the efforts that authors such as Durkheim made in order to 'free' sociology from its philosophical 'burden', this work has never been really achieved, and sociology has therefore always moved along diverse 'cultures', between science and literature.[10] Consciously or not, all sociological theories have always brought along philosophical views of order, change, subjectivity, history, 'good society', etc.

Community often appears, therefore, as a concept that is not only philosophically 'charged', but that is also an abbreviated expression of social life, a sort of 'zero degree' of every form of sociality. Each time community is mentioned, possibilities of life in common, illusions of collective life, and mentions of what could and should be shared, are working as background.

10 Cf. Wolf Lepenies, *Die Drei Kulturen*.

3 Textual evidences of the five uses of community in sociological theory (through the lens of the 'cultural semantics')

This section of the article will show some examples of forms through which the different dimensions of community appear strongly intertwined in the context of the work of some specially selected authors who are central to sociological tradition. As it will be seen, it is possible to simultaneously identify several dimensions in the same author, although sometimes one of them may mark the 'tone' over the others. It could not be otherwise, since sociology has always been all at the same time, regardless of the efforts made by some eminent sociologists to eliminate some dimensions of this intellectual enterprise (the historical, the typological-conceptual, the political-utopical, the socio-technological or 'engineer-style', and the ontological-philosophical dimension). Thus sociology has always and simultaneously maintained these orientations and purposes: to produce an explanation of long-term historical processes, or at least cooperate with such a kind of explanations; to establish a system of fundamental categories that can account for the forms that human interaction can take in every time and place; to provide incentives for the transformation of society in the desired direction; to propose concrete mechanisms, with a strongly practical imprint, in order to advance in that direction; and to offer a reflection, always philosophically loaded, about social life and, above all, about 'life in common' as such.

We will also see some examples of authors who are representative of particular cultural semantics, and others who make strenuous efforts to distance themselves from the cultural semantics that, at first, they 'belong'[11] to, given the space in which they grow up intellectually and in which they maintain the most of their dialogues.

---

11 The quotation marks do not seek other function than to relativize the suspicion of 'cultural determinism' that could fall on these reflections. It is important to stress this: to postulate the validity and effectiveness of certain 'cultural semantics' does not necessarily imply that the authors have to carry with them as if they were straitjackets. But there certainly are cultural semantics, and they have their own weight. Max Weber knew this better than anyone, as will be shown below.

3 a) Community as that which we are ceasing (or have already ceased) to be.

In sociology, it is almost inevitable (and this would be conceptually a serious error) to speak about community without making any reference to Ferdinand Tönnies. More incidentally cited than seriously studied, well known and appreciated by his contemporaries (although today almost nobody reads him), the work of Tönnies is a wonderful quarry of great intuitions that later would became almost common places in sociological discourse.

As early as 1887 he published his main work, *Gemeinschaft und Gesellschaft*,[12] indisputably one of the most important books of the intellectual and institutional history of sociology. Not only in this book, but also Ferdinand Tönnies' whole oeuvre is filled with references to the process of modernization as indicating the drawback of communitarian bonds and their replacement with societal relationships. In his main work, for instance, Tönnies propose a kind of 'Entwicklungsgeschichte', and in this way he takes a part – in a very strong form – in the evolutionary trends that were in vogue in his time. This is not the only dimension present in his work,[13] but it certainly marks the tone of the rest. The phrase found in the first pages of his book, which states that "Gemeinschaft is old; Gesellschaft is new as a name as well as a phenomenon", is one of the most eloquent manifestations of this orientation.[14]

Moreover, in Tönnies' tales there are a lot of images of decay, disintegration, degeneration, decline, etc., as hallmarks of societal present conditions. In many passages there are harsh words with which the modernization process that leads from the community to society is connoted as a "process of rapidly advancing disintegration".[15]

Tönnies is a magnificent example of what I have been naming in this text as the 'German semantics of the community'.[16] German semantics, even harboring

---

12 Three editions of Tönnies' *opus magnum* have been consulted for this article: the original German edition (*Gemeinschaft und Gesellschaft*), the English translation (*Community and Society*) and the Spanish translation (*Comunidad y Sociedad*).

13 In fact, we could also mention Tönnies in all other analytical dimensions that are deployed below in this paper, and in some cases we will do it.

14 Ferdinand Tönnies, *Community and Society*, 34.

15 Ibid., 204.

16 Obviously the word 'semantics' would require further specification, which for the moment I cannot give. I also know that in a text like this I take a risk by adopting a concept that, for example in the work of Niklas Luhmann, or Reinhard Koselleck, has more clear and precise definitions than those I can offer here. Provisionally, I would like to define 'semantics' as a set of senses more or less identifiable, which are endowed with certain attributes, specific to a particular area

within it very different versions and using a variety of intellectual sources, roughly has tended to give the community some affective (even directly irrational), culturalist, gregarious, romantic, natural, essential and emotional components. Consequently, it has tended to mobilize strong registers of a kind of 'immersion' of the individuals in a sort of collective fusion, based often on appeals to 'blood and soil', and conducting strict boundaries between an 'us' strongly distinguishable from another 'us'. Nothing could be more distant from the Anglo-Saxon semantics of the community, to which we will refer below, whose attributes can confront almost point to point with the above-mentioned.

Of course, it is not Tönnies the only author of those that we later would consider key authors for the sociological theory that had assumed the task to analyze a long historical process. As it is well known, Marx, Durkheim, Weber, Simmel, Parsons, Habermas, Luhmann, Giddens did this as well. But in none of these cases this process appears, as in Tönnies, as marked by a passage 'from community to society'. Moreover, in authors as Simmel and Durkheim the concept of community is used very incidentally, and when it is present, it takes more the form of a generic 'idea' of life in common and less the form of a fundamental sociological concept, like 'Wechselwirkung' in the work of the German and 'social bond' in the French sociologist. On the other hand, using different versions of a system theory, Parsons, Habermas and Luhmann also develop discussions about social change and the long-term evolution of societies, but in neither of them community appears as one of its milestones or key moments.[17]

---

of cultural influence. I have dealt with a little more detail about these problems of 'cultural semantics' in Pablo DE MARINIS, "*Gemeinschaft, community,* comunidad".

17 In order to avoid overwhelming the readers with a very long list of bibliographical references on community in the works of the authors mentioned above, I will cite just some secondary literature produced in our own research environment, where these and other authors (and other relevant secondary literature) are deeply recovered and discussed. On Durkheim, see Ramón RAMOS TORRE, "La comunidad moral", and Ana GRONDONA, "La sociología de Emile Durkheim". On Tönnies, see Pablo DE MARINIS, "Sociología clásica y comunidad" and Daniel ALVARO, "Los conceptos de 'comunidad' y 'sociedad' de Ferdinand Tönnies". On Weber, see Pablo DE MARINIS, "La comunidad según Max Weber", Victoria HAIDAR, "De la disolución a la recreación de la comunidad" and Emiliano TORTEROLA, "Racionalización y comunización". On Marx, Tönnies and Weber see the doctoral dissertation of Daniel ALVARO, *El problema de la comunidad*. On Parsons see Pablo DE MARINIS, "La comunidad societal" and Diego SADRINAS, "La comunidad societal". On Luhmann, see Mariano SASÍN, "La comunidad estéril" and on Giddens and Habermas see Alejandro BIALAKOWSKY, "Comunidad y sentido".

3 b) Community as one of the forms that interindividual relationships can empirically take

The history of sociological theory records several important efforts to build a set of basic sociological concepts. The well known example of the conceptual taxonomies plaguing Max Weber's Chapter I of *Economy and Society* stands out among these efforts.[18] Written in 1920, this kind of conceptual interventions are not the most frequent in a work like that of Weber's, especially interested in historical processes and individual genetic connections more that in a systematic, formal or categorial sociology. But anyway this chapter is very relevant in the context of his work, and it is even possible to detect an important precedent of it in the so-called 'Kategorienaufsatz' of 1913.[19] In both Weberian texts it is possible to find numerous references to the concept of community and, of course, also to society as well as to a number of related concepts, intricately constructed with prefixes, suffixes, and compound words with various adjectives.[20]

Schematically, I will only note here that the purpose of both Weberian texts is to achieve a formal typological system of categories, not a historical account, aimed at providing verstehende Soziologie with a strong categorial basis (of course this does not mean that Weber ignored the massive and unstoppable process of rationalization that had been taking place in the West for centuries). We need to emphasize now that what in 1913 was still called Gemeinschafsthandeln (community action) and was characterized as the primary object of the nascent comprehensive sociology, was replaced in 1920 by soziales Handeln (social action). Moreover, two new concepts take prominence in 1920 that were not deployed in 1913 in this way: Vergemeinschaftung and Vergesellschaftung. Both could be tentatively translated into English as 'communization' and 'socialization'.

In both concepts the primary focus of analysis is placed on the individual actor. Important Weberian scholars tend to agree in recognizing in such conceptual decisions the completion of the transition to a methodological individualism that Weber would have operated at the end of his life. In the first concept, communization, Weber underlines the preeminence of a subjective feeling of belonging to a group, together with other actors; in the second one, socialization, Weber emphasizes the idea that the attitude of the actor is based on the rational pursuit of his/her interests. In any case, Weber is talking about open, reversible,

18 Max WEBER, *Wirtschaft und Gesellschaft.*

19 Max WEBER, "Ueber einige Kategorien der verstehenden Soziologie".

20 I have developed elsewhere ("Las comunidades de Max Weber") a detailed analysis of these concepts, the transformations and substitutions that were operated on them between 1913 and 1920, and the obvious difficulties that have taken place in their translations into Spanish.

changeable processes, and not about consolidated entities. In sum, for Weber, the vast majority of social relations share some of the features of communization and socialization at the same time.

Both the replacement of Gemeinschaftshandeln through soziales Handeln and the coinage of concepts like Vergemeinschaftung and Vergesellschaftung (that rather refer to dynamic and reversible processes than to consolidated and stable entities) show Max Weber's efforts to introduce a strong 'desustancialización'[21] of the concept of community. In doing so, Weber took a clear distance from the meaning contents that were so usual in the context of the German semantics of community, and that from Tönnies onwards (and even before Tönnies) had reached a widespread diffusion and had aroused so many political-intellectual passions in that cultural field.

Indeed, the years of the First World War, and after them the whole period of the Weimar Republic, witnessed an impressive revitalization of community thinking in Germany and more generally in Mitteleuropa. It is precisely in this context that what I insist here to call the 'German cultural semantics of the community' took shape. The truth is that an enthusiastic 'communitarian climate' permeated the thinking of almost everybody, in social and political movements, in the wide cultural field and, more specifically, in the new Sozialwissenschaften and the (comparatively more established) Geisteswissenschaften.[22] A notable and lonely exception is offered by Helmuth Plessner in the middle of the Weimar Republic era, whose important book *Die Grenzen der Gemeinschaft* offered a strong criticism both to rightist and leftist communitarian radicalisms, as well as a defense of the liberal idea of society.[23]

3 c) The ills of society are cured with more community

All kinds of sociology intend to outline a 'good society'. However hard this discipline has worked since its founding days by adopting the usual standards of scientificity (that even today, with few exceptions, persist to prescribe objectivity

---

21 I can not find an English word for that. I use this word to describe a process by which something loses its character as entity, or loses substance and consistency.

22 As examples of the abundant literature that analyzed the profiles of that German communitarian cultural climate could be mentioned Winfried GEBHARDT, "'Warme Gemeinschaft' und 'kalte Gesellschaft'"; Domenico LOSURDO, *La comunidad, la muerte, Occidente*; Harry LIEBERSOHN, *Fate and Utopia* and Stefan BREUER, "'Gemeinschaft' in der 'deutschen Soziologie'".

23 Helmut PLESSNER, *Die Grenzen der Gemeinschaft*.

and Wertfreiheit), sociology has always flaunted normative commitments and has placed them in the center of its intellectual enterprise.

As we have seen in 3-a, references of negative tone about the process of modernization that led to our societal present were not infrequent in the classics of the discipline. Thus arose keywords like 'anomie' and 'moral mediocrity' in Durkheim, 'loss of meaning' in Weber, 'alienation' in Marx, 'depersonalization' in Simmel, and a plenty of other substantives, always in negative tone, in Tönnies, from hypocrisy to vanity through artificiality.

This critical attitude towards modernity (even a critique made 'from the inside' of modernity, recognizing the irreversibility of historical change and the undesirability of the return to the old 'solid' and stable times) has been dramatically reduced in Parsons. But it reemerges very sharply in a contemporary author, who owes much to Parsons (and to all the great classics of sociology, and not only of sociology): Jürgen Habermas. On one hand, this author does not hesitate to advance a description of the 'pathologies of modernity'. But on the other hand it absolves modernity itself when it states that the differentiation 'system-lifeworld' is not *per se* pathological (though it is pathological the way it assumes contemporaneously, which is conceptualized as 'colonization' of the lifeworld by the system, through the media of money and power).[24]

What is interesting for the purpose of this article is that practically all of these critical diagnoses of modern society (and the evils attributed to this condition) have been invariably accompanied by communitarian illusions. These illusions could take rather general and abstract forms, as desires and ideals of a 'good society' (notably different from what it is now), or more concrete forms, of specific community devices, which could serve to counter the adverse consequences of modernity, and to rehabilitate some form, although postsocietal, of the social bonds (and social order) eroded by modernization.

Obviously, Tönnies is the example that should first appear. He was definitely a skeptic critic of modernity, but he also had a utopian side, which imaginatively projected a future of postsocial rehabilitation of community. On one hand, his practical commitment to the German Social Democrats and his full support for communitarian experiences of the labor movement, as strike committees, unions, cooperatives, etc., are well known.[25] On the other hand, the sharp criticism of modern present times abounds in his scientific work. But it does not prevent him from proposing forward-projections of a future of equality and freedom, recovering at least some of that strong ethical dimension that in the past permeated

24 Jürgen HABERMAS, *Theorie des kommunikativen Handelns.*

25 See his references about the cooperative movement (*Principios de Sociología*, 73–75; *Comunidad y Sociedad*, 259s) and the workers' movement (*Principios de Sociología*, 72).

community relations, but also making use of the civilizational advancement of modern social relations. Because for Tönnies (and this is usually overlooked in the conventional reception of his work) progress, Aufklärung, development and civilization are "positive facts".[26]

In the epilogue added to the 1922 German fifth edition of *Community and Society*, Tönnies rebuilds and updates the main argument of the book, but giving it a historical mobility that previous editions lacked. After reviewing the historical process from the "age of the Gemeinschaft" (when family life and household are highlighted) to the "age of the Gesellschaft" (when commerce and life in the big city stand out), Tönnies observes that culture itself would succumb in this journey that goes from "original communism" (homely) to the individualism that arises from it (in the village and in the city), then independent individualism (proper of a big city or even universal) and then to (state dependent and international) socialism.

Unless (and here is his hope and his bet) "the essence and idea of Gemeinschaft" could foster "a new culture amidst the decaying one".[27] Thus, community does not appear as it has always been, as what was always there, or what has been lost, but as that which must be remade, rebuilt, recovered. In short, Tönnies is excited with a communitarian (but postsocial) future that will not come by itself or following a certain law of history. This future must be actively constructed by the tireless endeavors of those who can/want to impart the relationships between them an ethical tone and a 'human' character, which may go beyond the mere 'being next to each other', beyond the mere 'nebeneinander sein' that was so characteristic of 'the age of the Gesellschaft'.

All these Tönniesian approaches are at first sight very similar to those of Durkheim. In this French sociologist, the conceptual vocabulary of community is quite infrequent, and does not show an alignment to the German semantics. However, there is in his thought a very powerful 'idea of community', based on a diagnosis that was quite similar to that of his German contemporary. Several works by this author are very illustrative of what is proposed here. For example,

---

26 See, for example, this quotation: "I do not reject or ridicule the serious and radical reforms that are introduced to our ethical and social situation in society; rather, my intention was always to advocate for them. Nor I repudiate all the positive facts of progress, Enlightenment, development and free civilization, as if they had no value: my opinion was never a romantic one" (Ferdinand Tönnies, *Gemeinschaft und Gesellschaft*, XXXVI and XXXVII, my translation). Or the following one: "My 'pessimism' concerns not more than the future of contemporary culture, but not the future of culture in general" (Ferdinand Tönnies, "Zur Einleitung in die Soziologie") (quoted by David Frisby, "Soziologie und Moderne", 205; my translation).

27 Ferdinand Tönnies, *Community and Society*, 231.

the preface to the second edition of *De la division du travail social,*[28] where he called for the rehabilitation of professional corporations to counter anomic trends of modern society. His continuous appeals to the 'civic morality' in his *Leçons de sociologie* could also be mentioned.[29] Or, at a later stage of his work, in *Les Formes élémentaires de la vie religieuse*, where he appeals to collective rituals that could be useful in this "age of transition and moral mediocrity" in rehabilitating forms of "creative effervescence" to feed the moral reconstruction of society.[30]

3 d) The invention of 'community devices' as antidotes against modern disintegration

The practical stance that sociology possessed since its founding days has been already stressed above. Obviously, all the great sociologists that we know from their books, articles and scholarly activities have also allowed themselves to have certain 'interferences' with social practice, in particular with political activities. And in this practical side, almost invariably, the importance of community (or of different 'community devices') for the constitution or reconstitution of the social bonds spoiled by the processes of modernization is highlighted.

The sociologists of the Chicago School of Sociology embodied better than anybody this 'interface' knowledge developing between sociological discourses with scientific purposes and practical activities of social reform and social control. However, in the work of various representatives of this school, such as Robert Park, 'community-society' as a dichotomous conceptual couple can no longer be found in the same way that it appeared in Tönnies or, in his own way, also in Weber. Nor that sequential figure 'from community to society' can be found. For example, in Park, both concepts are often used synonymously.[31]

But for this section of the article we are interested above all on the specific 'communitarian devices' that sociologists proposed to 'solve social problems', in the orientation proposed by Durkheim (and already mentioned above) of a rehabilitation of professional corporations. In this context, the famous research project conducted by William I. Thomas and Florian Znaniecki, *The Polish Peasant in Europe and America*, published between 1918 and 1920, can be mentioned. Drawing on very novel methods for the time (as biographical analysis, or

28 Emile DURKHEIM, *La División del Trabajo Social*, 11–48.

29 ID., *Lecciones de Sociología*.

30 ID., *Las formas elementales*, 397–398 (my translation).

31 See Emiliano TORTEROLA ("Lazo social y metrópolis"), who articulates a comparative analysis between Simmel and the authors of Chicago.

the study of personal letters) and (although it sounds anachronistic) respecting Bruno Latour's suggestions to 'follow the actors', these sociologists accompany the Polish peasants in their immigration processes, first from small villages and towns to the big cities of Poland, and from there to the effervescent Chicago of the first decades of the twentieth century.[32]

In this way, they use the concepts of 'organization', 'disorganization' and 'reorganization' to account for the changes in the forms of interindividual linking of these peasants, who later would become workers, artisans, small traders, etc. Right there, at the moment when they established themselves in Chicago, is when all the 'social therapeutic' and reforming intentions of Chicago sociologists clearly emerged. Thus, they promote the creation of institutions (which I would like to call here 'communitarian devices') that allow individuals to maintain and strengthen ties with their countrymen groups, while at the same time not losing their sphere of individuality as citizens of a 'bigger society', this 'community of communities' according to John Dewey, precisely this American society which, willingly or unwillingly, had received them.[33]

Very typical arguments of the 'Anglo-Saxon semantics' of community, which were already referred to above, but which have not yet been clearly defined, are at stake here. In community, as in Gemeinschaft, reference is done to a collective entity, showing a group of individuals who live and act together with relative unity and cohesion. This, in turn, deserves strong 'value charges' and is usually designated as morally 'good', virtuous, ethical, full of positive connotations in the sense of the Baumanian 'good sensation' mentioned at the beginning of this article. So far, then, there would be no significant differences between Gemeinschaft and community.

But there are differences between Gemeinschaft and community, and they are important. Thus, in community, individuals do no longer appear immersed in a totality that ontologically precedes them, and that hardly recognizes traces of their identity and individuality, as in Gemeinschaft, but they are rather depicted as active, proactive and rational demiurges of the collective instance. Community is neither an undivided organic whole that does not support the recognition of the parts included in it, nor 'a set of parts that do not make a whole'. Rather, it should be referred to in another mode by those parties to proactively shape the whole. In this way, the totality involved in a community has less attributes of nat-

32 William E. Thomas and Florian Znaniecki, *El campesino polaco en Europa y América.*

33 See the articles of Ana Grondona "La 'comunidad' de Chicago" and Victoria Haidar "Una 'Comunidad de comunidades'". The first underlines the 'socio-technological' character of the Chicago School. The second reflects on the linking of this school with pragmatism.

urality, of necessity, authenticity and eternity, and has rather artifactual features, deliberately and voluntarily constructed by the activism of its own actors.

In community, as in Gemeinschaft, there may be affection, even intimacy, but in no case outbreaks of irrationality. In community, collective passions may well exist, but they certainly assume a comparatively restrained, moderate, temperate, and one might even say 'domesticated' tonality. Under no circumstances will community put the individual as fully (and perhaps sacrificially) subordinated to the priorities and (often intense) requirements of the collective entity, as is usually seen in (self) presentations of Gemeinschaft.[34]

In sum, the Anglo-Saxon semantics of community has taken more voluntaristic, proactive features, showing the conscious, rational and deliberate adherence of subjects to a group that includes them but, mostly, does not erase but rather can enhance their individuality. In that sense, its definition seems to include many of the features of other constructs within that tradition and that are also related to it, such as 'civil society' and 'civic sphere', full of contents that, in contrast to the German semantics of community, posses much more individualistic, contractual and liberal features.

However, this 'social engineering' approach to promote community devices is not an exclusive monopoly of Anglo-Saxon sociologists. We can also find it in Max Weber, at various times in his work. First, we could refer to the work of the young Weber, published in 1892, on the peasantry of East Prussia.[35] There, various superimposed reflections on community appear, from the dissolution of the "community of interests" between peasants and agrarian landowners produced by capitalist modernization, to threats experienced by the German "national community" due to the 'invasion' of the Polish peasants who replaced the German workforce that was migrating to the industrial cities of the West following "the powerful and purely psychological charm of 'freedom'".[36] On the

34 I will only mention here some of the many important texts about the semantics of community, which will be treated in more detail in another article. Suzanne KELLER ("The American Dream of Community") and Thomas BENDER (*Community and Social Change in America)* analyze the deeper meaning and contemporary relevance of the concept of community throughout the history of American cultural tradition. See also Cherry SCHRECKER (*La communauté)* for a comprehensive history of the concept of community in Anglo-Saxon sociology. Hans JOAS ("Gemeinschaft und Demokratie in den USA") and again Cherry SCHRECKER ("Community and Community Studies") face a deep comparative analysis between "community" and "Gemeinschaft". Some incidental but interesting references to this problem can be found in Hartmut ROSA et al. (*Theorien der Gemeinschaft zur Einführung*, 177–178), Dietmar WETZEL ("Gemeinschaft", 45–46) and Harry LIEBERSOHN (*Fate and Utopia*, 7).

35 Max WEBER, "Entwicklungstendenzen".

36 Ibid., 247; my translation.

other hand, the latest Weber[37] develops another interesting community figure, a countertendential device that, in his perspective, could stop the nonsense in which the German masses (defeated in World War I) were sunk: the 'plebiscitary democracy of the leader'.[38]

3 e) Community as 'zero degree' of sociality

All the great authors of the sociological tradition have strong philosophical backgrounds, and have developed their works in the context of discussions on the most diverse philosophical currents, from neo-Kantianism and historicism in the case of the Germans to positivism and utilitarianism in the case of the French, through pragmatism among Americans. Moreover, several of these authors have been considered by their colleagues rather as philosophers than as sociologists, as the cases of Simmel and Tönnies reveal.[39] What has been said about the sociological classics is also true for most contemporary authors in this discipline. It is known that Bourdieu made a strong reception of Heidegger and Sartre, and Luhmann did the same with the phenomenological tradition opened by Husserl. Habermas went even further, since the entire tradition of Western philosophy falls on the large table of his theoretical concerns.

However, beyond the undoubted philosophical contributions to sociology, a sort of 'antiphilosophical stance' can already be seen from the inception of the second, often overplayed to legitimize the effort to turn it into a 'science' based on observations and committed to empirical research. Talcott Parsons is a very good example of this perspective, beyond his epistemological commitments and debts with the 'analytical realism' of Alfred Whitehead. However, in the context of an article on community it is not necessary to refer to his whole fundamental work, but at least to the concept of 'societal community', with which we hope to show that even in an author as supposedly 'scientific' as Parsons, a strong normative claim with socio-philosophical roots is also present.[40]

---

37 Max WEBER, "Parlament und Regierung".

38 For a reflection on the first of the Weberian texts, see Victoria HAIDAR ("De la disolución a la recreación de la comunidad). For the second, see Pablo DE MARINIS ('La comunidad según Max Weber').

39 Sociology has a strange story, for it would later build up its pantheon of classics based on the figures of a neo-Hegelian philosopher who would become a political economist (Marx), an historian of culture (Weber) and only one sociologist strongly self-convinced of his status, and of the need of sociology as a science of the social (Durkheim).

40 I devoted an entire article to the problem of societal community in Parsonian work (Pablo DE MARINIS, "La comunidad societal de Talcott Parsons").

Societal community is a concept that Parsons introduced late in his work, after the adoption of the AGIL scheme of 4 functions, and is characterized as the structural core of society, as the integration subsystem of the social system. So it is defined by Parsons: "is the patterned normative order through which the life of a population is collectively organized. As an order, it contains values and differentiated and particularized norms (...) As a collectivity, it displays a patterned conception of membership, which distinguishes between those individuals who do and do not belong".[41] The most important property of the societal community "is the type and level of solidarity that -in the sense of Durkheim's term- characterizes the relations between its members".[42] In turn, solidarity is understood as "the extent to which (and the ways in which) it is expected, that the collective interest prevails over the interests of its members, provided that both are in conflict".[43]

Societal community appears to be some kind of trans-historical invariant that appears in every form of society and at every stage of social development, which guarantees that society members are essentially 'under jurisdiction' of it. It breaks the sequential-historical sense that the pair community/society has taken in other sociological discourses, paradigmatically in Tönnies. On the other hand, the Parsonian societal community does not possess an ideal-typical character. Social relations are not classified alternatively as community or society, each with its distinctive features, but as social relations in general. Thus, the concept of societal community is clearly located at another level, more generic and structural, not typological. In any case, it is noteworthy that for the definition of the integrative subsystem of the social system Parsons chose a noun as 'community' (even with the addition of the adjective 'societal') that such a strong semantic charge had had throughout the history of sociology. In short: in the Parsonian concept of societal community, the 'registers' which we have designated above as historical type and also as ideal type are not present.

But there seem to be strong utopian elements in this concept, which is striking in an author like Parsons, who in principle could be conceived as a sworn enemy of mixing claims about states of affairs in the world and normative judgments. Beyond his claim to have found a kind of historical invariant with his concept of societal community, Parsons needs to introduce some specifications in relation to the historical emergence of modern society.

In his view, what is characteristic of modern society is the presence of diverse forms of pluralism: pluralism of economic interests, of political groups, in the

41 Talcott Parsons, *Societies*, 10.
42 Id., "Sistemas sociales", 712.
43 Ibid.

cultural sphere (including the multi-denominational religious pluralism), pluralism of intellectual disciplines and ethical pluralism.[44] Parsons concedes that this pluralism can bring with it "a major problem of integration for a societal community",[45] subjecting the individual to a heavy pressure, because the membership of this individual in each group (or part) implies a set of requirements that might be inconsistent with that posed by society (or totality).

Thus, even in a context of pluralization of experiences, societal community has to develop and consolidate a number of basic and general values, as it has done at all times and places. But under the specific conditions of modernity they must be increasingly abstract, in order to be able to support, tolerate or at least not interfere on the effectiveness and validity of specific, concrete and particularized values related to different groups, sectors, strata, organizations, institutions in which individuals participate and in which an increasingly differentiated social system is divided. This process is called 'value generalization'.[46]

Thus, societal community takes the form of a normative, philosophically 'charged' base for 'good society', in which individuals can display their subjectivity without disrupting social order. Parsons emphasized the importance of this normative basis with an optimist attitude, which was dangerously close to innocence in regard to the alleged capacity of modern societies to integrate differences.

But Parsons was no innocent. Other works, in which he also heavily uses the vocabulary of societal community, show that he was fully aware that things that 'need to be' not often 'are'. The best example of this is an essay published by Parsons in 1965, in the middle of a country heavily convulsed by civil rights' struggles. There, in a long and intertwined historical and conceptual development, Parsons describes the progress in the integration of the diverse groups (Jews, Catholics, etc.) that converge in the American societal community and, above all, he wants to emphasize the tasks still 'sadly pending' in reference to black populations.[47]

It is precisely in this type of work where the normative 'slips' in Parsons acquire greater visibility. So if things are not going in the direction you would expect, you will have to manipulate them, through effective policies that allow social citizenship to perform effectively, also among black populations.

---

44 Ibid., 713.

45 Talcott PARSONS, *The System of Modern Societies*, 12.

46 Together with 'differentiation', 'inclusion' and 'adaptive upgrading', 'value generalization' is on the base of the evolutionary changes he studied in *Societies* and *The System of Modern Societies*.

47 Talcott PARSONS, "Full Citizenship".

The question (and the political project) in Parsons could be spelled out like this: what is to be done, then, to enable mechanisms that can allow – or to neutralize mechanisms that can obture – the generation of stable cultural norms and symbols that are motivationally binding for solidarity to prevail over self-interest, mechanisms that, at the same time, must be abstract and universalistic enough so that the various groups and subgroups that make up a complex and differentiated society like modern society can adhere voluntarily to them without feeling subjugated or overwhelmed in their uniqueness and particularity?

In addition to this strong normative vocation, an Anglo semantics of community (very similar to what we had seen operating in the authors of the Chicago School of Sociology) is also present in this argument. There are many and varied individuals and groups in society. And, normatively speaking, an attempt should be made for individuals to be actively involved in groups, to stand firm loyalties imposed upon them there, without interfering with or disrupting the commitments with an even larger entity, society. It would be better for this to happen in a 'free', 'autonomous' and 'voluntary' way. If it does not, society will deploy the necessary mechanisms of social control, which could put things back into place.

4 Conclusions

The conclusions of this work will be few and modest. They will be exhibited in the form of short thesis, and only intended to stimulate an endless debate.

- The legacy of sociology to the thinking of community is much richer and more varied than is usually supposed. We hope we have been able to display it here, even if in this schematic form.
- It is fruitful for theoretical reflection on community to take into consideration the variety of 'uses' and 'registers' that it has taken, and the diversity of cultural semantics that may have led to these uses.
- The problem of the variety of 'uses' of community clearly shows that sociology has always been, and remains so even today, a multifaceted intellectual enterprise. Explaining processes, coining category systems, designing political utopias, promoting specific techno-social devices, and normatively philosophizing, all at the same time, sociology does not stop to solve its endemic identity crisis, and perhaps precisely therein lies the wealth of its contributions.
- The problem of cultural semantics is just as relevant when you have to translate any text from one language to another. But even more so when you have to intellectually understand the 'travelling of ideas' from one context

to another, the 'reception' of a context into another, and the 'simultaneous problematizations' in different cultural contexts.

Bibliography

Alexander, Jeffrey, *Theoretical Logic in Sociology*, vol. 1: *Positivism, presuppositions, and current controversies*. Berkeley, California, University of California Press, 1982.

Alvaro, Daniel, "Los conceptos de 'comunidad' y 'sociedad' de Ferdinand Tönnies", *Papeles del CEIC* 52 (2010), 1–24.

Alvaro, Daniel, *El problema de la comunidad. Marx, Tönnies, Weber*. Buenos Aires, Prometeo Editorial, 2015.

Aron, Raymond, *La sociología alemana contemporánea*. Buenos Aires, Paidós, 1965 [*La Sociologie allemande contemporaine*. Paris, Alcan, 1935].

Bauman, Zygmunt, *Comunidad. En busca de seguridad en un mundo hostil*. Madrid, Siglo XXI Editores, 2003 [*Community: Seeking Safety in an Insecure World*. Cambridge, Polity Press, 2001].

Bender, Thomas, *Community and Social Change in America*. Baltimore/London, The Johns Hopkins University Press, 1982.

Bialakowsky, Alejandro, "Comunidad y sentido en la teoría sociológica contemporánea: las propuestas de A. Giddens y J. Habermas", *Papeles del CEIC* 53 (2010), 1–30.

Breuer, Stefan, "'Gemeinschaft' in der 'deutschen Soziologie'", *Zeitschrift für Soziologie* 31 (2002), 354–372.

de Marinis, Pablo, "Sociología clásica y comunidad: entre la nostalgia y la utopía (un recorrido por algunos textos de Ferdinand Tönnies)". In *La comunidad como pretexto. En torno al (re) surgimiento de las solidaridades comunitarias*, edited by Pablo de Marinis, Gabriel Gatti and Ignacio Irazuzta. Barcelona/México DF, Editorial Anthropos and Universidad Autónoma Metropolitana-Iztapalapa, 2010, 347–382.

de Marinis, Pablo, "La comunidad según Max Weber: desde el tipo ideal de la *Vergemeinschaftung* hasta la comunidad de los combatientes", *Papeles del CEIC* 58 (2010), 1–36.

de Marinis, Pablo, ed., *Comunidad: estudios de teoría sociológica*. Buenos Aires, Prometeo Editorial, 2012.

de Marinis, Pablo, "La comunidad societal de Talcott Parsons, entre la pretensión científica y el compromiso normativista". In *Comunidad: estudios de teoría sociológica*, edited by Pablo de Marinis. Buenos Aires, Prometeo, 2012, 231–263.

de Marinis, Pablo, "*Gemeinschaft, community*, comunidad: algunas reflexiones preliminares acerca de las variadas semánticas de la comunidad en la teoría sociológica", *Revista Argentina de Ciencia Política* 16 (2013), 87–104.

de Marinis, Pablo, "Las comunidades de Max Weber. Acerca de las tipologías sociológicas como medio de desustancialización de la comunidad". In *Max Weber en Iberoamérica. Nuevas interpretaciones, estudios empíricos y recepción*, edited by Alvaro Morcillo Laiz and Eduardo Weisz. México DF, Fondo de Cultura Económica, 2015, 293–320.

Durkheim, Emile, *Las formas elementales de la vida religiosa*. Madrid, Akal, 1992 [*Les formes élémentaires de la vie religieuse*. Paris, Alcan, 1912].

Durkheim, Emile, *Lecciones de Sociología*. Buenos Aires, Miño y Dávila, 2003 [*Leçons de sociologie: physique des moeurs et du droit. Cours de sociologie dispensés à Bordeaux entre 1890 et 1900*. Paris, Presses universitaires de France, 1950].
Durkheim, Emile, *La División del Trabajo Social*, Barcelona, Planeta-Agostini, 1985 [*De la division du travail social. Étude sur l'organisation des sociétés supérieures*. Paris, Alcan, 1893].
Frisby, David, "Soziologie und Moderne: Ferdinand Tönnies, Georg Simmel und Max Weber". In *Simmel und die frühen Soziologen. Nähe und Distanz zu Durkheim, Tönnies und Max Weber*, edited by Otthein Rammstedt (Suhrkamp-Taschenbuch Wissenschaft, 736). Frankfurt a. M., Suhrkamp, 1988, 196–221.
Gebhardt, Winfried, "'Warme Gemeinschaft' und 'kalte Gesellschaft'. Zur Kontinuität einer deutschen Denkfigur". In *Der Aufstand gegen den Bürger. Antibürgerliches Denken im 20. Jahrhundert*, edited by Günter Meuter and Henrique Ricardo Otten. Würzburg, Königshausen & Neumann, 1999, 165–184.
Grondona, Ana, "La sociología de Emile Durkheim. ¿Una definición 'comunitarista' de lo social?", *Papeles del CEIC* 55 (2010), 1–24.
Grondona, Ana, "La 'comunidad' de Chicago. Cuestión social, cuestión urbana y cambio social: una sociología de *lo* comunitario". In *Comunidad: estudios de teoría sociológica*, edited by Pablo de Marinis. Buenos Aires, Prometeo, 2012, 189–228.
Habermas, Jürgen, *Theorie des kommunikativen Handelns*, 2 vols. Frankfurt a. M., Suhrkamp Verlag, 1981.
Haidar, Victoria, "De la disolución a la recreación de la comunidad. Un contrapunto entre Max Weber y François Perroux", *Papeles del CEIC* 54 (2010), 1–28.
Haidar, Victoria, "Una 'Comunidad de comunidades': tras las huellas de una tradición liberal y democrática de pensamiento acerca de la comunidad en las obras de John Dewey y los sociólogos de la Escuela de Chicago". In *Comunidad: estudios de teoría sociológica*, edited by Pablo de Marinis. Buenos Aires, Prometeo, 2012, 141–187.
Joas, Hans, "Gemeinschaft und Demokratie in den USA. Die vergessene Vorgeschichte der Kommunitarismus-Diskussion" in *Soziale Gemeinschaften: Experimentierfelder für kollektive Lebensformen*, edited by Matthias Grundmann, Thomas Dierschke and Iris Kunze (Individuum und Gesellschaft, 3). Berlin, Lit Verlag, 2006, 31–42.
Keller, Suzanne, "The American Dream of Community: An Unfinished Agenda", *Sociological Forum* 3 (1988), 167–183.
Lepenies, Wolf, *Die Drei Kulturen. Soziologie zwischen Literatur und Wissenschaft*. München/ Wien, Carl Hanser Verlag, 1985.
Liebersohn, Harry, *Fate and Utopia in German Sociology 1870–1923*. Cambridge, MIT Press, 1988.
Losurdo, Domenico, *La comunidad, la muerte, Occidente. Heidegger y la 'ideología de la guerra'*. Buenos Aires, Losada, 2003.
Nisbet, Robert A., *The Sociological Tradition*. New York, Basic Books, 1966.
Parsons, Talcott, "Full Citizenship for the Negro American? A Sociological Problem", *Daedalus* 94 (1965), 1009–1054.
Parsons, Talcott, *Societies. Evolutionary and Comparative Perspectives*. Englewood Cliffs, N.J., Prentice Hall, 1966.
Parsons, Talcott, *The System of Modern Societies*. Englewood Cliffs, N.J., Prentice Hall, 1971.

Parsons, Talcott, "Sistemas sociales". In *Enciclopedia Internacional de las Ciencias Sociales*, vol. 9, edited by David Sills. Madrid, Aguilar, 1976, 710–721 [*International Encyclopedia of the Social Sciences*. New York, Macmillan and Free Press, 1968].

Pels, Dick, "Three Spaces of Social Theory: Towards a political geography of knowledge", *The Canadian Journal of Sociology / Cahiers canadiens de sociologie* 26 (2001), 31–56.

Pleßner, Helmuth, *Die Grenzen der Gemeinschaft. Eine Kritik des sozialen Radikalismus* (Suhrkamp-Taschenbuch Wissenschaft, 1540). Frankfurt a. M., Suhrkamp Verlag, 2002 [first Bonn, Cohen, 1924].

Ramos Torre, Ramón, "La comunidad moral en la obra de Émile Durkheim". In *La comunidad como pretexto. En torno al (re)surgimiento de las solidaridades comunitarias*, edited by Pablo de Marinis, Gabriel Gatti and Ignacio Irazuzta. Barcelona/México DF, Editorial Anthropos and Universidad Autónoma Metropolitana-Iztapalapa, 2010, 383–412.

Ritzer, George, "Metatheorizing in Sociology", *Sociological Forum* 5 (1990), 3–15.

Ritzer, George, "The Legitimation and Institutionalization of Metatheorizing in Sociology", *Sociological Perspectives* 35 (1992), 543–550.

Rosa, Hartmut, Lars Gertenbach, Henning Laux and David Strecker, *Theorien der Gemeinschaft zur Einführung* (Zur Einführung, 367). Hamburg, Junius Verlag, 2010.

Sadrinas, Diego, "La comunidad societal en la obra de Talcott Parsons: tensiones entre la inclusión y la exclusión". In *Comunidad: estudios de teoría sociológica*, edited by Pablo de Marinis. Buenos Aires, Prometeo, 2012, 265–305.

Sasín, Mariano, "La comunidad estéril. El recurso comunitario como forma de la autodescripción social", *Papeles del CEIC* 57 (2010), 1–35.

Schrecker, Cherry, *La communauté. Histoire critique d'un concept dans la sociologie anglo-saxonne*. Paris, L'Harmattan, 2006.

Schrecker, Cherry, "Community and Community Studies: a Return Journey". In *Transatlantic Voyages and Sociology. The Migration and Development of Ideas*, edited by Cherry Schrecker. Aldershot, Ashgate, 2010, 113–126.

Thomas, William E. and Florian Znaniecki, *El campesino polaco en Europa y América*. Madrid, CIS, 2004 [first 1918–19].

Tönnies, Ferdinand, "Zur Einleitung in die Soziologie", *Zeitschrift für Philosophie und philosophische Kritik* 115 (1899), 240–251.

Tönnies, Ferdinand, *Principios de sociología* (translated by Vicente Llorens). México, FCE, 1942 [*Einführung in die Soziologie*. Stuttgart, Enke Verlag, 1931].

Tönnies, Ferdinand, *Comunidad y sociedad* (translated by José Rovira Armengol). Buenos Aires, Losada, 1947.

Tönnies, Ferdinand, *Gemeinschaft und Gesellschaft. Grundbegriffe der reinen Soziologie*. Darmstadt, Wissenschaftliche Buchgesellschaft, 1979 [first *Gemeinschaft und Gesellschaft. Abhandlung des Communismus und des Socialismus als empiririscher Kulturformen*. Leipzig, Fues, 1887].

Tönnies, Ferdinand, *Community and Society* (translated by Charles Loomis). New Brunswick/ London, Transaction Publishers, 2004.

Torterola, Emiliano, "Racionalización y comunización en la esfera económica. Los matices del individualismo en la teoría de la modernidad weberiana", *Papeles del CEIC* 56 (2010), 1–24.

Torterola, Emiliano, "Lazo social y metrópolis. La comunidad en los orígenes de la sociología urbana: Georg Simmel y Robert E. Park". In *Comunidad: estudios de teoría sociológica*, edited by Pablo de Marinis. Buenos Aires, Prometeo, 2012, 109–140.

Weber, Max, "Ueber einige Kategorien der verstehenden Soziologie". In *Max Weber. Gesammelte Aufsätze zur Wissenschaftslehre*, edited by Johannes Winckelmann. Tübingen, Mohr Siebeck, [3]1968, 427–474 [first Tübingen, Mohr, 1922].

Weber, Max, *Wirtschaft und Gesellschaft. Grundriss der verstehenden Soziologie*. Tübingen, Mohr Siebeck, [5]1980 [first Tübingen, Mohr, 1922].

Weber, Max, "Entwickelungstendenzen in der Lage der ostelbischen Landarbeiter". In *Max Weber. Gesammelte Aufsätze zur Sozial- und Wirtschaftsgeschichte*, edited by Marianne Weber (Uni-Taschenbücher 1493). Tübingen, Mohr Siebeck, 1988, 470–507 [first Tübingen, Mohr, 1924].

Weber, Max, "Parlament und Regierung im neugeordneten Deutschland". In *Max Weber. Gesammelte Politische Schriften*, edited by Johannes Winckelmann. Tübingen, Mohr Siebeck, 1988, 306–443 [first München, Drei-Masken-Verlag, 1921].

Wetzel, Dietmar J., "Gemeinschaft. Vom Unteilbaren des geteilten Miteinanders". In *Poststrukturalistische Sozialwissenschaften*, edited by Stephan Moebius and Andreas Reckwitz (Suhrkamp Taschenbuch Wissenschaft, 1869). Frankfurt a. M., Suhrkamp Verlag, 2008, 43–57.

Williams, Raymond, "Community". In *Keywords: A Vocabulary of Culture and Society* (Revised Edition), edited by Raymond Williams. New York, Oxford University Press, 1983, 75–76.

José Itzigsohn
Community, Recognition, and Individual Autonomy

Community Figurations

The question posed by the organizers of the conference is to what extent the community presents a challenge to the autonomy of the individual. The thesis they propose states "that the single person needs the community and hence must live in community, which can, however, only be undergone under the influence of vast impositions upon his or her own individual autonomy, which in turn stipulate effortful acts of continual socialization, presents an immense challenge to human life." Implicit in this proposition is the assumption of an individual who needs a community to live but can think of him/her self as different and separated from the communities in which she or he is embedded and sees his/her individuality and autonomy as something separated and to be protected from those communities. The thesis states that living in community is a challenge and a burden to individual autonomy.

In this paper I argue that not only the individual needs the community to live, but that we cannot think of individuality and autonomy separated from community. In other words, individuals need communities in order to be autonomous. Modern society values individuality, autonomy, and self-expression. But these are the values and norms of our society and cannot be realized outside this type of society. Individuals are only individuals in society and membership in society implies membership in several communities. Being part of communities, to be sure, implies limits and constraints. Communities make demands on individuals, affect their identities, and limit what they can do. Communities can at times be oppressive. But in contemporary societies communities also enable people to think and act as autonomous individuals.

From a sociological perspective it is impossible to think of an autonomous individual disembedded from any community. As Durkheim[1] pointed out, individuality and individualism have become the dominant and unifying values of modern society, a religion of sorts. A society in which everyone does something different and pursues different life tracks is one in which individuality and individualism can soar. For Durkheim, the division of labor generated a form of organic solidarity that emerges from the complementarity of the activities of

1 Emile DURKHEIM, *The Division of Labour in Society.*

different people. But Durkheim realized that interdependence and some broad and distant shared values were not enough to sustain social solidarity and that anomie emerges in all realms of social action. For Durkheim anomie threatened the social order and it also threatened individuality and autonomy, as these are dependent on the social order and on the presence of solidarity. Durkheim saw the solution to the problem of anomie in the nurturing of communities that are closer to the everyday lives of individuals. In particular, he thought of those communities based on occupations, where people could identify more closely with others and adopt sets of norms that are closer to their individual experiences. Durkheim's point was that although organic solidarity was based on mutual interdependence and the belief in individualism, it still needed closer communities in order to sustain individuality.

But if the thesis of the conference presupposes an autonomous individual that stands separated from community and society, Durkheim's approach puts all the weight on the larger society as the maker of the individual, without leaving much room for autonomy or agency. The inevitable embeddedness of individuals in society was analysed in a more compelling way by Norbert Elias. Elias[2] argued that in order to conduct their lives humans enter into networks of interdependence with others. He called these networks figurations. Elias argued that the more interdependent individuals become the more they internalize the social constraints and accept the norms of the figurations they are part of. To that extent, Elias argument could be seen as supporting the thesis that community limits individual autonomy. But it is also this restraint, according to Elias, that allows individuals to interact as individuals and makes it possible for individualism to emerge.

Many of the figurations in which people participate are communities as defined by the organizers of this conference. Namely, a. their members see themselves as part of a collective; b. their actions depend on their shared identity; c. there is an emotional identification with the goals of the figuration; and d. individuals are willing to follow shared rules. Elias[3] points out that some figurations are close to the individual and can be seen directly in their everyday interaction (e.g. the family, local organizations, the workplace). Others are larger, more distant, and can be only analysed indirectly (e.g., the nation, the class, other forms of categorical memberships). Elias argument is that individuals are always members of figurations that have different degrees and forms of individuation and different demands and limits upon their members. All figurations have different divisions of power. And all also enable their members to act in society in

2 Norbert ELIAS, *The Civilizing Process.*
3 ID., *What is Sociology?*

order to fulfil some of their goals. There is no social life or individual autonomy outside of these figurations.

From Elias's perspective it is a mistake to think about individuals as if they were separate and autonomous from their figurations; as it is also a mistake to speak of these figurations without taking into account the actions and strategies of the individuals that constitute them. There are no individuals outside of society, and there is no society without individuals pursuing their strategies and goals, that in turn are affected by the properties of the figurations to which they belong. So rather than looking at society as a separate entity that determines what its members can do, Elias proposed to look at the characteristics of different figurations, with their different degree of individuation, different levels of interdependence, different power differentials, and different enabling characteristics. Furthermore, Elias[4] argues that different figurations are in competition with each other to establish the accepted social norms and habituses of the broader society they are a part of.

As Pablo de Marinis argued in his presentation in this conference, the centrality of community in contemporary social life was particularly emphasized by the Chicago School of Sociology. And as De Marinis points out the way the Chicago sociologists thought about community was different from the way European social theorists thought about it at the turn of the 20th century. It was the merit of the Chicago sociologists in the 1920s and 1930s to highlight the voluntary and instrumental character of communities and their role in promoting immigrant assimilation and in the construction of the urban space (Bulmer 1986).[5] The importance of community in the social life of the migrant, as de Marinis pointed out in his talk, is expressed in the description of the United States as a "community of communities." For the Chicago School, this mosaic of communities was what characterized Chicago (and other cities of immigration) in the United States in the first half of the twentieth century.

Nevertheless, the Chicago School was oblivious to the fact that communities develop in unequal social fields. The Chicago sociologists missed the systemic inequalities that marked community development. This point was stressed by W. E. B. Du Bois[6] two decades before the emergence of the Chicago School of sociology. Du Bois pointed to the centrality of the color line in the constitution of the individual and also the community in a racialized world. Although Robert Park was acquainted with the work of Du Bois, the Chicago School chose to ignore it. But Du Bois contributions are central to the analysis of communities and individ-

---

4 Id., *The Civilizing Process.*
5 See Martin Bulmer, *The Chicago School of Sociology.*
6 William E. B. Du Bois, *The Souls of Black Folk.*

uals. Du Bois argued that the color line is central to modernity and that it creates different experiences of the self and community. The experience of the racialized person and the racialized community is one of lack of recognition and invisibility. For Du Bois building community was central to withstand and to oppose racialization. Du Bois helps us think the field of inequalities that structures the formation of different selves and within which communities are constructed.[7] Du Bois was in fact the first theorists to emphasize the centrality of recognition in modern societies and his insights on racialization can be applied to other situations of subordination.

In this paper I rely on Elias' and Du Bois' insights to analyse how the community is intrinsically related to our ability to function as autonomous individuals. Individual autonomy is thought as the ability of the individual to make decisions about her/his life based on the individual's own motives rather than external social impositions. Critiques of capitalism and modernity have highlighted the numerous obstacles to individual autonomy that common people confront in contemporary societies. For Marx,[8] alienation and exploitation impeded the realization of the individual possibilities of creative autonomy. Marx argued that individual autonomy would only be possible in a community of associated producers that makes decisions together about what and how to produce. For Habermas,[9] it is the colonization of the lifeworld and the expansion of instrumental rationality that threatens individual autonomy. The latter could only be sustained by communicative rationality--that is, by the ability to make our voice heard in conversations between equals about our present and our collective future. Communicative rationality in turn depends on the one hand on an institutional order that would allow for undistorted communication, and on the other hand, on the values and norms present in the lifeworld's communities.

While these are powerful critiques of capitalism and modernity, it is still important to reflect about autonomy and community from the perspective of the everyday life of common individuals in contemporary societies. I argue that in everyday life individual autonomy refers to two things: first our capacity to make decisions about how we live our life, and, second, our ability to participate in the making of the collective decisions of the society in which we live. The first refers to our capacity to decide our lifestyle, how we use our leisure time, how and what we consume, whom we choose to relate with, and the like. The second refers to our ability to have our voice heard in the important social debates that affect our

---

7 See José ITZIGSOHN and Karida BROWN, "Sociology and the Theory of Double Consciousness", 231–248.
8 Karl MARX, *The Economic and Philosophical Manuscripts of 1844.*
9 Jürgen HABERMAS, *Legitimation Crisis.*

lives, to be part of the construction of the institutional, symbolic, and normative order of the communities in which we live. Yet, from a sociological perspective, as Durkheim pointed out, autonomy presupposes a figuration that allows for a large degree of individuation and that has a form of solidarity that can restrain the tendency towards destructive individualism. Furthermore, as Elias pointed out—and the organizers of the conference suggest—such a figuration implies a considerable degree of internalization of social control and the restraint of individual spontaneous behaviour. Only in a society that promotes individualism and in which individuals exercise a great degree of self restraint can individuals be autonomous in the terms described above.

It is necessary to explore then what are the preconditions for autonomy in contemporary societies. We need to establish what are the bases that allow individuals in contemporary societies to achieve a measure of control over their lifestyle and participate in in collective decision making. I argue that there are two preconditions that sustain individual autonomy in contemporary large social figurations: The first precondition is the need to secure an economic base. Freedom from want, as Franklin Roosevelt put it in his four freedoms speech,[10] is necessary to be able to make decisions about one's life. Without a secure economic base, we can certainly subsist, but we cannot shape our individual lifestyle as we would like. We cannot be fully autonomous individuals in modern society without access to some kind of work that would allow for our economic well-being and the realization of our individual goals.

Economic need is a barrier to individual autonomy. Being outside of the formal system of work, being a self-employed informal worker, does not make individuals more autonomous. On the contrary, most informal workers are marginalized in many different ways and they find it hard to shape their everyday lives. This does not mean that those who are on the margins of the employment world—those who toil in the vast realm of informal work—do not have agency. They indeed have agency and they deploy it in creative ways.[11] Their lifestyle choices, however, are subject to multiple social and economic constraints.

The second precondition that enhances individual autonomy is social recognition as full members of the national communities to which we belong. Without that recognition we cannot fully participate in the collective decision making of our national or local community. People who are not seen as full members of the community experience different forms of exclusion and marginalization

---

10 Roosevelt's State of the Union address, given on January 6 1941, is known as the four freedoms speech. In it Roosevelt articulates the four freedoms that should characterize contemporary democracies.

11 See Rina AGARWALA, *Informal Labor.*

that make them less equal individuals than those who are full members of their societies.[12] Exclusion from the national community means less opportunity for individual autonomous participation in the decision-making processes. As with informal workers, people excluded from their political communities still have agency, but they are not full participants in their political communities and therefore they have less of an ability to shape its present and future.

Of course, in conducting their lives people are subject to the norms, social pressures and more or less temporary fads of the figuration they inhabit. But within those limits individuals have different measures of autonomy to shape their lifestyles and participate in decision making. But they can do so only if they have a certain economic security and if they enjoy social recognition as members of their communities. Perhaps we ought to speak of autonomy as a matter of degree, not an absolute. The question then would not be whether individuals in contemporary society are autonomous but to what extent they can exercise that autonomy, that is, to what extent individuals can shape their everyday life and how much voice they have in the political and public sphere.

In the rest of the paper I present two cases, based on my own empirical research, that illustrate the argument that community is the foundation of individual autonomy—both in terms of being able to shape the individual lifestyle and in terms of individual participation in the decision making processes of the broader community. It is important to point out, though, that I am not arguing that through collective action people can improve their living conditions and/or affect the political communities in which they live. This is a well established argument. My argument is that only through community membership and community based collective action can people aspire to become autonomous individuals in our contemporary social figuration. The first case I discuss is that of the workers of recuperated enterprises in Argentina. The second case is that of Latino/a immigrants in the United States. In both cases the potential for being an autonomous individual is contingent on embracing membership in a community, with all the limiting and enabling elements that such membership may imply.

Community Action and the Defence of Individual Lifeworlds

First I examine the case of the workers of recuperated enterprises in Argentina. These are workers that appropriated the enterprises they worked for and chose

12 See Andreas WIMMER, *Nationalist Exclusion.*

to run them as self-managed cooperatives. They chose to do so in most cases because the enterprise was facing bankruptcy and they were facing a future of exclusion from the labor market. The decision to recuperate the enterprise was taken to protect the workers' lifeworld.[13] I have been doing research on how industrial democracy works and for that purpose I conducted interviews in six recuperated enterprises. I have followed two of these six enterprises for a period of eight years. The following is a synthesis of my findings. It does not correspond to the specific situation of anyone of the particular enterprises I have looked at, but it summarizes what I have learned from the combination of their experiences.

In our socioeconomic figuration people need jobs that makes possible a certain standard of living to be able to exercise individual autonomy in shaping their lives. Individual autonomy depends on access to economic resources and that depends, in part, on the institutions and norms of the local and national communities that individuals inhabit. In looking to secure their economic well-being individuals face a labour market that is a construction of the national community they live in. No economy is autarchic and every national or local economy is tied to global economies, but people confront labour markets that are local and subject to the institutions, regulations (or lack thereof) and the cultural expectations of their national communities.[14] Of course individuals can try to migrate, but if they are successful in doing so they will be facing a different labour market subject to the institutions and conventions of a different local and national community.

The experience of the workers of recuperated enterprises show the importance of community for securing the autonomy of the individual. These are workers who decided to appropriate the enterprise they were working for and run it as a workers managed cooperative. The workers of recuperated enterprises, however, were not utopists or believers in cooperativism. They recuperated their enterprises to defend a certain way of being, certain notion of individuality and rights prevalent in Argentine society.[15] The form of individuality and autonomy that workers were defending was based on the presence of relatively well paid manufacturing jobs and strong unions. Argentine manufacturing workers enjoyed for decades—since the first Peronist government (1946–1955)—a relatively high standard of living and a great pride and social recognition in being workers.[16] They derived their identity from their condition as industrial workers, and, until the 1990s, they enjoyed a standard of living that allowed them considerable

13 See Julián REBÓN, *Desobedeciendo al Desempleo.*

14 See José ITZIGSOHN, *Developing Poverty.*

15 See José ITZIGSOHN and Julián REBÓN, "The Recuperation of Enterprises", 178–196.

16 See Daniel JAMES, *Resistance and Integration.*

leeway on shaping their everyday lives. The closing of the workplace in a context of the mass unemployment produced by the 2001 economic crisis threatened that lifeworld. The recuperation of the enterprise was a Polanyan response aimed to defend the workers' lifeworld and to protect their identities and autonomy.[17]

The recuperation of enterprises was possible due to the participation of the workers in different figurations—that is, in different types of community. The first community of relevance was the workplace. The workplace is a figuration, with certain authority lines and forms of organization of work. It is doubtful that the whole workplace was a community—by the definition proposed by the conference organizers—before it went bankrupt. There might have been many communities within the larger figuration that was the workplace: based on where people worked, what kind of work they did, whether they participated in the union or on leisure activities together, etc. But once the enterprise went bankrupt the workers had to constitute themselves as a community—as workers of a certain workplace—in order to be able to appropriate it. And indeed, not all the workers who were employed in the enterprise became part of the community of workers that recuperated the enterprise.

The recuperation of enterprises challenged existing bankruptcy laws and norms concerning property rights and led to different levels of conflict and confrontation between the workers, the previous owners, and the state. The levels of conflict in different recuperation processes varied, but all of them required organization and community. The possibility of fighting to preserve the workers' lifeworld depended on creating numerous forms of solidarity based on different communities.[18] In addition to the workers becoming a community for the purpose of recuperating the workplace, they also enjoyed the support of the local communities in which the workplaces were located. The local communities sometimes mobilized in support of the workers, or sometimes individuals simply contributed with foodstuff donations for the struggling workers. Workers also received the support of other organized groups of workers and social movements—communities of struggle—that mobilized to support them politically or to provide them with financial or technical assistance.[19] In the process of recuperation workers also appealed to a general societal consensus that work is a social value and that the employers were breaking their social obligations to their workers when they

17 See José ITZIGSOHN and Julián REBÓN, "The Recuperation of Enterprises", 178–196 and Beverly SILVER, *Forces of Labor.*
18 See José ITZIGSOHN and Julián REBÓN, "The Recuperation of Enterprises", 178–196 and Julián REBÓN, *La Empresa de la Autonomía.*
19 See Julián REBÓN, *Desobedeciendo al Desempleo.*

closed the workplace. Indeed, the support for the recuperations is rooted in a broad societal support for the right of people to have decent employment.

In other words, the recuperation of enterprises was possible because the employees of the bankrupted enterprises enjoyed social recognition as industrial workers, as individuals who have contributed to the enterprise they worked for, and as individuals who have a right to work. Public opinion supported the recuperation of enterprises because it saw them as a struggle to defend the right to work and the right to a dignified life based on the individual's work. It is the recognition of the worker and the legitimacy of the claim that everyone deserves the opportunity to be employed that was the base for the support for the practice of enterprise recuperation. Without this normative commitment of the larger national figuration to the rights of workers, the workers' actions would not have enjoyed widespread support. The workers of the recuperated enterprises could defend their individual autonomy and their lifeworld by forging their own community, appealing to the solidarity of different communities, and appealing to the values and recognition of the larger national figuration to which they belong.

Once the workers succeeded in recuperating the enterprise, however, they confronted the need to run the cooperative in an economically efficient way. This is the most basic rule of business, and although the workers of the recuperated enterprises see themselves as workers, they are also in fact a collective of entrepreneurs. If they did not succeed economically, they would not be able to preserve their autonomy in shaping their lifeworld. The workers decided to run the workplace in a democratic way, through community self-management. They did not need to do so; they could have hired experts, outside managers. But their previous experience with bosses and managers was so bad that they decided to rely on their own capacities as a community to run the enterprise. At this point the task the workers confronted was to create rules for their community. These rules had to do first of all with the organization of the work process. Workers had to decide about issues such as showing up for work, arriving on time, organizing the work process. These are the everyday issues that contribute to a smooth running of an enterprise, whether a privately owned or a self-managed one.

The first thing the workers—and the researcher—realize is the need for order. The challenges for the workers' community were very clear. First, what to do about absenteeism—e.g., members of the community who repeatedly got medical certificates to stay at home. Will all the sick days be paid or should the cooperative establish certain limits? Second, what to do with workers who arrive late to work? Will this lead to reductions in wages? Is it a good idea to establish incentives for showing up everyday to work or for arriving on time? How are tasks on the workplace going to be assigned and what to do with those who put little effort into their tasks? The first task of the workers then was to establish rules and a certain discipline for their community. That is, they had to discuss what kind of impositions, constraints, and limits they were to put on themselves. In many places these decisions consumed a lot of assembly time and caused bitter debates. The new cooperatives went back and forth, experimenting with different combinations of incentives and penalties: from instituting awards for coming to work regularly, to discounting sick-days when they were too recurrent or not properly documented, from appealing to personal responsibility to trying different ways of dividing the work between the members.

That is, the first order of business for the community of workers is to establish rules that limit the freedom of its individual members. This is obvious as it is impossible to run any organization in a different way. Recuperating the workplace is what allows the workers to maintain their identities, their autonomy, and their lifeworld. But in order to do that they need to subject themselves to the discipline of their own community. This is what the organizers of the conference point out: the community imposes limitations and it demands work to socialize people into its norms. And the self-imposed discipline of the workers can feel as onerous constraints on many of them, limiting their ability to act as they would like. On the other hand, without these impositions, the workers could not run the cooperative effectively; and therefore they cannot maintain their autonomy and individuality. The community of workers then is both a source of limitations to what people can do and a base for their autonomy in everyday life.

Yet, the decisions described above point to another dimension of worker autonomy in the recuperated enterprises, that is, the ability of the workers to participate in making the decisions about how to run the enterprise. The rules and norms the workers impose on themselves are absolutely necessary also in order to run the place in a democratic way. The workers chose to create self-managed enterprises run through assemblies, and, like in any community that makes decisions through deliberation, the workers had to establish rules and procedures for decision making: rules about creating agendas, process of debate, time and lengths of meetings, and what decisions to bring to the collective and which ones to delegate to the elected management. They had to choose what decisions to delegate on the leadership of the cooperative, they had to agree on how to operate the business aspects of the enterprise and how to use the cooperative resources and surpluses: should those be used for investment, savings, or wages? All these rules are community impositions on the individual workers. But those rules also make it possible for people to participate autonomously and actively in shaping their workplace community. Workplaces that failed to create these rules or in which the rules collapsed became very contentious communities in which autonomous individuality could not thrive or communities in which the workers delegated most of the decisions on the leadership of the cooperative, giving up the option of participating actively in running their workplace.

The workers that recuperated their enterprises were a very small fraction of the Argentine workers who confronted unemployment and exclusion from the labor market. Recuperating an enterprise requires a strong commitment to a struggle with an uncertain future: workers first faced the struggle of appropriating their workplace, then they faced the task of organizing as a self-managed cooperative, and finally they faced the challenge of making the cooperative economically sustainable. Most workers chose not to participate in that project. Many joined other communities of struggle, such as unemployed people movements. Others went through the fracture of their lifeworld isolated from other workers and many were pushed to structural marginality. All of them experienced poverty and constraints on their ability to shape their lives. Many of them gained part of that ability back when the economy recovered. As a result of their decision to create a community based on the enterprise and through the collective action based on that community, a minority of workers was able to defend their lifeworld and autonomy and gained the opportunity to shape the everyday life of their workplace.

Community Organization for Autonomous Political Voice

As shown in the previous section, in the process of defending their lifeworld and their identity, the workers of the recuperated enterprises won the ability to decide collectively on the everyday management of their workplace. This is an instance of the second aspect of individual autonomy that I want to address: the ability to participate in the making of collective decisions in the communities in which we leave, particularly in the political realm. One way in which we participate in collective decision making in democratic polities is through the vote. The vote is an individual form of conveying our preferences. But the vote presupposes a political figuration that allows for the vote and the existence of options from which we can choose. Furthermore, in order to be an autonomous participant in collective decision making an individual has to be recognized as a full and equal member of the political community. If an individual is not recognized as an equal member, then that person can not express her or his voice, and her or his preferences may not be part of the public agenda. Other ways of participation in the collective decisions of the national or the local community is involvement in political groups, social movements, or community or grassroots organizations. Voluntary participation in political or social movement organizations imply identification with a particular type of community--the community constituted by the organization and perhaps, with a larger community that the organization aims to represent. Again, to be an autonomous participant in these type of organizations a person need to be recognized as an equal, otherwise, she or he cannot participate in setting the agenda of the group. Immigrants and ethnoracial minorities are often in the margins of the definition of the national identity and are not seen as full and equal members of the political community. Therefore, they cannot participate in the political sphere as fully autonomous individuals.

Latino/a immigrants in the United States exemplify the problem of lack of recognition. The Latino/a identity in the United States is constructed against the background of ethnoracial categorization and exclusions. Latinos/as are seen and defined as a panethnic group but also as a racialized group, as part of the U.S. non white population.[20] The racialization of Latinos/as is different from the racialization of African Americans that Du Bois[21] described. Latinos/as complicate the working of the color line. Latinos/as are a multiracial group and many are perceived by the institutions of U.S. society as white. Bonilla Silva[22] has argued that light skin Latinos/as may achieve the status of honorary whites in U.S. society. But as immigrants Latinos/as are not full members of the national community and many experience racial discrimination and exclusion.

Those Latinos/as who are are undocumented are thoroughly excluded from the political sphere. This is a clear case of exclusion and non-recognition. But, even when they have documented status and they can participate formally in the political process of the larger political community, their ability to have their voice heard in the public sphere is limited. In a democratic polity individuals can participate in all sorts of political organizations, organizations based on embracing specific issues or a particular ideology. Being a democrat, an environmentalist, or a defender of gun owners' rights is a choice that autonomous citizens can make. Latinos/as of course can choose those identifications and participate in those organizations as individuals, and they indeed do so. But within those organizations they may encounter a variety of forms of exclusions and they may find that their experiences and concerns are not recognized or validated. The subtle exclusion of Latinos/as in the political sphere takes many forms, such as ignoring their specific concerns, ignoring them as a constituency, discouraging or suppressing their vote, a ceiling in their their ability to join the upper ranks of political organizations or the political system, and even ignoring them for political patronage. As a result, whereas Latino/a immigrants often have formal rights and can par-

20 See Linda ALCOFF, *Visible Identities.*

21 William E. B. DU BOIS, *Dusk of Dawn.*

22 Eduardo BONILLA SILVA, *Racism without Racists.*

ticipate in politics as individuals, they cannot bring their agenda to the fore as autonomous individuals as they are often not seen as equal participants in the political process.

The only way Latino/a immigrants can acquire an autonomous voice and presence and be considered full participants of the U.S. political process is through organizing around their identity as an ethnoracial group and through their organization in panethnic political, community, and social movement organizations. Only through embracing Latino/a identity and organizing on that base can the group make their concerns public and be recognized as a constituency with its own interests. Latino/as mobilize on the base of their ethnoracial identity to get a foothold in the political arena and get recognized as actors. Through this process Latinos manage to put some issues in the agenda and become a coveted voting group. It is only through making collective demands as a group, constructing organizations, and mobilizing around their ethnoracial identity—all processes that demand the collective effort of building and identity and imply impositions and the erasure of difference —is that Latinos/as manage to achieve recognition and realize their individual right to participate in politics as autonomous individuals.

This does not mean that there is one unified Latino/a agenda as Latino/as have different interests and points of view. But it does show that it is only by embracing community, with all its limitations and enabling features, that people manage to achieve autonomy as individuals. This is in fact no different than the historical trajectory of other groups and social categories. Workers, women, African Americans all have gained a voice as individuals in politics through embracing their group identities and building communities and communities organizations. Furthermore, when group organizations weaken, as is the case of unions or labor parties, individuals of that group—in this particular example workers-- loose part of their ability to make their voice heard in the public sphere. As a result, the individuals of that group lose part of the ability to participate in politics as autonomous individuals.

Collective action as Latinos/as, however, implies identifying as part of a community that is defined by the state. Latino/a as a form of ethnoracial classification and identification is a construction created by the U.S. state bureaucracy and imposed on a group of immigrants and their descendants. People from Latin America don't identify with this label outside the United States and they do not necessarily do so once in the United States. In the United States immigrants are categorized around the different accepted categories of identity available in the larger national community. Those categories are used in the census and in multitude of instances in private and public life. People are constantly asked to fill forms stating their ethnic and/or racial identity. People are also treated by mainstream institutions according to those categories. Identifying people as Latinos/as is a classificatory devise of the state, a label that has racializing effects.[23] Adopting and mobilizing around this label means accepting the racializing classifications of the U.S. state and society. Individuals of Latin American origin do this because, as Bourdieu[24] argued, naming a group means bringing it into existence. Given that they are classified and racialized as Latinos/as, individuals of Latin American origin can only have their specific concerns as immigrants and as a racialized group addressed in the public sphere—and therefore can only become autonomous individuals—by embracing and organizing around this label.

I could observe this process in my own research on the process of incorporation of the Dominican first and second generation in Providence, Rhode Island.[25] The experience of incorporation for the majority of Dominicans is affected by their encounter with racialization and the color line. Dominicans encounter the racializing and discriminatory practices with which the U.S society and state relate to people it considers black. This is indeed reflected in the perceptions of many Dominican immigrants and their children who equate being American—in the sense of being a full member of U.S. society—with whiteness. Very few Dominicans can claim whiteness in the United States or achieve the status of honorary whites—although an important percentage self-identifies as white in the census.

The story of incorporation of Dominicans in Providence is one of self-organization for recognition and access to resources. As the Dominican community in Providence grew and Dominicans became citizens they could participate in city politics as individuals. They could vote for the mayor and city counsellors and they could write and call the offices of the mayor or their council representatives. Yet, in spite of the numerical growth of the Dominican community, for a long period the city authorities did not pay attention to their particular needs, such as improving the public schools and their neighbourhoods, and recognition as a constituency. As a result, they also did not participate much in the political process. This changed when some individuals decided to gather together to participate in politics as Latinos/as. Dominicans organized with other people of Latin American origin who were seeking to have their issues and their presence recognized. They did so because no separated group had the numbers to be effective in the electoral arena by itself and because there is an available classificatory label that unified all the groups of Latin American origin.

As they organized as Latinos/as several Dominicans got access to elected positions in the city council, in the state chamber of representatives and the state

23 See Cristina MORA, *Making Hispanics* and Linda ALCOFF, *Visible Identities*.
24 Pierre BOURDIEU, *Language and Symbolic Power*.
25 José ITZIGSOHN, *Encountering American Faultlines*.

senate, and even a former mayor was of Dominican origin. Today Dominicans and Latinos/as have a strong presence in city hall and the city council. They have gained resources for their neighbourhoods and associations, and a voice in many urban public debates. They have not changed much their socio economic situation because many of the issues that affect them—such as joblessness or low paying jobs and immigration concerns—are beyond the scope of city politics. Also, the question of recognition and full membership in the national community transcends the city. The color line is part of the social structure and Dominicans are indeed still racialized in the larger field of U.S. sociocultural classification.

But at the level of local city politics they have achieved a measure of recognition. And it is this recognition that allows them to participate as autonomous individuals in city politics because they can influence the public debate and because they can make choices as into what capacity to participate—e.g., whether to organize and participate as Latinos/as or based on other identification. Achieving full recognition is a complex and protracted process. Becoming a recognized actor at the level of city politics—and also at the level of national politics—is a step in the struggle for full membership and for as much individual autonomy as it is possible in contemporary democratic polities. Through their panethnic political organization Dominicans and Latinos/as became recognized as actors and as a constituency in the city and have gained a voice and access to some resources. In this process they have gained also autonomy as individuals.

Community, Recognition, Autonomy

Contemporary society values individuality and individual autonomy. There is not much room in the self-imagination of such a society for community. Within this understanding of modernity the community may be seen as oppressive or atavistic. It is the merit of the organizers of the conference to emphasize the centrality of community for human life, but their premise is that although indispensable, the community threatens individual autonomy. This may indeed be true in some cases, but in this paper I have argued that in our contemporary figuration the values of individuality and autonomy can only be realized in the frame of community membership and community based action. I have shown this through the description of the experiences of the workers of recuperated enterprises in Argentina and Latino/a immigrants in the United States.

Prior to the economic crisis of the turn to the 21st century, the lifeworld and the autonomy of manufacturing workers in Argentina was built on the presence of strong unions, —a figuration that emerged from Peronism and recognized workers as full and valuable members of society. When that particular historical figuration and its communities broke down with neoliberalism, some workers chose to rely on their workplace communities to appropriate their workplace and protect their lifeworld. In this way they defended their identities and autonomy as individuals and they also gained autonomy in the management of their everyday work life.

Similarly, participation in politics as autonomous individuals depends on the presence of a national figuration in which institutions secure the ability of individuals to have their voice heard and express their preferences. Yet, national membership always excludes some segments of the population, and, furthermore, in multiracial/multi-ethnic figurations—like most contemporary political communities—individual participation is mediated by ethnoracial belonging. Those who are not recognized as full members of the national community have less of an ability to affect the public sphere. The only way excluded individuals can gain equality—and therefore full autonomy—in the political sphere is by identifying and organizing along community lines. In this way they can become autonomous individuals and have their concerns considered in the public sphere.

It is the different networks of interdependence that people belong to—Elias' figurations—that constitute the base for individual autonomy in everyday life. These community figurations indeed make impositions on their members. Figurations, as Elias argued, have a power structure and demand compliance with their norms. Moreover, as Elias remarked, in order to accommodate those disciplines and norms, individuals have to internalize self-restraint. Yet, as Du Bois argued, given the existing forms of exclusion and lack of recognition, it is only through memberships in the different communities to which they belong that individuals can achieve recognition and become full and autonomous members of the larger figurations in which they live. Only those individuals that enjoy economic, racial, or gender privilege can think of themselves as being autonomous without membership in a community. And that is possible only because the institutions of the larger figuration sustain their privilege.

Bibliography

Agarwala, Rina, *Informal Labor, Formal Politics, and Dignified Discontent in India*. Cambridge, Cambridge University Press, 2013.

Alcoff, Linda Martín, *Visible Identities. Race, Gender, and the Self*. New York, Oxford University Press, 2005.

Bonilla-Silva, Eduardo, *Racism without Racists.Color-Blind Racism and the Persistence of Racial Inequality in America*. Lanham, Rowman & Littlefield Publishers, 2006.

Bourdieu, Pierre, *Language and Symbolic Power*. Cambridge, Harvard University Press, 1993.
Bulmer, Martin, *The Chicago School of Sociology*. Chicago, University of Chicago Press, 1986.
Du Bois, William E. B., *The Souls of Black Folk*. London, Longmans, Green and Co. Ltd., 1965 [reprint of 1903].
Du Bois, William E. B., *Dusk of Dawn*. New Brunswick, Transaction Publishers, 2012 [reprint of 1940].
Durkheim, Emile, *The Division of Labor in Society*. New York, NY Free Press, 1997.
Elias, Norbert, *What is Sociology*. New York, Columbia University Press, 1984.
Elias, Norbert, *The Civilizing Process*. Malden, Blackwell Publishing, 2000.
Habermas, Jürgen, *Legitimation Crisis*. Boston, Beacon Press, 1975.
Itzigsohn, José, *Developing Poverty. The State, Labor Market Deregulation, and the Informal Economy in Costa Rica and the Dominican Republic*. University Park, Penn State University Press, 2000.
Itzigsohn, José, *Encountering American Faultlines. Race, Class, and the Dominican Experience in Providence*. New York, Russell Sage Foundation, 2009.
Itzigsohn, José and Karida Brown, "Sociology and the Theory of Double Consciousness. W. E. B. Du Bois's Phenomenology of Racialized Subjectivity", *Du Bois Review* 12 (2015), 231–248.
Itzigsohn, José and Julián Rebón, "The Recuperation of Enterprises. Defending Workers' Lifeworld, Creating New Tools of Contention", *Latin American Research Review* 15 (2015), 178–196.
James, Daniel, *Resistance and Integration. Peronism and the Argentine Working Class, 1946–1976*. New York, Cambridge University Press, 1988.
Marx, Karl, *The Economic and Philosophical Manuscripts of 1844*. Amherst, Prometheus Books, 1988.
Mora, Cristina, *Making Hispanics. How Activists, Bureaucrats, and Media Constructed a New American*. Chicago, University of Chicago Press, 2014.
Rebón, Julián, *Desobedeciendo al Desempleo*. Buenos Aires, PICASO, 2005.
Rebón, Julián, *La Empresa de la Autonomía*. Buenos Aires, PICASO, 2007.
Silver, Beverly, *Forces of Labor. Workers' Movements and Globalization since 1870*. New York, Cambridge University Press, 2003.
Wimmer, Andreas, *Nationalist Exclusion and Ethnic Conflict. Shadows of Modernity*. New York, Cambridge University Press, 2002.

Karl Siegbert Rehberg

Community as Point of Origin and as Reason for Yearning

Reflections from Anthropological, Sociological and Political Perspectives

I

Mapping to the topical frame of this conference, with its series of comparative studies of world interpretations and reality constructions in historical, anthropological and cultural study perspectives, we can assume an – usually unacknowledged – consensus. A consensus in spite of the most diverging interpretations of human and cultural history: we agree and act on the assumption that the culturally diverse modes of human life – from earliest tribal communities that were determined and regulated by blood relationships to advanced civilizations – originated from forms of communal life. The phantasm painted by (the French philosopher) Jean-Jacques Rousseau of the savage who forlornly wanders about in the woods only describes a fictitious pre-societal state. The cultural anthropogenesis is a later state, achieved by peasants and herdsmen through communal life without rule or oppression.[1]

The term "community" is widely used in common (!) speech; in scientific terminology however "community" characterizes specific socio-historical stages of development on the one hand, on the other hand it is being utilized as a specialist term for marking structure in sociological and historical comparisons.

It is remarkable that in Antiquity there has been no clear distinction between the concepts of community (Gemeinschaft) and society (Gesellschaft). Reading medieval translations of the works of Aristotle we find that often they used the terms „communio" or "communitas" while Thomas Aquinas and others utilized the term "societas" instead. There are numerous data to proof the synonymous usage of both terms[2]; at least we can regard this as terminological interlacing that is to found with Jean Bodin, for example. Still (in the early phase of Enlightenment) Gottfried Wilhelm Leibniz speaks of an "unrestrictedly equal society"

1 Cf. Jean-Jacques ROSSEAU, *Discours sur l'origine*.
2 Cf. Manfred RIEDEL, "Gesellschaft, Gemeinschaft", esp. 803–805.

as well as of larger communities that already know the distinction between ruler and subject.[3]

German authors of the classical period in Weimar (Johann Wolfgang von Goethe, Friedrich Schiller) and, even more, of Romanticism show a different usage: At the end of the 18th century, the differentiation between community and society was determined by the difference between political and personal relations and the political society was usually applied to the state. Friedrich Schleiermacher pointed out the dualism of "mechanical" community and "community of unmediated understanding".[4] This heralds new fusions (and fluctuations) in the usage of the terms, with the German poet Heinrich von Kleist wishing for a societal alliance or rather a bonded society ("Gesellschaftsbund"[5]).

Imaginations about the first forms of social life as well as the concepts of community that are related to systems of family relationships always imply a substantial presupposition of the intimate relations between the people involved. Mind that these interrelations did not necessarily have to be informal; not long ago children in Europe used to address their parents formally. The communal model of sociality can be broadened, it may include the "community" of all members of an ethnic group or followers of a religion. Consequently quasi-family or imaginary forms of cohabitation evolve, like those in monastic communities. The Church embodies a social institution that is being determined by "complexio oppositorum" (Carl Schmitt)[6]; it wants to embody union with God and therefore at the same time 'community' of all human beings.

The medievalist Peter von Moos formulated aptly: "The church is, following common sense, a prototype of institutionality". At the same time it actually does not want to be an institution "but *communion sanctorum*, God given, not founded by humans, a spiritual community incorporating all generations of deceased, contemporary and future Christians until the Last Judgement".[7]

Interestingly even the culturally advanced rule of elite warriors relies on the suggestion of community which is established through the dynastic legitimation of the nobility. A strict borderline between internal and external relations determines this group. The basic model of a highly intimate conjunction on the basis of blood relation, patronage, consecration etc. enables a wide overstretching of

3 Cf. Gottfried Wilhelm LEIBNIZ, "Vom Naturrecht", cit. in: RIEDEL, "Gesellschaft, Gemeinschaft", 809.

4 Cf. Friedrich SCHLEIERMACHER, "Versuch einer Theorie des geselligen Betragens ", cit in ibid., 832.

5 Cf. Heinrich VON KLEIST, "Gewerbefreiheit", cit in ibid., 831.

6 Cf. Carl SCHMITT, *Roman Catholicism and Political Form.*

7 Cf. Peter VON MOOS, "Krise und Kritik der Institutionalität", 301 sq.

the original communal life. The result is an institutional fiction, quite possibly connectable with even more abstract processes of socialization, but nonetheless and exactly for that reason very effective.[8]

Friendly "folksiness" now acts as medium in order to build community, folksiness in the sense of free sympathetic relations between human beings who are familiar with each other. In his Philosophy of Law ("Rechtsphilosophie", 1821) Georg Wilhelm Friedrich Hegel distinguishes the concept of community which is based on love and familiarity – with family as prototype – sharply from the "system of needs" ("System der Bedürfnisse"), the capitalist economic sphere which is based on individual interests, and additionally both phenomena from the constitutional state that abrogate all of those differences.[9]

II

Lewis Henry Morgan's description of "prehistorical society" gained a lot of influence (in the scientific world). Morgan contrived an evolutionary ladder, beginning with earliest ways of cohabitation, passing on to the subdivision of society due to differences in blood relation or gender, until the development of the monogamous family. For the earliest companionships like in Polynesia or Australia he assumed a kind of "communism" that was linked to agricultural production. He also already perceived the complicated systems to classify kinship and recorded and tabulated them as did Claude Lévi-Strauss much later in his studies on Elementary Forms of Family and Totemism.[10]

Morgan's influence on Karl Marx and his historical-philosophical sketches often has been mentioned with the latter having taken over Morgan's assumption of native "communist" manifestations of community has often been described. Japanese poststructuralist philosopher and interpreter of Marx' philosophy Yoshimito Takaaki however explicated very accurately the difference between these two classical authors: In no way Marx adopted Morgan's evolutionary determinism for his own history based model of human evolution.[11] That only was implemented later by doctrinaire Soviet invention of "Marxism-Leninism".

Other scenarios of human origin, like the theory of matriarchy published in 1861 by Johann Jakob Bachofen added to the variations on models of community

8 Cf. Karl-Siegbert Rehberg, *Symbolische Ordnungen.*
9 Cf. Georg Wilhelm Friedrich Hegel, *Elements of the Philosophy of Right*, 807–816.
10 Cf. Lewis H. Morgan, *Ancient Society.*
11 Cf. Manuel Yang, "Specter of the Commons".

in the context of early history of civilization. Bachofen claimed that the archetypal community had been dominated by mothers.[12]

III

The founders of the new academic science of sociology shared this motif of evolution. Hence Émile Durkheim contrasted the early "segmental" societies and their strong, quasi automatic binding force with the highly differentiated industrial society, defined by division of labour, where a communal unity had been destroyed. He presumed a remaining "organic" – that is to say: tied to functional specializations – "solidarity" that always is fragile and susceptible to crises. The binding force of the early societies Durkheim named "mechanical solidarity". Solidarity in this case means "integration". In order to re-stabilise the modern type of society he wanted to draw upon civil religious rituals that originated from totemism of tribal communities.[13]

Max Weber – whose study of social action resulted in a theory of understanding that is to be used as basis for sociological explanations –assumed in his "Basic Sociological Terms" "social action" that is individual and at the same time related to other human beings. In so far social relationships are the subject of Weber's analysis. Coming from this assumption he developed categories to capture larger social unities, like for example the state or the church. Surprisingly after thorough analysis of his first drafts of "interpretive sociology" (Verstehende Soziologie) it transpired that Weber had originally assumed "collective action" (Gemeinschaftshandeln) by primal household units, ethnical communities, the market community, political groups, and – especially for the development of the medieval European city – by "oath-bound communes" (Schwurgemeinschaften), the institutionalization of corporative fraternizations. From these corporative unions emerged more complex political institutional nexus like those built in Florence and Venice in recourse to the ancient Roman Republic.[14]

It is true that communal action is also possible under the conditions of rationally organized polities. But with modernization it obviously suffers a certain debilitation. This can be clearly seen when Weber writes about religious communities with strong belief in redemption and the consecutively accrued genuine "ethics of fraternity" (Brüderlichkeitsethik) that might be inner-worldly

12 Cf. BACHOFEN, *Mother Right*.
13 Cf. Émile DURKHEIM, *The Division of Labour in Society*.
14 Cf. Max WEBER, *Economy and Society*, esp. 3–62 and 901–940.

entrenched as well: In relation to war he still stresses the "pathos" of the sense of community and of "the warriors' dedication and unconditional communal willingness to sacrifice"[15] (as we can see it with suicide attackers these days). Only after including models of political economy Weber arrives at a firm methodologically individualistic base in his sociology.[16] Now he avoids the term "community" (Gemeinschaft) as well as the term "society" (Gesellschaft) by choosing the more dynamic processual terms "communification" (Vergemeinschaftung) and "socialization" or (more appropriately) "societification" (Vergesellschaftung).

IV

The German Ferdinand Tönnies is probably the most important author on our topic on account of his major work "Community and Society" (Gemeinschaft und Gesellschaft), conceived in 1880/1881, elaborated further in 1887 and 1912.[17] This book has often been misunderstood, it is in fact difficult and requires a lot of prerequisites, most of all it actually is misinterpretative.

Remarkable the philosophical background Tönnies imbibed his oeuvre with ancient ideas of "ordo", traditions of natural law, the works of Baruch de Spinoza, Arthur Schopenhauer and Friedrich Nietzsche as well as Karl Marx and Friedrich Engels etc.

Tönnies approach reflects the deep crisis of political order that arose after the destruction of monarchic and clerical legitimism in the French Revolution. The same influence shaped the emergence of the science of sociology in France and Germany and is now imprinted in our current debates about communality and a "civil religious" foundation of democratic autocracy (or rather: self-government).

Resulting from this historical framework emerged the focal topoi of conservative as well as progressive positions, of positions that support the political system as well as of those that wish to break it, all of which are still discussed today (see part VI). Tönnies strongly argued in favor of sociopolitical reforms and the development of social democracy. But his argumentation was based on the deep feeling of insecurity – especially amongst the intellectual elites –, that resulted from the implementation of central state bureaucracy and industrial capitalism.

Tönnies, observing the cultural tendencies and achievements of modernity, was no fatalist and no despiser. Seeing the ambivalent processes of moderniza-

---

15 Cf. Id., "Religious Rejections of the World", 323–359.

16 Cf. Karl-Siegbert Rehberg, "Handeln, Handlung".

17 Cf. Ferdinand Tönnies, *Community and Society*.

tion he deduced the self-commitment to strengthen opposition with willingness to install reforms and to support more evolved forms of communality, that he presumably found in 19th century cooperatives.

Modernization, especially occidental processes of rationalization, ruptured communal and traditional contexts of living; this finding is the self-evident premise of all theorems about the evolution of social life and it is the most important background foil for any typological analysis of capitalist industrialism. This is not to be confused with a real disappearance of early or elementary forms of sociality. There is rather a concurrency as we always live in communal relations as well.

Tönnies used an ideal-typical method in order to compare historical states: He contrasted family based agrarian production and its basal power interrelations with a political and economic world governed by means of contracts. His "theory of society"[18] is (one could say) Marx, this "most profound social philosopher" according to Tönnies, but Marx being read through the lenses of Thomas Hobbes and Arthur Schopenhauer.[19]

This is not the place to report on the multi-layered social, psychic, spatial, legal or economic dichotomies. Tönnies himself got by with the help of schemes like the following:

"Community" allows this string of associations: 1. family life – unity – convictions – people (Volk), 2. village – custom – disposition (Gemüt) – commonwealth (Gemeinwesen), 3. city – religion – conscience – church.

In contrast "society" is explicable with 1. large city – convention – endeavor – "society as such", 2. nation – politics – calculation – state or nation, 3. "cosmopolitical life – public opinion – self-consciousness – republic of scholars".[20]

Conceptions of justice and ideals of community were part of an assumed "primordial communism", a concept Tönnies shared. Differentiation processes later decomposed its social equality. Tönnies' benchmark for the evaluation of practical politics was community spirit, including when discussing ways to abolish national wars. He experienced the power constellations created by the Treaty of Versailles in 1919, including the establishment of the League of Nations, which was understood as institutionalization of the victor. Disapproving this approach he considered chances of pacification of humanity via a "federation of world states" (Weltstaatenbund) or a "world government", and he even pondered possibilities of a European union.[21]

---

18 Ibid., part 1, Nr. 19–40.

19 Ferdinand Tönnies, *Gemeinschaft und Gesellschaft*, XXIV, XXXII sq. and XLVIII.

20 Id., *Community and Society*, part 5, Nr. 7.

21 Ibid., part 3, Nr. 29, part 4, Nr. 5, part 5, Nr. 9.

In spite of his typological method Tönnies' descriptions of community often remain substantialistic in an unchecked way. This can be seen for example in his ideas about gender relations.[22] Following Tönnies' descriptions, the isolated male, whether in a Mediterranean *kafenion*, in a provincial club or in metropolitan trade union or in political party activities, in industry and commerce and in other public or semi-public spheres of activity, could regard himself as stand-alone mover of the world, as existing outside the home, that is: outside of community, just a pride (sometimes a bit lost) social atom – much in the sense of Max Stirner as "The Ego and his Own" (Der Einzige und sein Eigenthum).[23] At the same time Tönnies was a social democratic supporter of reforms who also promoted – in a mildly patriarchic way, one could say – demands by the bourgeois women's movement.

V

In contrast to effects of the industrial revolution at the end of the 19th century romantic ideals of merging and semantics of love, which often were additionally – especially in a 'secular religion' (Säkularreligion[24]) of arts – loaded with religious content contributed to new longings for communitization. This can be seen for example in the German youth movement. Very soon right wing as well as left wing thinkers developed comprehensive concepts of community – reaching as far as the totalitarian ideologies of unity like the communist class community or the fascist national community.

Philosopher Helmuth Plessner – whose concept of "perceived body" (Leib) I presented here last year[25] – opposed such "social radicalism" when he published his writing "Limits of Community" in 1924. Philosophical anthropology, a science whose major authors were Max Scheler, Arnold Gehlen and Plessner himself[26], offered a counter-position against any one-sided argumentation or radicalization. He wrote: "Nature is the enemy of radicalism".[27] Inherent in a human beings personality is not only her or his openness towards others, not only the „urge for disclosure", but also the desire for distance and reserve, much needed by a being

22 This was very clearly highlighted in: Michael T. GREVEN, "Geschlechterpolarität" and Bärbel MEURER, "Die Frau in 'Gemeinschaft und Gesellschaft'".

23 Max STIRNER, *The Ego and his Own*.

24 Cf. Karl-Siegbert REHBERG, "Roma capitale delle arti".

25 Cf. ID., "Self-reference and Sociality".

26 Cf. ID., "'Philosophical Anthropology'".

27 Helmuth PLESSNER, "Grenzen der Gemeinschaft", esp. 14.

that essentially defines itself in terms of external bodily borders.[28] In order to avoid total subjection or the individual to a community Plessner argued in favour of a protection of the personality with the means of "form" and "dignity", of ceremonial and prestige, which would make the individual unassailable. This shows very clearly Plessner's bourgeois-liberal repulse against every kind of radicalism of his time, even against an absolute pacifism, absolute in a way that Max Weber would have called "ethical by attitude" (gesinnungsethisch). He equated the anti-sociality and the pathos of community of "internationalistic humanity-communism" with "nationalistic communism" as he called the National Socialism.[29] In his concerns during the 1920s he put the main focus nonetheless on the possibilities of a socialist revolution and the feared loss of individuality.

VI

In the 1980s communitarians in the USA reformulated the reminder that the freedom of the individual increasingly destroys the conditions for life in urban pluralistic societies of mass consumption – as I would call them – because a society without communal cohesion would lead to a collapse of life together.

Michael Walzer, one of the exponents of communitarianism, applied this criticism even on intellectual achievements, including the scientific approach, although we tend to be very proud of these achievements in our epoch of functional differentiation, beginning with Durkheim and not ending with Niklas Luhmann. Talcott Parsons for example, when analyzing important professional groups such as doctors, scientists, judges, pointed out the suppression of "traditional affectivity" and its replacement by "affective neutrality" that appears to be indispensable in order to deliver peak performance in such professions.[30] Karl Mannheim, the main author of the sociology of knowledge, demanded that the "free-floating" intellectuals who are not bound to any status, analyze different world views and their embeddedness within differing motivations in the light of modernity.[31]

We know the "masters of critique", such as Jean-Paul Sartre and the authors of Critical Theory, whereas Pierre Bourdieu tried to diminish his role and to

28 Cf. ibid., 63.
29 Cf. ibid., 49.
30 Cf. Talcott PARSONS, *The Social System*, esp. 102–105.
31 Cf. Karl MANNHEIM, *Konservatismus*, esp. 25–27; see also ID., "The Problem of the Intelligentsia".

present himself – as consequence of "1968" – simply as "collective intellectual".[32] That definitely was a – maybe coquettish – misunderstanding as his spectacular actions with groups of unemployed, like the occupation of the École Normale Superieur in Rue d'Ulm, gained so much attention by the media basically because this intellectual, being the first sociologist with a chair at the Collège de France, had initiated the occupation.

Walzer argued that in every case the influence of such grand intellectuals was based on a distance from society, including to those that they wanted to provide with a public voice. Appellatively he contrasted their appearance with Jewish prophecy that had always been entrenched *in the* community because the prophets spoke as messengers of their Covenant God Jahveh, not because of any spiritual superiority.[33]

For public consumption and understanding Amitai Etzoni summarized liberal thinking on community in his book "The Golden Rule" and his "Global Communitarian Manifesto".[34]

Robert Bellah also invoked most effectively the threat to liberal individualism by its transformation into ego-centered self-reference.[35] Bellah implicitly was connected to cultural critique of the 20th century that was driven by complaining about the fall of the "great" individuality or on the contrary by celebrating this very fall in depersonalizations of structuralism and postmodern deconstructivism. Politically opposing authors such as Theodor W. Adorno and Arnold Gehlen[36] only saw adjustment where there had been "personality" before. Bellah expresses these thoughts using a citation of Maxim Gorki. According to Gorki the contemporary person is "inwardly used up" and "unstable". I keep a critical distance to generalizations like that; an analytically detached view from outside seems indispensable for sociology – René König even said sociologists had to be a bit like a Jew.[37] Still Gorki's statement that human beings were driven by a "pathologically enhanced sensitivity" evokes the latest races for political correctness, where the caution about the usage of terms - as important as it is - replaces far too often the necessity for action. Just the act of avoiding certain terms seems to grant moral justification and dispenses of any further activities.

---

32 Cf. Pierre BOURDIEU, *Counterfire*.

33 Cf. Michael WALZER, "Der Prophet als Gesellschaftskritiker" and ID., *The Company of Critics*.

34 Cf. Amitai ETZIONI, *The New Golden Rule* and ID., *The Spirit of Community*.

35 Cf. Robert N. BELLAH, *Habits of the Heart*.

36 Cf. Arnold GEHLEN, *Man in the Age of Technology*, esp. chapter 7 (127–141) and chapter 9 (159–166).

37 Cf. René KÖNIG, "Die Juden und die Soziologie", esp. 333, 339 sq. and 342; also Ralf DAHRENDORF, "Soziologie und Nationalsozialismus", esp. 114 sq.

Like other communitarians Bellah essentially does not doubt the importance of an individuality that is able to develop freely, but he opposes the resulting loss of community and common welfare as sign of the times. Dissident-religiously motivated individualism of responsibility from the times of the founding of the United States of America would be more and more replaced by an expressive self-centredness – that can be studied with the role models of successful managers as well as with the booming therapy "culture" – this is the core thesis Bellah and his team illustrated with ideal-typical biographies. Particularly in neoliberal capitalism and with the downfall of the welfare state, the consequences of such an undermining of communality and consequently of society can be seen.

When Alexis de Tocqueville analyzed the "Habits of the Heart" in his report about the "Democracy in America" from 1835/1840 – this title, by the way, is the expression that Bellah used as title for his own book – Tocqueville did this on the foil of a in French Revolution reanimated ancient republicanism. On the American side of the Atlantic he mainly applied those habits to the very important family life and after that to the prevailing parish religiosity and to the importance of local politics and of free, that is non-governmental institutions.[38]

Since Aristotle's times the middle classes were regarded as most important social group representing this phenomenon, people who ask for freedom to develop and reach personal advancement while at the same time distancing themselves from "those below".

US-american major cities in the 1980s were determined by a retreat into private dwellings and personal contacts whose inside world seemed to offer some remaining controllability while public spaces were abandoned and left to anonymous devastation; personal security generally appeared as threatened. In this particular context the claim against a "culture of singulation" is understandable. Bellah opposes these trends of self-isolation with hopes for a re-foundation of the freedom spirit that once founded the US-society. He aimed for a culture of active cohesion and civil religious loyalty to the constitution, aided by new "communities of memory" (Erinnerungsgemeinschaften).[39] The term was coined by Maurice Halbwachs, the French sociologist, murdered by the Germans in the concentration camp Buchenwald. He regarded "collective memory" as crucial requirement for social integration.[40] Since Rousseau's work "Contrat Social" fear emerged, especially in France, in front of possible dissolution of social cohesion due to the victory over the personal reign of monarchs and their support by the Roman Church.

---

38 Cf. Alexis DE TOCQUEVILLE, *Democracy in America*, 287.
39 Cf. BELLAH: *Habits of the Heart*, Chapter 11.
40 Cf. Maurice HALBWACHS, *La mémoire collective* and Heike DELITZ, "Gemeinschaft".

Communitarians did not yet take into consideration the threat of fundamentalism – currently Islamic – as a form of terror by the collective. Notably – and Etzioni explicitly pointed this out[41] – fundamentalist Protestantism gained public importance in the US precisely during the time when the major works of the liberal call for community were published. Obviously progressive as well as conservative powers within the deeply divided American society fought against the *inner* destruction of social coherence after *exterior* threats by various kinds of state totalitarianisms had vanished.

VII

There had been consensus on communities as defining the primal forms of human co-existence. In the same manner the dissolution of communities in the wake of modernisation, accompagnied by an elated individualism, is taken for granted. Self-empowerment of the person was inspired by the renaissance man as a new type of mankind. This model of personality developed due to a 19th century "hysterical renaissance"-thinking – as Thomas Mann called it[42] – that had been influenced by Jacob Burckhardt's broadly received work "The Civilization of the Renaissance in Italy"[43], and by the following excessive increase in relevance in the works of Friedrich Nietzsche. Here bourgeois phantasies about "good old times" were fostered – although especially in the *Quattrocento* communities were of decisive importance.

Fears of losing social bonds lead to drawing scenarios of decline. All this belongs to a repertoire of cultural critique that moans the loss of community. As early as 1884 Karl Marx and Friedrich Engels predicted this loss in their "Communist Manifesto": The victory of bourgeois activism which in the wake of capitalism ruled the world, would lead the downfall of all traditions and communal forms of living; especially the family would disintegrate.[44] Correspondences between such pictures of alienation and hymnal motives of deep intimacy in romantic friendship are evident. Community as early stadium in the development of complex societies became a medium to heal the self-destruction of society. At the same time it served as objective of yearning for something unachievable.

Just recently the term "home" – "Heimat" – comes into vogue again, a term long time tabooed especially in Germany after 1945 because of Hitler's political

41 Cf. Etzioni: *The New Golden Rule*, XVIII.

42 Cf. Thomas Mann, *Betrachtungen eines Unpolitischen*, 586.

43 Cf. Jacob Burckhardt, *The Civilization of the Renaissance in Italy.*

44 Cf. Karl Marx and Friedrich Engels, *Manifesto of the Communist Party.*

crimes – as well as the terms "community" and "nation". Instead of "home" terms for community were in use, similar to the word "Führer" that was replaced by the English expression "leader"; and even in museums the English translation "guide" was used instead of "Führer". "Heimat", often related to the childhood feeling of security, turns out to be indispensable, although it cannot be found in reality, at best it is imaginable. The most natural aspect of communal life becomes unachievable; only thinkable as part of the past that, according to Theodor W. Adorno[45] or Ernst Bloch, assists in the critique of the existing society. In his major work "The Principle of Hope" (Das Prinzip Hoffnung), written 1938 to 1947 in US-american exile and first published in three volumes 1954 to 1959 in the GDR (German Democratic Republic), Bloch states:

> The real genesis is not at its beginning but it is running on empty, and it only begins to start, when society and existence begin to be radical, that means take to its roots. [...] Once [the human] is based in real democracy, without dissolution of the self or alienation, then something arises in the world, that reminds everyone of childhood and that no-one has met yet: Heimat.[46]

At the beginning of the 20th century there existed comparably symbiotic models of communal life in different "dialogue-philosophies" and "meeting" or rather "communication philosophies". Martin Buber for example regarded the "I-you-relationship" as necessary for a community of persons with love being its highest form of realization.[47] In the same context Herbert Marcuse criticizes the reifying domination over humans in capitalism and bureaucracy and develops a counter-project with his work "Eros and Civilization".[48] His political demand of a general "erotic" relationship between humans – not at all limited to sexuality – was picked up and tried out by hippies and other protest cultures on the eve of "1968": "Make love, not war". All of these endeavours focused on healing the society with community.

Such a 'Search of Lost Time' points at communities as precondition for the formation of one's own person. It is not about the disappearance of the person, but about its place in the lives of emancipated people who do not only consider their own personal interests.

Especially in postmodernism and its explanation of a social evolution, that involves plurality, absolute subjective choices and combinations of diverse realms of meaning, we find a lot of different expressions of longing for commu-

---

45 Cf. Theodor W. ADORNO, *Minima Moralia*, aphorisms 1 u. 2.

46 Ernst BLOCH, *The Principle of Hope*, esp. 165–179.

47 Cf. Martin BUBER: *I and Thou*.

48 Cf. Herbert MARCUSE, *Eros and Civilization*.

nity. It is not any longer the 'community in blood and lineage' or the 'community of land and goods' that binds people to each other, but something like a 'community of spirit' or, in simpler terms, of common values und their realization. Paradoxically speaking: "communities without community".[49] The longing reaches for "post-traditional" communities – this resembles communitarian positions – which is no coincidence as we live in a 'post-heroic' society.

Excursus: A side note about the experiences of German re-unification after the end of the Cold War: After the breakdown of the Soviet system, that hardly anyone foresaw, and after the following end of the GDR immediately the loss of community was bemoaned. Beside the shock of omnipresent acceleration this was part of the most important interpretations of the liberating, but also irritating changes of life. The "system" of the old German Federal Republic appeared after 1990 as "society par excellence".

East Germany instead was regarded as a sunk refuge of communal life. Very often so-called communities by distress (Notgemeinschaften) gave comfort in difficult times and social change. For this reason people tended to forget the intensity of control and the anxiousness of the ruling elite which had let to the state socialist control system. Instead they liked to remember house festivities, and forgot the "house books" that were used to register each and every visitor.[50]

In mass consumption societies with highly functional differentiations collective action appears to exist only at the edges of individual existence, maybe only as residue for emotional balance. Another phenomenon of collaborative condensation seems to have vanished as well: the huge actual "masses" that were taken for granted in the age of predominating factory work and endless queues of unemployed in front of welfare institutions. The masses are replaced by "singularization" on a *massive* scale. Mass media produce parallel experiences for millions of "lonely" or, in any case, 'solitary' people. 'Substitute masses' make an appearance only in soccer or baseball arenas.

For a quarter of a century, especially with the *linguistic turn*, cultural theories have been developing descriptions of a world that can be "read" like a "text". Consequently additional concepts recently evolved like the discovery of the "figurative nature" of things – the *iconic turn*. Chris Dercon, the former director of the Tate Modern, recently said in a lecture: "Pictures are more important than words".[51] In our times of a new interest in material objects we can also observe a new form of materialization of social relationships, a revitalization of *ad hoc* communities, at events like *public screening* or facebook mass get-togethers of virtual

49 Cf. Slavoj Žižek, zit. in: Dietmar J. Wetzel, "Gemeinschaft", esp. 54 sq.
50 Cf. Karl-Siegbert Rehberg, *Ost – West*.
51 Lecture on "The Future of Cultural Institutions" by Chris Dercon at the Staatliche Kunstsammlung Dresden on 23rd of October 2015.

"friends". Thus we see: the longing for community is inescapable even when the individual competition intensifies and aggravates.

Translated by Heike Friauf

Bibliography

Adorno, Theodor W., *Minima Moralia. Reflections from Damaged Life* [first German in 1951], transl. by Edmund F. N. Jephcott. London, Verso, 1999.

Bachofen, Mother Right = Bachofen, Johann Jakob, *An English translation of Bachofen's Mutterrecht (Mother Right). A Study of the Religious and Juridical Aspects of Gynecocracy in the Ancient World*, transl. and abridged by David Partenheimer. Lewiston, Mellen, 2003.

Bellah, Robert N., *Habits of the Heart. Individualism and Commitment in American Life*. Berkeley, The Regents of the University of California, 1985.

Bloch, Ernst, *The Principle of Hope*, transl. by Neville Plaice, Stephen Plaice and Paul Knight. Oxford, Blackwell, 1986.

Bourdieu, Pierre, *Counterfire. Against the Tyranny of the Market*, transl. by Chris Turner. London, Verso, 2002.

Buber, Martin, *I and Thou* [first German in 1923], transl. by Ronald Gregor Smith. London et al., Scribner, 2000.

Burckhardt, Jacob, *The Civilization of the Renaissance in Italy* [first German in 1860], transl. by Samuel G. C. Middlemore. London, Phaidon, 1945.

Dahrendorf, Ralf, "Soziologie und Nationalsozialismus". In *Deutsches Geistesleben und Nationalsozialismus¨ Eine Vortragsreihe der Universität Tübingen*, edited by Andreas Flitner. Tübingen, Wunderlich, 1965, 108–124.

Delitz, Heike, "Gemeinschaft". In *Handbuch der politischen Philosophie und Sozialphilosophie*, vol. 1, edited by Stefan Gosepath, Wilfried Hinsch and Beate Rössler. Berlin/New York, De Gruyter, 2008, 376–379.

Durkheim, Émile, *The Division of Labour in Society*, transl. by Wilfred D. Walls. Basingstoke, Palgrave, 2002.

Émile Durkheim, *The Elementary Forms of the Religious Life*, transl. by Joseph W. Swain. New York, Free Press, 1968.

Etzioni, Amitai, *The Spirit of Community. Rights, Responsibilities, and the Communitarian Agenda*. New York, Crown, 1993.

Etzioni, Amitai, *The New Golden Rule. Community and Morality in a Democratic Society*. New York, Basic Books, 1996.

Gehlen, Arnold, *Man in the Age of Technology* [first German 1957], transl. by Patricia Lipscomb with a foreword by Peter L. Berger. New York, Columbia University Press, 1980.

Greven, Michael T., "Geschlechterpolarität und Theorie der Weiblichkeit in 'Gemeinschaft und Gesellschaft' von Tönnies". In *Hundert Jahre "Gemeinschaft und Gesellschaft". Ferdinand Tönnies in der internationalen Diskussion*, edited by Lars Clausen and Carsten Schlüter. Opladen, Leske + Budrich, 1991, 357–374.

Halbwachs, Maurice, *La mémoire collective*. Paris, Presses universitaires de France, 1950.

Hegel, Georg Wilhelm Friedrich, *Elements of the Philosophy of Right*, edited by Allen W. Wood, transl. by Hugh B. Nisbet. Cambridge, Cambridge University Press, 1991.

Kleist, Heinrich von, "Gewerbefreiheit", *Berliner Abendblatt* (1810), 216.
König, René, "Die Juden und die Soziologie". In Id., *Soziologie in Deutschland. Begründer, Verächter, Verfechter*. München, Hanser, 1987, 329–342.
Leibniz, Gottfried Wilhelm, "Vom Naturrecht". In *Leibnitz's Deutsche Schriften*, edited by Gottschalk Eduard Guhrauer, vol. 1. Berlin, Veit, 1838, 414–419.
Mann, Thomas, *Betrachtungen eines Unpolitischen*. Berlin, Fischer, 1919.
Mannheim, Karl, *Konservatismus. Ein Beitrag zur Soziologie des Wissens*, edited by David Kettler, Volker Meja and Nico Stehr (Suhrkamp-Taschenbuch Wissenschaft, 478). Frankfurt a. M., Suhrkamp, 1984.
Mannheim, Karl, "The Problem of the Intelligentsia, an Enquiry into its Past and Present Role". In *Essays on the Sociology of Culture*, ed. by Bryan S. Turner (Routledge Sociology Classics). London/New York, Routledge, 1992, 91–170.
Marcuse, Herbert, *Eros and Civilization. A Philosophical Inquiry into Freud*. Boston, Beacon, 1955.
Marx, Karl and Friedrich Engels, *Manifesto of the Communist Party* [first German in 1848], transl. by Samuel Moore, authorized by Friedrich Engels. London, Reeves, 1888.
Meurer, Bärbel, "Die Frau in 'Gemeinschaft und Gesellschaft'". In *Hundert Jahre „Gemeinschaft und Gesellschaft". Ferdinand Tönnies in der internationalen Diskussion*, edited by Lars Clausen and Carsten Schlüter. Opladen, Leske + Budrich, 1991, 357–391.
Moos, Peter von, "Krise und Kritik der Institutionalität. Die mittelalterliche Kirche als 'Anstalt' und 'Himmelreich auf Erden'". In *Institutionalität und Symbolisierung. Verstetigungen kultureller Ordnungsmuster in Vergangenheit und Gegenwart*, edited by Gert Melville. Köln/Weimar/Wien, Böhlau, 2001, 293–340.
Morgan, Lewis H., *Ancient Society, or, Researches in the Lines of Human Progress from Savagery, through Barbarism to Civilization*. New York, Holt, 1877.
Parsons, Talcott, *The Social System*. New York, Free Press, 1951.
Plessner, Helmuth, "Grenzen der Gemeinschaft. Eine Kritik des sozialen Radikalismus" [first 1924]. In *Gesammelte Schriften*, vol. 5: *Macht und menschliche Natur*, edited by Günter Dux, Odo Marquard and Elisabeth Ströker. Frankfurt a. M., Suhrkamp, 7–133.
Rehberg, Karl-Siegbert, "Ost – West". In *Deutschland – eine gespaltene Gesellschaft*, edited by Stephan Lessenich and Frank Nullmeier. Frankfurt a. M./New York, Campus, 2006, 209–233.
Rehberg, Karl-Siegbert, "Roma capitale delle arti. Transzendenzen und Kunstkonkurrenzen". In *Transzendenz und die Konstitution von Ordnungen*, edited by Hans Vorländer. Berlin, De Gruyter, 2013, 66–93.
Rehberg, Karl-Siegbert, *Symbolische Ordnungen. Beiträge zu einer soziologischen Theorie der Institutionen*, edited by Hans Vorländer. Baden-Baden, Nomos, 2014.
Rehberg, Karl-Siegbert, "'Philosophical Anthropology' as an Interpretation of Human Life Forms". In *Life Configurations*, edited by Gert Melville and Carlos Ruta (Challenges of Life. Essays on Philosophical and Cultural Anthropology, 1). Berlin/Boston, De Gruyter, 2014, 25–45.
Rehberg, Karl-Siegbert, "Handeln, Handlung". In *Max Weber-Handbuch. Leben – Werk – Wirkung*, edited by Hans-Peter Müller and Steffen Sigmund. Stuttgart/Weimar: Metzler 2014, 58–63.
Rehberg, Karl-Siegbert, "Self-reference and Sociality. The Differenciation of 'Perceived Body' and 'Corpus' in Philosophical Anthropology". In *Thinking the Body as a Basis, Provocation and Burden of Life. Studies in Intercultural and Historical Contexts*, edited by Gert Melville

and Carlos Ruta (Challenges of Life. Essays on Philosophical and Cultural Anthropology, 2). Berlin/Boston, De Gruyter, 2015, 19–32.
Riedel, Manfred, "Gesellschaft, Gemeinschaft". In *Geschichtliche Grundbegriffe. Historisches Lexikon zur politisch-sozialen Sprache in Deutschland* [Basic Concepts in History: A Dictionary on Historical Principles of Political and Social Language in Germany], vol. 2, edited by Otto Brunner, Werner Conze and Reinhart Koselleck. Stuttgart, Klett-Cotta, 1975, 801–862.
Rousseau, Jean-Jacques, *Discours sur l'origine et les fondements de l'inégalité parmi les hommes. Discours sur les sciences et les arts*. Paris, Flammarion 1992.
Schleiermacher, Friedrich, "Versuch einer Theorie des geselligen Betragens". In *Ausgewählte Werke*, edited by Otto Braun and Johann Bauer, vol. 2, Leipzig 1913.
Schmitt, Carl, *Roman Catholicism and Political Form* [first in German 1923], transl. by Gary L. Ulmen (Contributions in Political Science, 380). Westport/London, Greenwood, 1996.
Stirner, Max, *The Ego and his Own. The case of the individual against authority*, transl. by Steven T. Byington, edited with annotations and an introduction by James J. Martin. New York, Dover, 1973.
Tocqueville, Alexis de, *Democracy in America*, transl. by George Lawrence and edited by Jacob P. Mayer. New York, Doubleday, 1969.
Tönnies, Ferdinand, *Gemeinschaft und Gesellschaft. Grundbegriffe der reinen Soziologie* [reprint 2nd ed. 1912]. Darmstadt, Wissenschaftliche Buchgesellschaft, 1963.
Tönnies, Ferdinand, *Community and Society (Gemeinschaft und Gesellschaft)*. With a new introduction by John Samples. New Brunswick/London, Transaction, 2004.
Walzer, Michael, *The Company of Critics. Social Criticism and Political Commitment in the Twentieth Century*. New York, Basic Books, 1988.
Walzer, Michael, "Der Prophet als Gesellschaftskritiker". In Id., *Kritik und Gemeinsinn. Drei Wege der Gesellschaftskritik*, transl. by Otto Kallscheuer. Berlin, Rotbuch, 1990, 83–125.
Weber, Max, *Economy and Society. An Outline of interpretative Sociology*, transl. by Guenther Roth and Claus Wittich. Berkeley/Los Angeles/London, University of California Press, 1978.
Weber, Max, "Religious Rejections of the World and Their Directions". In Id., *Essays in Sociology*, transl., edited with an introduction by Hans H. Gerth and Charles Wright Mills. New York, Oxford University Press, 1946 323–359.
Wetzel, Dietmar J., "Gemeinschaft. Vom Unteilbaren des geteilten Miteinanders". In Poststrukturalistische Sozialwissenschaften, edited by Stephan Moebius and Andreas Reckwitz (Suhrkamp-Taschenbuch Wissenschaft, 1869). Frankfurt a. M., Suhrkamp, 2008, 43–57.
Yang, Manuel, "Specter of the Commons. Karl Marx, Lewis Henry Morgan, and Nineteenth Century European Stadialism", *borderlands* 11 (2012) [e-journal: www.borderlands.net.au].

Alejandro Grimson

Heterogeneity, Community and Cultural Configurations

Introduction

We live in impressively heterogeneous "communities", with migration, gender, ethnic, class and aesthetic tensions. It is also obvious that we live in societies with fragmentation, violence and inequality. The question is why to continue thinking and talking of "society" or "community".

If we try to use some classical concepts, like culture, we can generate a misunderstanding. If culture supposes homogeneity and equality between its members, it is easy to view contemporary societies as the opposite of any culture in this sense. Porosity, globalization, hybridization, transnationalism, translocalism, contrasting cosmopolitanism, are they truly the opposite of culture?

The postmodern turn in social anthropology discards units, frames, borders and underlies these "new processes". Apparently, communities, as territorial units, are things of the past. Now, we live with cultural otherness in the same city and the same neighborhoods. We have atomization, social networks and transnational communities.

On the one hand, there are old essentialists who talk about culture as a space without heterogeneity, sometimes including their praise to preserve homogeneity. On the other hand, we have the option to discard culture, because communities have exploded.

How should we conceive social spaces? Can we have networks and flows without frames? There is a general confusion on how to think the relationship between homogeneity and heterogeneity in our contemporary world.

The classical notion of culture in its various definitions has a strong problem, because it overlooks five crucial issues: heterogeneity, power, conflict, historicity and inequality.

Should we preserve a concept that does not recognize these five dimensions? Nevertheless, on the other hand, there is another question: what does culture allow us to think on? It allows us to think on frames of meanings, on borders of practices and rituals, on real or virtual territories, on common languages or common symbolic codes, on horizons of social and political imagination, and so on.

Against essentialism, which takes culture and rejects the idea that within cultures there is conflict and inequalities of power, postmodernism proposes to discard not only culture but also generally class, territories or borders. It assumes heterogeneity, discarding culture. My aim is to reconsider theoretical articulations among culture, heterogeneity, power, conflict, inequality, and historicity. This analysis is necessary to understand contemporary communities.

The aim is not to find an arithmetic formula to indicate amounts of culture and conflict. The point is that conflict is constitutive of culture, and there are no social conflicts without cultural meaning. For this reason I have proposed the concept of cultural configurations. A cultural configuration implies that there is a set of meanings and practices, a common symbolic language, certain limits of social and political imagination, and also heterogeneity, inequality and conflict within a frame.

What is new?

Is migration a new human process? Has our world got a special inequality or heterogeneity? From my point of view, the visibility of migration, difference and exclusion has grown rapidly in the last decades, especially because of the revolution of communication. Europe has a new migration process, but looking at the whole world only one century ago the migration process was from Europe to the Americas, Australia and other places. Migration, in a broad sense, as displacement, is indispensable to understand very old texts, including the Old Testament. One of the Pentateuch books is entitled "Exodus". The history of migration began together with the history of humankind.

My argument is that this visibility changed our view and conceptualization of contemporary heterogeneity, but we need to apply this new way of interpretation to every society, including the past ones. Inequality and heterogeneity are constitutive of all societies.

Marx, Rousseau and others built the idea that in the past or in other parts of the world there are societies without inequality and heterogeneity. I am not discussing the political implications of this kind of imagination in their own worlds. Nevertheless, I think that all ethnographic and historical evidence shows clearly that humankind has never known homogeneous societies, communities or tribes. This is obvious for class societies: slavery, feudalism, capitalism or post-capitalist societies.

There were and there are societies without class or racial classifications. So, to discuss this general thesis it is necessary to consider societies without economic stratification, ethnicity or racial differentiation. Is there any ethnographic or historical evidence that there are or there were societies without differences and inequalities related to sex-gender and age?

The knowledge on the sometimes called "simple societies" is clear: there are not societies without age and gender heterogeneity and inequalities. At the same time, obviously age and gender relationships are different in each social context. There are not societies without power inequalities, but this constitutive feature of any society varies deeply in time and space.

From the contemporary point of view, small societies usually seem homogeneous. However, little and daily things such as birth, hunting, travel, a storm, death, sexuality or food have different meanings for their members. Exceptional events like war, a natural disaster, the arrival of foreigners create tension and differentiation lines.

New ways of social imagination are a crucial dimension. However, empirical relationships are not less relevant. In any society there are human beings without some rights, at least by biological reasons: children have not the same political rights than adults. In any society there is an established differentiation between men and women. Women, men, ancient people, young people, give different meanings to similar phenomena.

Communication

Now, if heterogeneity is constitutive of the community as a whole, this challenges us to rethink theories of communication. The term communication, originally, had no technical specificity, but it meant to share. Communicating was linked to generate community or even communion. Similarly, publishing meant to publicize an event or a story. A number of works of what was denominated as the "new communication"[1] found its anchor on this etymology.

Now, the assumption of heterogeneity implies that the code difference is found in both the generation as well as in the interpretation of symbolic processes. Therefore, two classical forms of defining communication are not necessarily synonyms in the XXI century: to publish or to make public do not always involve putting in common. Publishing, as we can see, can be a part of - and generate – hermeneutics battles.

It is therefore necessary to consider two concepts related to the concept of communication. On the one hand, the idea of *contact*, that is the fact that two individuals or groups do not enter codeshare relationship, either physical or

1 Yves WINKIN, *La nueva comunicación.*

virtual. At least in the initial stage, with full lack of understanding, it is a process of communication. For example, the moment in which Columbus observes Caribbean people and wonders whether they are human or not, as he believes they do not speak any language. This is a moment when there is some contact, as we have seen, but still there is minimal interaction, since one of them do not even know if the others are human. Years later the Aztecs themselves also questioned whether Cortes was human.

History shows that if the situation prolongs contact time, by necessity or desire, growing understanding situations are generated, even if they were situations that led to the extermination or domination of one group by another.

Contact implies the existence of the tiniest imaginable frame we can imagine. Goffman[2] extensively developed the notion of frame, continuing Bateson's distinction[3] between reality and fiction (or game). Bateson holds meta-communicative frames that allow us to properly interpret whether a word or message must be treated or not as part of a game.

In his *Frame Analysis*, Goffman notices a context of rapid changes and the coexistence of disparate frames in specific situations of interaction. Contact is at least one card, a wink of the eye, a ' hi ' in a chat written in a language that may be completely unknown for the recipient. For Goffman, frame originates in a social interaction. There is no frame before the interaction because the zero degree of communication is the contact.

At one extreme is the pure contact is, two people in the same scene who do not understand each other, they know nothing of the language or codes of the other person. On the other hand, the complete understanding between two people or two groups is at the opposite extreme. At least, on a theoretical level, it is necessary to distinguish the understanding as a hermeneutical task of social research that seeks to capture a world -other, from the understanding that arises in the everyday processes of social interaction.

It is possible to rely on the knowledge deployed by the authors who were devoted to clarify the interpretative activity of philosophy, humanities or social sciences[4], for questioning about the interpretative processes that are generated in the most banal or dramatic situations of human beings interacting in their daily lives.

The absolute understanding is an ideal horizon, in the sense that it is empirically nonexistent. It is a perfect commensurability of interpretative code that generates identical understandings, point to point, in every fact, gesture, word or

2 Erving Goffman, *Frame Analysis*.

3 Gregory Bateson, *Pasos hacia una ecología de la mente*, 205–222 sq.

4 E.g. Hans-Georg Gadamer, *Verdad y método*; Paul Ricoeur, *El conflicto de las interpretaciones*.

story. As we know, there is never a complete understanding between two human beings which can be maintained over time. Not even in the most intimate relationships. Differences in gender, age, language, class, path, different subjectivities, suggest that heterogeneity is constitutive of human communication. If we analyze cooperative relationships over time, we can see that the idea of perfect commensurability, point by point, may be a utopia or an illusion, but never a lasting fact.

This contrast between contact and understanding clearly expresses the greatest challenge we have when studying communication processes. That is to say, processes of sharing, wondering when to generate community or communion. The challenge is how to conceptualize heterogeneity. At one extreme, communication presupposed homogeneity when it was assumed that it generates common sense and community. At the other end, communication seems impossible, because the differences constitute insurmountable borders. In fact, communication is a symbolic process of intersection between dissimilar but not immeasurable prospects. Communication requires more than contact, it implies multiple levels of understanding, although the absolute understanding is only an illusion. In certain moments of collective effervescence, in special rituals or celebrations of peculiar intensity, it can be produced what anthropologists call *communitas.*[5] *Communitas* refers to an unstructured state in which all members of a community are equal allowing them to share a common experience. It is about an illusion becoming performative and is lived by multiple subjectivities. This is a limited suture time, where the existence and imagination of the community is celebrated and remanufactured.

Communication occurs in contexts that require something shared, even to disagree. If there is communication there are "language games"[6]. The key point is that even disagreement requires some degree of understanding, something shared. We can only think differently if we assume that the other thinks and understands something, albeit limited, from their point of view. We could not become aware of our disagreements if we did not understand something of what others affirm, say, imagine or feel.

We have thus come to a central point. If in the traditional paradigm there could only be communication where the encoding of the message was identical to the decoding, the kingdom of the alleged homogenous culture was the favorite space of that imagination. We can see, in a sense, the opposite. Social situations where there is full identity code are quite bounded, restricted and simple. They are not useful as a metaphor for highly complex communication processes.

5 Victor TURNER, *Dramas, Fields, and Metaphors.*

6 Ludwig WITTGENSTEIN, *Investigaciones filosóficas*, 40–41.

Rather, in contemporary society it is evident that in all processes of circulation of sense there is a difference of meaning.

If certain theoretical moments that difference was intended as a possibility, we should point out that today the difference is rather constitutive of communication processes.

Generally, interpretative differences are ascribed to the existence of conflicting interests. Power inequalities, which are linked to any of the above dimensions, cause dissimilar perspectives and from there, some of the contrasting interpretations of the same facts are derived. Or even emerge divergences about which are the facts themselves, which could be object of interpretation.

At the same time, it is clear that there are multiple social, macro and micro situations in which differences of meaning cannot be allocated to different interests. Perspectives for interpreting reality are not a mechanical consequence of supposedly objective interests. They are also a result of incorporated common sense, wishes or instituted forms of imagination. When a teacher or a doctor understands the needs of children of middle classes and does not interpret successfully to those from indigenous children, this may be sometimes ill will, but sometimes it shows only the sedimentation of a mode of communication that does not incorporate into its practice the registration of heterogeneity and inequality.

Anyone has known couples who want to continue together but, however, they fail to understand each other. Something similar can be seen in misunderstandings between parents and children. To believe that understanding is a derivation of shared interests would suppose, for example, that if two people undertake a common action and all rules are established there would only be space for a communicative transparency. That perspective cannot but be struck by misunderstandings when it would be expected the same surprise at partnerships involving full understanding by the participants.

Focusing exclusively on the interests is the consequence of a perspective that sees the understanding modes as merely instrumental issues of means-end relationships. However, the constitution of desires and forms of imagination is a highly complex process. There are deposits of perceptual matrixes and common senses which do not keep any instrumental connection with an alleged interest defined instrumentally.

Cultural configurations

Our main obstacle is that when we refer to "culture" the image of homogeneity reappears before us. In a culture, the classic anthropologist would say, people speak the same language, believe in the same gods, eat the same food, share a worldview ... They have everything in common, culture is community.

The difference is constitutive of all social processes, as it was showed by several authors[7]. This implies that the cultural unity of society is simply a political illusion homogenizing with performative capacity. However, certain conceptions which radicalized postmodern postures even claimed that we must give up on any notion of culture, because what lie on earth is simply chaos, fragmentation and chance.

Now, on the one hand, we cannot conceive social processes and processes of movement without a sense of heterogeneity that does not necessarily imply a chaotic dynamics. It is quite evident that most babies born in Japan learn Japanese as their first language and not Aymara. Examples like this, which can be transported to religion or music, imply that there are cultural borders in the contemporary world. Only what is on one side and the other side of a frontier are not homogeneous entities, uniforms inside. What is on one side and the other are different frames of joint heterogeneity.

When we ask ourselves which are the cultural boundaries of the contemporary world, we should not look for homogeneous groups. We must look for groups or societies or movements which establish a frontier of significance in such a way that heterogeneities and conflicts acquire different meanings to both sides of those borders. If we take any country in Latin America or Europe, different cultural settings are addressed. Not because there are national essences that make all equal in every country, but because the differences are processed differently in one context and in the other. At the same time, in many countries there are regions, provinces or communities that are, on another scale, cultural configurations. A town and sometimes also a neighborhood can be thought of as different cultural scale settings.

The "cultural configurations" notions refer to this social spaces where there are: (1) shared languages of conflict (racial, political, class, gender, geographical, etc.); (2) instituted horizons of the possible (genocide, political violence, kind of social protest), (3) sedimented logic of disputes (negotiation, confrontation, destruction, terrorism), (4) and borders of meanings.

Every configuration is a frame, but any frame is not a configuration. Within a configuration there is even a system of possible and impossible frames. "Frame" is a broader concept. We can find different kinds of frames. In the scene of contact between two persons who does not understand the language of the other you have a frame, but there is not a common sense, there is not a shared experience, there is not a historical process and a sedimented logic of the relation. When the Span-

7 Homi BHABHA, *El lugar de la cultura*; Nancy FRASER, "La justicia en la era".

iards arrived to America there was a frame of interaction, and different frames before heterogeneity and inequality of power became in a cultural configuration.

A cultural configuration is a social space in which heterogeneity and inequality are organized in a way of life, languages games, and alterity games.

Within a configuration there is even a system of possible and impossible frames. The setting is historical, whereupon the possible and impossible change, and in this process the configuration itself is transformed.

Cultural configurations are not things that exist in the world, such as mountains or seas, but they are lenses through which we can read certain processes more adequately. Classical anthropology assumed that there are many cultures in the world and that our job is to describe them[8]. The notion of configuration does not imply that there are real configurations in the world, but that there is a heuristic tool, a concept that can be useful in specific cases.

A cultural configuration is a social space in which there are shared languages and codes, instituted horizons as possible, sedimented logics of the conflict. The concept could be applied to a school, to different institutions, aesthetic movements, migrant groups or various territorial spaces. Unlike culture, it always implies the existence of disputes and powers, of heterogeneity and inequalities, and changes.

The main issue is the relationship of the parts and the whole. If we add up each of the points of a painting by a great artist, the result is always much lower than the whole, which was to distribute those same points in a specific way. Therefore, "the whole is greater than the sum of its parts". Because "the whole" is a figure, the figure enables or disables that a point should become a part and the "whole" regulates forms of interaction of the parts. Think of any city. We can distinguish the existence of parties, class, gender differences or generations. These parts interact in a specific way. A cultural configuration has "parts" and defines how a social actor could be a part. Political parties, social movements, ethnic claims, unions, provinces. Obviously, the whole cannot do it all, but we will reconsider this point later.

These settings are not just there, waiting for someone to register them. What is "out there", objectively so to speak, are points, parts, joints. Figures are the ones we, as researchers, can see when observing these realities.

A crucial question is about what is the logic of constitution of the parts, what are the criteria enabling discussions and tensions. We know configurations that give highly variable relevance to gender, generation, class, ethnics, the racial, the political, the territorial, the provincial, the regional.

8 Lila Abu-Lughod, "La interpretación de las culturas", 57–90; Christoph Brumann, "Writing for Culture", 1–41; Alejandro Grimson, *Los límites de la cultura.*

In turn, we know configurations that avoid explicit conflicts and other ones that prevent other conflicts from reaching brutal explanation. One dimension is the intensity and the other is the verbal or physical explanation of conflicts. We could add the negotiation (the “brega” in the Puerto Rican sense) to the epic, seduction extermination, dissent to confrontation.

At the same time, we know of configurations that tend to be structured numerically in two, three or many, as the parts override identification with the whole. That is, there are historical formations in which the parts tend to the dichotomization, and others in which languages of shifting alliances may turn to get structured. These variations result in a constitutive heterogeneity that is not chaotic because it tends to organize themselves into historical contexts and specific power relations.

When transnational processes are discussed, it often seems that the *transnational* would be *trans contextual*. However, there is nothing human outside the contexts. The new layer is a transnational context which is entangled with other layers of unpredictable and changing modes. If we apply this to migration processes and the transnational outside “trans contextual” we could think of a being-in-the-world Bolivian, Mexican, Paraguayan, Brazilian, Mapuche, Chilote, Mixtec, Aymara, which would have the following peculiarity: their cultural history, categories of identification, language, their ways of perception, would move from one world to another, which would result in a being-outside-the-world. A trans contextual essence. But if the transnational is a constitutive dimension of all these contexts, these inhabited worlds are multiple and interconnected at the same time.

Classifications

Homogeneous societies are only possible if there are spaces with only one meaning for each symbol, one sense for each word. One meaning for each gesture, without tensions, without conflicts. The question is not if there are social contexts without inequality and heterogeneity, because there are not. The question is their specific features in each social context, in each community, in each society.

Also, the question is not if slavery or capitalism is only a variation in inequality, because they are not. The question is why inequality is legitimate and tolerated in each specific place.

Obviously, to understand this it is necessary to move to a methodological relativism. It is not an ethical relativism, but a way to understand if there is a common perspective and rationality on contextualized inequalities. A not egalitarian consensus could be, in a second possibility, also the result of the convergence of different rationalities. In this case there is a contingent and less stabilized suture. Although it crosses stresses, it avoids major conflicts. The third possibility is obviously an open social conflict.

This is the *Chakrabarty problem*: how to localize universal terms like “class”, “race”, or “inequality”. His book[9] begins with the story of when he failed to find social classes in India. He was not looking on social differences, but as all a Marxist generation (or more), he was looking on social classes as Marx has defined them for Europe one century before. His project began from that impossibility: provincializing the Metropolis. There are not universal concepts without a historical context in their meanings and implications. Locating thinking, historicizing concepts, is a great challenge of our times. To contextualize community and equality is extremely difficult.

“Ethnic politics” is one way to speak on politics. But it is common sense in a specific cultural configuration. So it is not possible to tell one from another. In which historical and configurational conditions heterogeneity became ethnic diversity? Or racial difference? In which configurational conditions inequality became racial inequality? Or class inequality? Or territorial inequality?

An example from Argentina. In the political space of Argentina, ethnicity is invisible in politics. So, at a national or urban level, there are very different classificatory features from countries where race or ethnicity are weapons of political conflict. Nevertheless, when doing fieldwork in a large shantytown of Buenos Aires, I have found the political relevance of Paraguayan and Bolivian migrants in local politics. In local disputes, it is impossible to win the political representation of this shantytown without the support of the Paraguayan or Bolivian people. So, there are different visibilities and relevance of ethnicity at different levels of analysis in the same country.

What is society and what is community? The are levels of analysis. In each country, city, or level there are different ways in which the puzzle works. Different pieces, different articulation between pieces and different figures. On the other hand, they are claims for identity. Sometimes, the simple assertion of a right derives from belonging to a certain community. Sometimes, the fundamentalist claim of superiority includes the right to destroy the rest of the humankind.

As a level of analysis, cultural configurations could be neighborhoods, cities, provinces, countries, transnational movements, and so on. In other words, there are frames where heterogeneity and inequality have different meaning. Within those frames there is a certain historical logic of relation between parts, between social space categories, between groups and persons.

---

9 Dipesh Chakrabarty, *Al margen de Europa*.

Social anthropology demonstrated a long time ago that the notion of person varies between these social frames[10]. In any community, in any social group there exists an identity toolbox, a set of classification categories. There are kinship categories, with gender, descent and alliance differences. There are different kinds of ethnic, class, gender, age, generation, music, dressing, national, regional classification, and so on.

The main ways in which societies divide themselves, classify their parts, vary from one urban, provincial or national context to another. For example, in a qualitative research we asked social leaders in seven cities of Brazil and other seven cities in Argentina how Brazilian/Argentinean society was divided. In both countries, the people interviewed talked about a division between rich and poor people. But while in Brazil they underlie the ethnic and racial division, such concept did not appear in Argentina; in this country they underlie political divisions[11]. So, in each country you do not find homogeneity: you find a language to talk on heterogeneity and inequality. To justify, to celebrate or to question there is a specific language game. Alterity games have different social classifications, terms, and meanings.

The fact that Argentineans do not talk explicitly in racial terms does not imply that race and racism are not problems. The question is if race plays a role in the alterity games, because there are problems that societies fail to recognize.

An example: are there more indigenous people in Brazil or in Argentina? In every place where I did this question people agreed that it was obvious that there were more indigenous people in Brazil. Nevertheless, in the 2010 National Census there were more than 850,000 people with self-identification as indigenous in Brazil, and 950,000 in Argentina. This meant 0.4% in Brazil and 2.4% in Argentina. Why do Argentineans, Brazilians and others have this misperception? Basically, because the national imaginary was built in both countries talking on the "melting pot" of races, but such "races" were defined in extremely different ways. This classic national imaginary says that the Brazilian is the result of the mixture of three races: black, white and indigenous. Argentinean classical national imaginary says that the Argentinean is the result of a mix of several races: Spaniard, Italian, French, German, British, and so on. All of this "races" are European nationalities. In this official story in Argentina "there is not racism because there are not black people". A racist belief... Depending on how we understand the question, it is possible to give two different answers. Brazilians consider indige-

---

10 Clifford GEERTZ, *Conocimiento local*; Marcel MAUSS, "Sobre una categoría del espíritu humano", 303–333.

11 Alejandro GRIMSON, *Pasiones nacionales*; Pablo SEMÁN and Silvina MERENSON, "¿Cómo se dividen brasileños y argentinos?", 189–210.

nous people as their ancestors, and therefore perceive that there are more indigenous people in Brazil. It is, obviously, a reasonable statement. However, there is another way to understand the term “indigenous”: that is why in the Census only 0.4% Brazilians are personally linked with the category.

False cognates

So, in the contemporary world there are borders. Borders do not separate homogeneous cultures, but heterogeneous configurations. A configuration is a frame with a sedimented logic of relation between the parts and the whole.

The problem of “community”, like the problem of “culture”, is that the concept could be useful to analyze one part or the whole. In addition, some parts perceive themselves as a “community” and in some cultural configurations some parts perceive themselves as the representation of the whole. If the national imaginary is defined by whiteness or any other racial feature, any person or group with other features is imaginarily excluded from that definition of “community”. As belonging in the imagination is performative, any border could make an actual exclusion.

How can we localize borders? It is crucial here to distinguish another aspect of cultural configuration and traditional culture concepts. In classic anthropological theory, cultures exist objectively. In the world there is cultural diversity, there are different cultures. Each culture is an entity and the aim of anthropology is to describe each culture. At least during the first half of the twentieth century, anthropologists looked only on non- Western communities, trying to find homogeneity. The problem is, ironically, that anthropologists arrived always after the colonizers and the priests did. So, if they observed objectively the social spaces, they could not see homogeneity without considering inequalities of gender, age or power. The trick, described by Leach[12], was to work on and to analyze the society *as if* it did not have any intercultural contact, *as if* it did not have any historical change or any heterogeneity. This trick has a scientific and political aim: to preserve and know cultures before their western “contamination”. On the one hand, the problem is the anthropologist was imagining a society, a culture, a community, by only observing supposedly non-contaminated areas of life. It is difficult to describe this as “objectivity”. On the other hand, the problem is which part of the history of humankind could be described without referring to intercultural communication, contamination, diffusion and conflicts.

---

12 Edmund Leach, *Sistemas políticos de Alta Birmania*.

It is clear why anthropologists have a strong reaction against this theoretical tradition. However, social life is not pure chaos and fragmentation. Because, as I have just said, there are borders at different levels and scales of analysis. How can we localize borders of cultural configurations? A methodological way to find borders is to find "false cognates", "false friends" and their flows within society and between societies.

Let us take two languages, English and Spanish. There are a lot of similar words in both languages. Some of them have the same meaning. For example, television/televisión. These are true cognates. Nevertheless, other words are very similar, but their meanings are absolutely different.

English word	Spanish word	English meaning of Spanish word	Spanish meaning of English word
Actual	Actual	Current	Verdadero
Advertise	Advertir	Warn	Anunciar
Contest	Contestar	Answer	Concursar
Billion	Billón	Trillion	Mil millones

The simple idea of false cognates is when you think that you understand, when you are certain of it, you have actually misunderstood. My point is that race, class, nation, equality, diversity are all false cognates. In all these cases we suppose that we understand what people are saying when they use these words. But we do not, as they are false cognates. Obviously, we always have the possibility of understanding. However, we first need to consider an unknown language as a challenge, and then we need to learn to understand.

For example, nationalism could be associated with war or xenophobia in some contexts. However, Ghandi was also a nationalist leader in his own context, without war or xenophobia. So, to assume that nationalism has a clear and unique meaning in every context is a usual but terrible mistake.

Race has different meanings in different configurations. Obama was the first Afro-American president in the United States. Nevertheless, people like Obama, with a white parent and a black one, is considered in Brazil a "mulato", not a black or afro person. In Brazil and in the United States there are a large number of people with afro ascendancy. The Brazilian system of classification establishes a great number of intermediate categories between white, black o indigenous. "Mulato" is only one of these. So, this cultural configuration underlies some differences that are invisible in the cultural configurations of the United States, because "race" is associated there with a specific "colour", and "mulato" and "afro" are considered to have the same colour. There are contrasting alterity games. Any white person born in Latin America belongs to the hispanic *race*. This is an extremely peculiar notion of race.

A configuration establishes a border to the false cognate. It is the physical or virtual place where a term, a practice, a ritual, changes the meaning. And other configurations can exist within a given configuration, the space of communication can be studied by the dynamics of false cognates: how, where, when they move, and who moves them.

There is a crucial difference between the border where the false cognate becomes meaningless, where it simply is the end of a meaning regime and the beginning of another, and when the border is found where the false cognate disguises basic social dispute, where differences of meaning do allude to heterogeneity but also to unbalanced power and inequality. In the first case, the false cognate points to border settings. In the second, it underlines the border between the parts within the whole.

There are terms that are self-evident landmarks at first sight, and therefore they do not seem to be false cognates. This is especially true with proper names; as if I point to a country or a person, then that seems obvious and we seem to understand. However, as in any language today, names such as "Che Guevara", "Bush" or "Putin" open a load of heavy divergent connotations.

As shown, to provincialize language is a movement of contextualization. Such contextualization will not be equal to translate uniformly from one language to another. A more complex, but essential challenge is to try to translate from one configuration to another. Thus, each term may not only have a meaning, but in some cases it may be the epicenter of a dispute over meanings. How can we translate to English language terms as "Tango", "folklore", "cabecita negra", "Peronist", "gorila"? What cannot be understood in literal translation, sometimes need a full range of accents and connotations, as there are terms which meanings are accessible only by opening the black box of specific degrees of heterogeneity.

Hegemony

What is hegemony in relation to these notions of cultural configurations? It is the institution of common sense, of a certainty, a framework of interpretation. Hegemony never sits in one place, but it is the social process that produces consensual meanings. One of the main features of the concept of hegemony by Gramsci is linked to its relational character. Hegemony presupposes a link between parts, which can be understood as powerful and subaltern.

One of Gramsci conception is the relational character between power and subaltern actors. If our main question is how subalterns react to power, which their tactics and strategies of resistance are, we will analyze social processes from the point of view of power. There is a problem if we move to hegemonic-centrism, because the subaltern's points of view could be impossible to understand in the language of power. On the other end, we have the theory of the popular culture, absolutely self-generative, which denies the relational feature of identities, power, conflict, and so on.

Today we see that this postulate, indeed crucial, has also posed a problem. The hegemonic-centrism has become one of the major epistemological obstacles to the understanding of cultural hegemony. The main question is how the hegemonic-centered matrix subordinates the reaction against the strategies and actions to the devices of power. The range of options seems wide. Subalterns can naturalize the hegemonic common sense and be subjected to it, they can be seduced by sophisticated frames, they can resist openly, they can stand silently in a restricted area and they may establish local negotiations and other similar variants.

These options are based on the assumption that power is a given thing, against which the subalterns act in a certain way. The power is thus reified, it is an apparatus, an arranged device more than a process of relational assemblages, more than a contingent and situated plot. But to observe the social and cultural processes from the point of view of power, and to wonder how the subalterns react may be a problem. Thanks to Gramsci, it was known that power was not somewhere, and thanks to Foucault, that power was not at one end of the relationship. But a central-hegemonic matrix incorporates processes to account for the "answer" and incorporates the power of "the subaltern" as a readable and intelligible action from a given power.

For sure, we do not need to go the opposite end. A self-generating or an extremely autonomous vision can lose sight of existing relationships, in which the powers are unevenly distributed. In that regard, the conceptualization of the subaltern processes must be distinguished as if they were purely self-constituted on the one hand and as if they were "in relation to" a reified power, on the other. There are theoretical intermediate points which can be discussed, but the analysis can also be approached from empirical research, to rethink the question of which are the borders of hegemony. That is to say, whether the subaltern can speak and how. In a vision where a space for dialogue is already regimented, a voice can only be pronounced in a place of hegemonically constituted enunciation. In that sense, the audibility of voices would be a guarantee of non-subalternity. On the one hand, in the most dichotomous pattern, the subaltern always speaks truly when they are counter-hegemonic. Besides, in the other theoretical end the subaltern cannot speak, because they can only take the word from a place of enabled enunciation and constitute hegemony[13].

However, our societies are full of murmurs, of eventual shouting, divergent melodies, of multiple sounds which no theoretical assertion can mute. Nor any ethnocentric or centric-hegemonic matrix can reduce percussion and harmony, tears and applause, laughter and insult to a line between acceptance and rejection of power. A great deal of the social and the cultural happens outside the field of vision of that centrism, and also with irreducible meanings of it.

We must ask ourselves which are the ways in which different social groups perceive themselves, conceive their otherness, imagine and practice relationships. The groups, therefore, should not be analyzed in relation to hegemony as if it were a given thing rather than an open process. To de-center ourselves, our relationships with specific others must be analyzed and be related to the languages we do not understand or do not share or the ones we reject, which is far from being equivalent.

In that sense, in Latin America such an approach not only leads us to question ourselves about confrontations and negotiations. It also leads us to question about the symbolic systems and the cracks, erosion and undercutting in such schemes. And it also leads us to question about the indifference, real or potential, for hegemonic confrontations and the alternative interventions at the border.

Heterogeneity and hegemony are constitutive of the community, they are conditions of communication. Flows of false cognates come across social life. In proposing an anthropological twist about misunderstanding we are postulating that it can be read ethnocentrically in two different ways. On the one hand, as an aberrant interpretation, the direction on which all differential readings would imply a crushing qualifier. But not necessarily as "diversion" or "resistance" as this would imply a hegemonic-centrism where preferential hegemonic reading would be proper and any difference would be read in relation to it.

An anthropological intervention in this regard is expected to be read as an attempt to pluralize the recognition of groups, actors and languages. The proposed run-out certainly involves taking the actually existing, any irreducible heterogeneous hegemony. Not just ruling-out *per se* the possibility of establishing a comprehensive hegemonic language of such heterogeneity. We claim, however, that the ethical and political option should not to be drawn on the basis of turning dumbly to speak languages we do not understand. Or on the basis of translating in a wrong way, relying on our own meanings for the words of those groups that use similar terms to our senses. If these languages are learned, if these voices are properly interpreted, the counter-hegemonic project will no longer be exactly

---

13 Cfr. Gayatri Spivak, "Can the Subaltern Speaks?", 271–313.

what it was before such cosmopolitanism. Far from an offset that will result in a flight to great conflicts, it will understand the different definitions of conflicts or situations, to consider languages from elsewhere and to foster the potential of emancipatory projects.

We should ask about borders of hegemony by looking at the ways in which the actually existing subaltern speaks. Sometimes, they speak within hegemonic borders. And sometimes they speak from the border or moving borders of cultural configurations. My point is that, when social actors play the game of the sedimented hegemony, they are naturalizing a cultural configuration and its common sense. Only when they have a reflexive view on cultural configuration, when they see how their own common sense define their social imagination and their social actions, they can try to change the rules, to change the ways of struggle, to change the alterity games. If they are successful, the hegemony should be transformed, the features of cultural configuration will move up to a new common sense on heterogeneity and inequality.

So, from this perspective, what can community be? From an ethical or political point of view we can define communities from certain values, rules, consensus that makes life possible. From an ethnographic and empirical point of view, the world is full of fragmentation, segregation, violence; extreme inequalities are not the opposite of community. Flows, hybridization, porosity, cosmopolitanism are not the opposite of language. There are different ways of heterogeneity and inequality. However, the cultural configuration community is the limit of the social imagination, of the bounds of possibility; it is the frontier of visibility as it implies a border where false cognates begin. The question is which kind of inequalities a society can tolerate, understand, legitimate, reject, and hate. On the other hand, which kind of equalities a society does tolerate. For example, child malnutrition could be unacceptable for the same cultural configuration that accepts other remarkable inequalities. Different classical inequalities, like slavery or male domination, have moved to the intolerable level. In that process, within a cultural configuration, there is not a homogeneous time, but heterogeneous temporalities. When did inequality between white and afro people end in the United States? Did it end with slavery, with the civil rights? Will they come to and end when no afro young man could die in the hands of the police because of his "race"? The question is a trap, because inequalities do not end from one day to the other. Equality is a historical process, a disputed one.

Equality, a huge term used for more than two centuries, has multiple meanings within multiple frames. Challenging hegemony and common sense implies, necessarily, to challenge the sedimented meaning and the social relations existing in them.

Bibliography

Abu-Lughod, Lila, "La interpretación de las culturas después de la televisión", *Etnografías contemporáneas* 1 (2005), 57–90.
Bateson, Gregory, *Pasos hacia una ecología de la mente*. Buenos Aires, Ediciones Carlos Lohlé, 1976.
Bhabha, Homi, *El lugar de la cultura*. Buenos Aires, Manantial, 2002.
Brumann, Christoph et al., "Writing for Culture: Why a Successful Concept Should Not Be Discarded", "Comments" and "Reply", *Current Anthropology* 40 (1999), 1–41.
Chakrabarty, Dipesh, *Al margen de Europa*. Madrid, Tusquets, 2008.
Fraser, Nancy, "La justicia en la era de las 'políticas de identidad': redistribución, reconocimiento y participación", *Apuntes de Investigación* 2/3 (1998), 21–43.
Gadamer, Hans-Georg, *Verdad y método*. Salamanca, Sígueme, 1984.
Geertz, Clifford, *Conocimiento local. Ensayos sobre la interpretación de las culturas*. Barcelona/Buenos Aires/México, Paidós, 1994.
Goffman, Erving, *Frame Analysis. Los marcos de la experiencia*. Madrid, CIS and Siglo XXI, 2006.
Grimson, Alejandro, *Pasiones nacionales*. Buenos Aires, Edhasa, 2007.
Grimson, Alejandro, *Los límites de la cultura. Críticas de las teorías de la identidad*. Buenos Aires, Siglo XXI, 2011.
Leach, Edmund, *Sistemas políticos de Alta Birmania*. Barcelona, Anagrama, 1977.
Mauss, Marcel, "Sobre una categoría del espíritu humano: la noción de persona y la noción de 'yo'". In *Sociología y antropología*, edited by Marcel Mauss. Madrid, Tecnos, 1991, 303–333.
Ricoeur, Paul, *El conflicto de las interpretaciones. Ensayos de hermenéutica*. Buenos Aires, Fondo de Cultura Económica, 2008.
Semán, Pablo and Silvina Merenson, "¿Cómo se dividen brasileños y argentinos?". In *Pasiones nacionales* edited by Alejandro Grimson. Buenos Aires, Edhasa, 2007, 189–210.
Spivak, Gayatri, "Can the Subaltern Speaks?". In *Marxism and the Interpretation of Culture*, edited by Lawrence Grossberg and Cary Nelson. Basingstoke, Macmillan, 1988, 271–313.
Turner, Victor, *Dramas, Fields, and Metaphors*. Ithaca/London, Cornell University Press, 1975.
Winkin, Yves, *La nueva comunicación*. Barcelona, Kairós, 1987.
Wittgenstein, Ludwig, *Investigaciones filosóficas*. Barcelona, Crítica, 1988.

Santiago Gonzalez Casares

Community and Eventfulness

The community *happens*, the communal phenomenon can only be understood in terms of its eventfulness. The organized community appears as an authentic possibility for man to realize himself, offering meaning and a promise of a just and legitimate existence. The notion of community has accompanied all forms of human social organization throughout History; it precedes individual existence and gives purpose to every human endeavor. Nevertheless, it can deceive and disperse all possibility of a genuine subjectivity and dilute human determination in a neutral and undetermined mass (*Das Man*). Community can elevate the lonely and isolated individual subject to a collective reality in which he can: fall in love, work, militate (inspire), play – Create. But in a community, man can also hate, murder, refuse - Destroy. What *does* constitute a community? How does it come about? Who is responsible for it? Do all gatherings of human beings constitute a community? We will utilize the phenomenological method to address some of these questions while attempting to establish, at the same time, a definite presence of the problem throughout the efforts of phenomenology as a whole.

The problem of community has haunted the efforts of the phenomenological method and conceptual undertakings since its birth. Each and every phenomenological reduction addressed, sooner than later, a certain form of the communal phenomenon. Some of these efforts were clearer than others, some more developed. All of them agreed upon the fact that, in order for the phenomenon of community to appear phenomenologically, some form of an intersubjective "we" must intervene. In order to facilitate the emergence of a proper and authentic (*Eigentlich*) community, we will utilize the term "Nos-otros" instead of the "we". This term better illustrates the fundamental components of the phenomenon: the *nos* as that which is in common, and *otros* maintains the otherness that makes each one of the *nos* his own man, irreplaceable and irreducible: Unique. In order to maintain a phenomenological reduction functional, both of these elements must remain identifiable. This subjective distinction, or as Jean-Luc Marion would say, this ontical difference (Who?), is the condition of *possibility* for the emergence of a such a thing as a community. It becomes vital in order to achieve phenomenological clarity, in order for the community to appear, to happen. The analysis will then need to describe to the highest extent possible the phenomenological make-up of the *nos* as much as the alterity that constitutes the *otros*. We will then explore the phenomenological attributes of Husserl's transcendental reduction, the ontological analytic of the *Dasein*, Levinas' ethical attempt, and finally Marion's erotic phenomenology. Phenomenology, to a certain extent, is a failed attempt to reach this common phenomenon. From Husserl's intermonadic-cultural community, to Heidegger's existential care of the *Mitdasein*, from Levinas' universal community of just selves to Marion's community of lovers: all of these are attempts to reach intersubjectivity in its phenomenon.

It will be the purpose of this essay to analyze the condition of possibility for the full phenomenological manifestation of the community as a fundamental disposition of the self. The community emerges as an authentic outlook for the reducing subjectivity. It is the opportunity for the individual to become a collective self, thus appropriating its existence through the realization *in*, *with*, *for* and *by* the community, an ethical and erotic realization. It is only through the community that the individual can find meaning beyond the autistic certainty of being, and the self-evidence of life. We can structure the history of the phenomenological method through a genealogy of a series of figures of the community, subjectivity constituted *in (as)*, *with*, *for* and *by* the "appropriating we" that gives birth to a community. The communal phenomenon is thus dependent upon its eventfulness, the bond has to be created and re-created ad-aeternum; it will *happen* over and over again. The appropriating eventfulness of community (*Ereignis*) is this constant transformation of each and every one of the subjectivities included in it.

The transcendental *Nosotros*

Let us begin with the first phenomenological attempt to address the selfhood of the communal phenomenon: the transcendental reduction to the inner workings of the constituting subject. In his courses "First Philosophy" Husserl calls his reduction the "philosopher being born into philosophy"[1]. It is the reducing subject who re-directs his gaze from the banality of the surrounding world, through reasoning (thought), to his conscious states. These states are immanent to the operating transcendental subjectivity and belong to its own sphere of consciousness. This "we" (*nosotros*) appears in Husserl's transcendental reduction under the modality of the *alter ego* in the inactual mind states of conscience. It is the first phenomenological attempt to elucidate the stranger *in* me, who is not me but is *as* I am. The "transcendental we" appears as the second term of the first reduction where the "I" becomes another, and both together become "we", first instance of the *inter*: reduction to the stranger and constitution of a "transcendental we". The "philosopher being born into philosophy" is Husserl's initial name for his reduction, that is, the transcendental subject that accomplishes the

1 See HUSSERL, *Philosophie Première*, 3–36.

intersubjective reduction to the "sphere of the own" by the means of the "pairing analogy". It is in this manner that the subject operating the reduction arrives at the *alter*, ultimately reaching the "transcendental we", that would spread out into the whole of the community and *communities*. The question of intersubjectivity in Husserl's phenomenology could be resumed thusly: "We" *in* Me *as* you. The merit of this reduction is to place the problem of the "We" at the center of phenomenological consideration. He understands man as a communal being and attempts to describe it with the utmost effort as to how this occurs. Nevertheless, he does not quite attain the self of the common, nor the "We" in itself. Husserl ends up limitating the alter ego to the caprices of the constituting subject and loosing thusly the irreducible alterity of the other. The "We" looses itself in the mind states of the subject and the selfhood of the community fails to fully manifest. It does not account for the exteriority of the other even if it maintains to a certain extent the selfhood individual subjectivity.

What isn't community

The heideggerian Da-sein awakens the verbality of Being through the phenomenon of the world (*worldliness* of the world) and its diverse modalities of being. The Dasein *opens* a world of *possibilities*, of utensils ready-to-hand disposed in a circumlocution of signs and signals. He is thus ejected in this world that he *comprehends affectingly* (*Verstehen*) through different dispositions (Curiosity, Fear, Ambiguity) as the fundamental anguish (*Angst*) in which the totality of the world falls into meaninglessness and reveals nothingness, opening the possibility for meaningful existence[2].

This ontological reduction (to the *ontological difference* between Being and being) attains in this manner the *Mitdasein* and the community as Solicitude (*Fursorge*), which allows the *Dasein* to reach the other in itself, from my own destiny towards the other's irreplaceable purpose. This expression of the "Nosotros", one-with-the-other, acknowledges the uniqueness of the other, for he too *opens* a world (his own) as I do, lives as I do and dies as I do but not from the same death, each one bears his own. The solicitude (care-for) implies the irreplaceability of each member of the nos-otros, for each one of "us" encounters their own inalienable and destined crossroads.

The interpreter as we could call this ontological reduction reaches an authentic "nosotros" (Mitdasein, Fursorge) and an inauthentic nosotros (*Das Man*).

2 See HEIDEGGER, Martin, *Sein und Zeit*, 167–200.

Community is not an undetermined ensemble of individuals co-existing side by side in everyday life at any time or place. It is hard to imagine such a thing as a community of the living, as Michel Henry would suggest. This notion seems to fall under the universal neutral that, by phenomenological criteria, constitutes Heidegger's *Das Man* in the existential-analytic of the *Da-Sein*, *ens realisimum*, possessor of all the qualities of Being (of "life" in Henryan terms[3]), that which is common to all beings. It seems to suffer the same fate as all ontologies, the indetermination of Being. The poverty of its categories and the horizon of meaninglessness prevent any appropriation of self, neither individual nor collective. In the anonymity of the Common-to-all, all individual and collective selves who would want to know something (any-thing) more about his (their) life (ot only what, but *who* he is), will find him (their)-self(ves) undermined. Heidegger's *Man* crushes all *possibility* and, therefore, the community's capacity to occur, its eventfulness. *Das Man* is incapable of naming; it ignores any possible manifestation of a particular subjectivity by subduing it to ontological indetermination. Every-one is a living *thing*, no *one* is, every-one lives, no community *happens*. For every community transforms the existence that surrounds it into something new, it differs from the totality of living-beings through its *occurrence*, exceptional to any other, from the newness *(Novedad)* of its *happening*. Every community is unique and irreplaceable, original and unprecedented. It would then seem contradictory to define a community's particularity only in reference of its participation to the universality of living beings. It is evident that any community is ultimately a community of life, but that does not tell us anything about that particular community, the characterization of a community of living beings threatens to deny each community its difference and condemn it to anonymity. As Buber would put, it would fall to the "common abyss"[4], where nothing really *happens*.

The ontological reduction leaves us then with this formula: *Nosotros* as Me *with* you. The merit of this reduction is that it does in fact reach the other as such, in its uniqueness and specific modality of being. His otherness remains intact as it appears in its irreducible authenticity, in the openness of his personal and irreplaceable world. There is no possible substitution: each Dasein manually opens his own destiny. Another merit is alerting us, trough the analysis of the *Das Man*, what does not constitute a community. The problem that remains unresolved is the mysterious ontological nature of the other, his ethical manifestation.

3 Michel Henry argues for a "community of life" in Henry, *Phenomenologie Materielle*, 160 sq.
4 Buber, *Between Man and Man*, 38.

Impossible Community

The just-one (Levinas' ethical reduction to the "trace of the other"[5]) reaches an "asymmetrical nosotros" (responsible we) emerging from the original an-archy of the dual relation from one-to-the-other. From the face of the third person (justice), in which all other faces are watching, Levinas insures an "ethical community" founded in the infinite responsibility "for-the-other" of the rapport of the initial sociality (dual).

Nos-otros as Me *for* You.

The just-one is also a dynamic incarnation of the question (*Sache*) for it appears as the instauration of ethics as a first philosophy, beyond all possible ontological explanation. The ontological attempt to reach inter-subjectivity fails, as it remains captive of the avatars of Being and its insufficient categories. Once obtained the phenomenon of the world (being-in-the-world), it is internal functioning, its affective disposition and its openness, the possibility of reaching the otherness of the other becomes clearer. If the husserlian attempt remains at the stadium of intentional conscience, and consequently, the world and otherness are reduced to states of consciousness of the constituting subject's intention; the possibility of encountering the other as such – him also opening a world of possibilities – becomes problematic. Levinas attempts to restructure the phenomenological analysis and description, based upon the "verbality of being" (Heidegger) in order to discover the phenomenon of otherness. In fact, the levinasean reduction to the trace of the other, first attempt to get at the "We" through the otherness of the other, accomplishes itself in the verb. It will not be the ancient Being but a certain otherwise and beyond, that will not be contained by the conscious categories nor by Being's possibilities. This verb beyond and "otherwise than being" comes from a "Pré-dire originaire"[6] that ravages all ontological "dit" (said) to immemorial and an-arquic Time. The imposition of the ethical interpellation of the other's face hinders all subjective operation of an ego as the central pole of intentional aspirations. The face contra-dicts and de-centers the captive subject who inhabits beond and otherwise: "over there" (là-bas). Subjectivity is no longer confined to finitude (the cage marked by the impossibility of all my possibilities,

5 See Levinas, Emmanuel. *La trace de l'autre*, 261–182.
6 Originary pre-saying see Id., *Autrement qu'être ou au-delà de l'essence*, 78–100.

being-toward-death, Da-sein) and will now be obliged to respond to the otherwise than being – the infinite relationship that makes ethics possible.

Infinity escapes all possible ontological descriptions, implies an abyss (*Abgrund*) that cannot be contained in any sort of metaphysical foundation (*cogito, conatus, Hypokeimenon, Vorstellung, Geist, etc.*). Infinity as the an-archic responsibility for-the-other ends up enforcing *substitution*. Substitution implies a hyperbole of responsibility where the subject is the warrant of the other, perhaps to the point of not leaving place for its *own* destination. The Other counter-acts, imposes his action over mine, I am the hostage of his whim and pleasures – even to his murder ("Thou shall not kill"). It is from the other's counter-intention, counter-intuition and counter-experience that the other obliges me to the point of substituting my own actions to the other's faults. The responsibility of the ethical universality of the an-archic Good, makes the subject responsible to the fifth degree: of his own faults, of his own faults in front of the other, of the others faults (in his place), for the other unknown and even for my persecutor. This hostage-situation is an attempt at a definition of subjectivity in which he is neither the beginning nor the end of philosophy. Subjectivity is no longer the *arche* of thought, it gives way to an-archy. Levinas argues that we can indeed surpass the metaphysical notion of *essentia*, through substitution, toward the Good and responsibility.

But ethics does not manage to shed some light on the "Who?" of the We, for the universal face remains in-attainable because of its neutrality and indetermination: all the faces, no faces. *Das Man*.

Community of two (Lovers)

The lover as a phenomenological reduction incarnates the question-for-love (M'aime-t-on?)[7], no-thing matters to him more than the answer to said question, he goes through the road sketched by the just-one but demands now the certainty of knowing "who" is *nosotros*? (De qui s'agit-il?). The lover wants to detach an individual face from the universal responsibility of levinasean ethics ("d'autrui à l'individu") to love.[8] That is, following the modality of thinking-the-other, the lover requires the ontical distinction of demanding to know who to love. He follows in this manner all the requirements of the phenomenological method,

7 It is the initial question of Marion's erotic reduction. See Marion, Jean-Luc. *Le phénomène érotique*, 111–143.

8 "*Who* is it?" Marion attemps to go beyond levinasean ethics, from the ethical universal to the individual, the loved one. See Id., "D'autrui à l'individu".

for he puts forward an "erotical reduction" anchored in a phenomenology of donation. The lover, as he incarnates the erotic question, goes from the ethical response to the erotic *adonné* (given-one). He embodies the question-for-love that precedes all metaphysical, ontological and ethical questioning, "L'amour avant toute chose"[9], love as first philosophy.

Marion analyzes de possibility of substituting levinasean ethics erotic reduction. To love respecting the ontological exigency of leaving-the-other-in-liberty (irreplaceable) as the ontical constraint to know "who?" is the recipient of my love. Therefore being able to surpass the ethical setback of leaving the other undetermined and anonymous. There is no substitution possible, the other is no longer only a face: she is her scent, her eyes, her mouth, her hands and her voice. She is the one and only, irreplaceable; and so am I. I now know my purpose, her love; love is able to *give* me my own *ipseity*, *interior intimo meo* (Augustin)[10].

To reach the erotic phenomenon, Marion must apply his phenomenology of giveness (*donation*), the reduction to the self of its phenomenon – the gift (le Don). Marion distinguishes between the given in the economic ex-change, haunted by reciprocity and the gift, that which is given by giveness in loosing ("à perte"[11]) and without reservations. The gift is always innocent, easily identifiable, and nevertheless of a phenomenological depth without limits. For the gift seems to appear in its own absence, that is, the absence of the given because it is not an object (anonymous commodity) and has no possibility of being ex-changed. It *is* simply *not* there. The gift is of no consequence, it is not like being awarded the Oscar for the gift does not accept any recognition; it is gratuitous, without reason ("sans pourquoi"[12]).

In order to verify the phenomenological coherence of these claims, Marion goes on to analyze the gift by its three components: the giver, the givee and the gift itself (donateur, donataire, don[13]). He will corroborate through phenomenological reduction the possibility of giveness by placing in parenthesis the three instances. Initially, we could do without the giver, that is, givenness can take place without any-one to claim having actually given any-thing. The gift of life after a difficult disease, for example, as well as the gift of freedom after a prolonged prison sentence. Secondly, we can do without the givee (receiver of the

---

9 See Id., *Le phénomène érotique*, 111–143.

10 See Id., *Au lieu de soi*, 29–70.

11 Following love's internal logic in the erotic reduction, all that is given lovingly does not expect anything in return. To give love is to loose all hope of reciprocity. See Id., *Le phénomène érotique*, 133–143.

12 "Without because". There is no reason for loving, love is beyond reason, it follows de principle of insufficient reason, see ibid., 125–133.

13 See Id., *Etant Donné*, 121–124.

gift) for it is him that will re-introduce the gift in the economical exchange (market logic). Giveness can take place absent of any sort of reception whatsoever. If the givee is not present, or even suggested (offering to Mother Earth) it is even more a gift when it remains loyal to its gratuitous nature. Lastly, giveness can take place without an object given (gift) for the gift can often be immaterial. It can be a kiss goodbye to a loved one, even a hand given in marriage. Marion consequently succeeds in reducing the phenomenon to its giveness as well as establishing love as a privileged phenomenon worthy of phenomenological efforts. Love is in its own right the condition of possibility for solidarity and solidarity is what lies at the foundation of any *given* community.

Nos-otros

Community is not impossible, au contraire! It is the actual explanation of human possibility. The nos-otros of a given community always comes before the ego and its reduction; it has the ontological precedence of a "retard orginaire du moi par rapport à moi-même"[14]. Such as ethics and erotics precede ontology and metaphysics, the nos-otros ante-cedes the I and, moreover, constitutes a possibility of ensuring a just and legitimate existence.

But the community depends on the full occurrence of this Nos-otros. In order for the ties of solidarity necessary to allow the instauration of the communal phenomenon, the phenomenon of nos-otros must be well described and understood. No community can *organize* itself happily and fruitfully without being founded in an ethical-erotical root. The nos-otros obtained ethically constraints the reducing ego to obey the nobility of the Other's illeity (*Hauteur d'illéité*). Erotically, it carries out the necessity to know Who? -to-love. In an "organized community" each member (companion) must share something profoundly intimate with each other through the appropriating eventfulness (*Er-eignis*) that characterizes every authentic community. Communal *Er-eignis* implicates and singularizes each companion in his or her "passibility" (*passibilité*). Passible means for Romano that eventfulness always precedes all possibility for the appropriation of subjectivity, all that is possible is previously determined by eventfulness and condemns the subject to an originary passivity, hence its passibility[15].

As the eventfulness of community manifests itself, it automatically names its members in their uniqueness and originality. Each member of the community

---

14 "Originary delay of me to myself". ROMANO, *L'événement et le monde*, 146.
15 See ibid., 134.

is *related* in a sense to one another. There is something that each of us share with each other that at the same time singularizes us and makes us who we are. One can easily supply, when meeting someone new, our affiliation to a certain community (proximity, ideological, aesthetic). This association implies a partial definition of our selves for it says something that each of us chooses and reaffirms each time. The affiliation to a community can never be an imposition for we have to choose it at every moment and every place. We can be born in France and decide to move to Argentina and acquire the Argentinian citizenship.

Nevertheless, at the same time, the happening of a community is unpredictable, irreversible and im-possible (passible). Romano gives the example of the experience of mourning the death of a loved one, in which every person implicated has a distinct reason for grieving, for being there. We can apply the same phenomenological structure to the eventfulness of a soccer team, in which not only each player is implicated, but also has a very specific function in the mechanism and eventual development and outcome of the game. The eventfulness of the actual game exceeds all possible interpretation demanding an infinite hermeneutical analysis as it "saturates"[16] all Kantian categories of experience. Its donation implies a counter-experience that transforms each present companion (team supporter) through a counter-intention and counter-intuition that subverts all possibility of constitution. It is appropriating in that it converts the participating subjectivity into a collective *nosotros*.

Politology does not attain the self of the communal phenomenon, for it lacks method and the theoretical tools to get at the concept of community in itself. As do all the secondary sciences, such as anthropology, ethnology, sociology, etc. It is not about the identification of the members of the community amongst themselves as equals, it is about the realization and appropriation of each individual in the differance[17] that constitutes every community. A community is that which differs, there isn't a chance for a community to survive if such a community is organized through the identification of its participants. There is a rare possibility (if any) of such a community to prolong itself in time for it does not allow the alterity constitutive of the nos-*otros* to express itself, condemning the collective to

16 The eventfulness inherent to any given community defines it as a saturated phenomenon. This expression, coined by Jean-Luc MARION, explains the occurrence of certain phenomena that subvert the Kantian categories of experience. These phenomena overcharge (saturate) with intuition all categories destroying the possibility of detaining a concept. Community as a phenomenon seems to saturate all four categories, quantity, quality, modality and relation, for the subject is de-centered, ex-ceeded and counter-experienced by the collective of the community. See ID., *De Surcroît*.

17 This term coined by Jacques Derrida implies a "Sameness which is not identical", which illustrates quite well the self of the communal phenomenon.

eventual self-annihilation by suffocation. The community is defined by an open structure that embraces de difference for this is what constitutes it. Different *Daseins* (human realities) with distinct hopes and longings are at the core of the communal origin, which is defined by an irreducible exterior. A community, in order to endure time, should remain always an open entity following its ethical and erotic mandates.

It is for this reason that not every grouping of individuals with common characteristics implies a community, it may not impose on others under no circumstances for it must obey the principles that gave birth to it, ethical and erotic givenness. Love cannot be imposed as ethical responsibility and respect interdict any sort of violence or submission of the Other. It is here that Levinas misunderstands the importance of the Hauteur de l'Autre as he tries to *substitute* the actions of the Other for my own.

The community can only manifest itself properly if its members recognize themselves and the other in an ethical-erotic collective creation. Creation implies involvement, as it occurs it carries with it the hands of both the single one (to use Kierkegaard's terms) and the Other, the companion (the co-being sharing the community). There are certain degrees of companionship, higher degrees, such as friends and family, and lower ones, such as the union companion, the supporter, and the militant. Even lower ones that should be analyzed in depth for it is not certain that they truly constitute an accomplished community. For example, the neighbor, with whom we share a territory, certain amounts of time and dialogue, does not necessarily integrate a community. In this case, we can kindly wave hello to the fellow man from across the street and not be involved at all in that action. That is, the bare minimum of my world is involved waving hello as it could having a polite conversation about the weather with the doorman in the elevator. These situations would easily fall into the neutral chitchat where no one says anything about anyone (*Das Man*). The community hurts, you may not remain indifferent in front of its happening. It can never be polite, for its creation involves, at least, a bit of our humanity, we leave something of ours in the communal exchange, something that is no-thing. It is every-thing; at least temporarily, we risk the totality of our world, our worldliness. My companion has also to risk his world and a new world will be created, a common world appertaining to each one of us, we are part of the community as much as we risk our world in it and for it.

The time of the community is Now! The Community creates and is created in the gesture of its creation. It is in this manner that it remains open (*toujours-à-venir).* The gesture of creation always implies its inevitable recreation. As it is *toujours-à-venir*, the communal phenomenon remains open and in need of continual re-creation.

What is "common" in the community is *nos-otros*. It is not life that links us together in the communal phenomenon for life cannot determine a community as it looses its phenomenological privilege in the neutral and universal void of all living beings. It is the fact that the ethical and erotic origin manifests itself only in the appropriating appropriation of the many, a specific "us" that guards a definite "fidelity" (Badiou) to the eventfulness of the community. It could be said that we are diverting the phenomenological privilege to the entity to the detriment of Being, as would Levinas in "Totality and infinity". This would undermine all the efforts of the whole of the said French radical phenomenology (counter-phenomenology). But the fact is that the "We" is

not an entity, it is not a single I, you, nor He, it is a singular collective phenomenon that is not affected by the idiosyncrasies of the individual subject : the "We" does not die as a it is clearly beyond being and life (Henry).

The fact that we are indeed a community of living beings is undeniable, but to certify that does not tell who I am, what my purpose is nor says anything about my possibility (A quoi bon?). In order to get at a meaningful existence for subjectivity, for nor being neither life tell anything about myself nor my companion, it is necessary to establish the collective subjectivity as operating the reduction. We will thusly discover that at the origin of the collective phenomenon, hides eventfulness with all the power of its unpredictability and irreversibility.

Bibliography

Husserl, Edmond. *Philosophie Première (1923–1924). Deuxième partie: Théorie de la Réduction Phénoménologique*. French translation by Arion L. Kelkel, Paris, Presses universitaires de France (Epiméthée), 1972.

Heidegger, Martin, *Sein und Zeit*. Tübingen, Max Niemeyer Verlag, [19]1993.

Henry, Michel, *Phenomenologie Materielle*. Paris, Presses universitaires de France (Epimethee), 1990.

Buber, Martin. *Between Man and Man*. London/New York, Routledge, 2002.

Levinas, Emmanuel, "La trace de l'autre". In Id., *En Découvrant l'existence avec Husserl et Heidegger*. Paris, Vrin, [3]2001, 261–282.

Levinas, Emmanuel, *Autrement qu'être ou au-delà de l'essence*. La Haye, Editions Martinus Nijhoff, 1974.

Marion, Jean-Luc, *Le phénomène érotique. Six méditations*. Paris, Grasset et Fasquelle (Figures), 2003.

Marion, Jean-Luc, *Etant Donné. Essais d'une Phénoménologie de la donation*. Paris, Presses universitaires de France (Épiméthée), 1997.

Marion, Jean-Luc, "D'autrui à l'individu". In Emmanuel Levinas, *Positivité et Transcendance. Suivi de Lévinas et la phénoménologie*. Sous la direction de Jean-Luc Marion. Paris, Presses universitaires de France (Épiméthée), 2000, 288–308

Marion, Jean-Luc, *De Surcroît. Études sur les Phénomènes Saturés*. Paris, Presses universitaires de France (Perspectives critiques), 2001.
Marion, Jean-Luc, *Au lieu de soi. L'approche de Saint Augustin*. Paris, Presses universitaires de France (Epimethée), 2008.
Romano, Claude. *L'événement et le monde*. Paris, Presses universitaires de France, (Épiméthée) 1998.

Ariel Wilkis

Community and Money: An Approach from Moral Sociology

In the 19th century, sociologists presented money in opposition to community bonds. In the classic text by Ferdinand Tönnies "Community and Society"- from 1887, money plays a key role in the connections among "society" in contrast to those based on the affect and morals of the "community." At the time, this narrative showed how social life relies on money in a dynamic that unraveled traditional ties and replaced others such as impersonality, abstraction, rationality and neutrality. Karl Marx and Georg Simmel made important contributions to this narrative, analyzing a type of money that was disengaged from affect or moral ties. During most of the 20th century, sociology maintained this distinction. However, starting in the 1990s, a new generation of sociologists began to reconsider this interpretation, forging a new agenda for sociology, one that focused on the morality and affect inherent to money exchanges.

Today I hope to show how my own work contributes to this new sociological narrative of money. I would like to tell you how the core ideas of the perspective known as the moral sociology of money reveal sociology's ability to reconnect money with community bonds.

The moral sociology of money

My work challenges two of the fundamental paradigms of approaching the moral value of money. One of these defends money, sustaining that access to it favors freedom and empowerment. The second one condemns money, arguing that it serves only to corrode moral ties and convey individualistic values while being used as a source of exploitation. In spite of their apparent contradictions, both paradigms have something in common: they are unilateral visions that do not provide a broad enough approach to the moral value of money. I argue that money does not have an absolute moral value that allows it to be either praised or condemned. Money participates in relationships based on solidarity and competition; it comes into play in the construction of power and domination, and it helps to build both reciprocal obligations as well as egoism. My purpose is to provide a narrative of money to reveal how money forges connections in the social life as people negotiate its shifting and contradictory moral meanings.

Between 2006 and 2012 I carried out an ethnographic study in the most impoverished neighborhoods of Greater Buenos Aires to understand the social uses of money among the poor[1].

There is extensive literature showing that the way the poor relate to money is a critical aspect of social, political and economic dynamics today. Knowledge of these dynamics allows us to interpret the current conditions of integration and subjection of those most disadvantaged in the face of processes like globalization, financialization and neoliberalism. Most of the literature concentrates on only a certain aspect of how the poor relate to money (retail, migration, consumption, debts, etc.) My work proposes looking at money as a way to connect these heterogeneous spheres of social life. According to Wright Mills, knowledge is guided by an imagination that seeks connections where there apparently are none[2]. Money, in my work, embodies this sociological imagination. Through it, we can travel into spheres of social life that are generally analyzed separately such as family, politics, religion and the marketplace. Instead of providing fragments of knowledge, this proposal aims to highlight the continuity among these spheres. In my perspective, money appears as a conceptual and methodological tool to restore a unity to social life that appeared lost.

Inspired by Viviana Zelizer's work[3], I argue that money can be split into puzzle pieces that reveal the myriad opinions and feelings associated with it. Filling in the puzzle of social life required that we examine each piece. Without these different pieces of money, it is not possible to complete the puzzle of individual and collective lives. Only looking at one kind of money would leave the puzzle incomplete. Some of these pieces appear in my fieldwork.

To show my perspective "in action", I will present a few scenes in the life of Mary and her family.

Scenes of Money

Scene 1

Mary is a fifty-eight-year-old woman who lives in Villa Olimpia, a *villa miseria* (slum) in Greater Buenos Aires west of the country's capital city. Mary's own work begins at *La Salada*, an enormous street market that extends along the bank of

1 Ariel WILKIS, *Las sospechas del dinero.*

2 Wright MILLS, *Sociological Imagination.*

3 Viviana ZELIZER, *The Social Meaning of Money.*

El Riachuelo (the most polluted river in the area), where she buys clothes at a low price and then resells them. Sometimes opportunities come looking for Mary. When her sons arrive home from the meat processing factory where they work, they take several pounds of stolen beef out of their bags. Before they change out of their blood-stained clothes, they package the meat in smaller portions; the clients begin ringing the bell shortly after they finish. They negotiate the price for each chunk. Money and meat are exchanged in front of Mary's attentive eyes. Once the deals are over, she demands that her sons share part of the proceeds. 'They know they have to give me the money because I do my share as well!' she says to me. Mary imposes this principle not only on the money that comes from stolen beef, but also on the salaries her sons earn. For this household head, her family's strength and unity depend on this standard of equal distribution. Mary believes that money must be *safeguarded* because that is the only way to take care of her family.

Scene 2

Mary's oldest daughter Sandra and her husband Daniel needed US$ 300 to move to a new house in Villa Olimpia. She asked her uncle Jorge to lend her the money. They reached an agreement with Sandra's uncle to wait until they had saved up the full amount owed before repaying him.

Two weeks after Sandra's uncle had lent them the money, the family got together to celebrate the birthday of one of Mary's other daughters. The dinner ended at around midnight. Mary was cleaning up the kitchen with her daughters when she heard a loud argument coming from the patio, where the family was playing cards. The card game had been interrupted.

Quit cheating.

Who're you calling a cheater? You're the cheater and you always have been.

Jorge was shouting above the others. He was the one who had started the argument. In just a few seconds, he had taken the deck of cards and tossed it in the faces of his opponents. Sandra's husband had remained aloof, not wanting to get involved. But he couldn't hold back when he saw the cards thrown in his wife's face. He challenged Jorge.

Come out onto the street and fight like a man, instead of going after a woman.

What's your problem? Everyone knows you're a deadbeat.

His words pushed Daniel over the edge; he left the house and headed home for his gun. Sandra saw that look in his eye and ran after him. She rushed inside their new house, followed by her siblings. While they all tried to stop Daniel from going outside with the gun, Uncle Jorge, drunk, came up to the house shouting, "I want my money. You're never going to pay me back because you're a deadbeat."

Sandra remembers this dramatic situation all too well. It went on for hours. Her uncle was saying that they would never return the loan; he had humiliated them before their relatives and neighbors. The long night ended at around four o'clock the next morning; several of Sandra's siblings and Daniel managed to come up with the US$300. With the money in hand, they went to the uncle's house. They counted out the money and handed it over—and haven't spoken to him since.

Scene 3

While I was finishing up my fieldwork, a debate was sparked by a press statement made by a politician from one of the opposition parties: "Beneficiaries of welfare plans use the money to buy drugs and gamble." As a result of his statement, a new debate began on a theme that has come up from time to time since the State began to develop a policy of paying the poor for complying with certain requirements (getting their children vaccinated, presenting school attendance certificates).

The last time I visited her, Mary was excited over some very good news. Her grandchildren had begun receiving scholarships from the government; her younger son had also been approved for a government job plan and would begin work at a construction co-op. For Mary and her family, the ability to relax and enjoy life was connected to the welfare money they received, money which was often used to finance things they otherwise could not have afforded.

Scene 4

Mary gets paid for her work as an activist, 'a political salary,' she likes to clarify. In Villa Olimpia, Mary isn't the only one who received money in exchange for her work to expand Luis Salcedo's political career, the local political boss. A lot of local residents receive a political salary, since the higher up the local leader goes, the more people are needed to consolidate this growth. At the same time, more and more financial support is provided by the national government to ensure that Salcedo and his people continue to express their support for the administration's initiatives. For Mary and other residents of the neighborhood involved in politics in exchange for payment, this *militant money* brings its own uncertainties: it is rarely clear how much they will get or when they will be paid. Nevertheless, Mary does what is expected of her for this "job", visiting her neighbors, resolving their problems on behalf of Salcedo and inviting them to rallies and demonstrations. Then she waits.

It would be hard to describe Mary and her family's everyday life without explaining her relationship to multiple pieces of money. How can her suffering, concerns, dreams and hopes be understood without it? By excluding money from this narrative, a portion of Mary's inner world would be relegated and silenced. At the social level, excluding money would also leave important gaps. It would also be hard to understand her relationships with her family, neighborhoods, church, political network and her employers if we do not understand her relationship to money.

Through the story of Mary and her family, money can be seen as a moral test for people and their social ties, and we will see how the hierarchies and decisions of the family are negotiated based on the meaning attributed to money. One example is how Mary demanded that her sons contribute part of the money they get from selling meat taken from the factory to the household budget. They initially refused to hand over the money as they wanted to mark their financial independence. However, Mary managed the household's money to guarantee the family's wellbeing and reinforce her authority as the head of the household. Mary thus depended on receiving payment for her work from the neighborhood's political leader. Her level of uncertainty grew when this *militant* money wasn't paid. This meant that the leader was not acknowledging her efforts and her trust in him diminished. As we saw earlier, Mary's daughter Sandra and her husband had received money *lent* by Mary's brother. The accusation that they would not pay back that debt constituted a public affront. Mary's brother used this *lent* money to question the morality of the borrowers. This in turn led to a violent encounter and two options: pay the debt or resort to a weapon. The couple opted for the first. Another type of money that circulated is *suspicious* money, mainly government welfare. As seen in the press statement by the politician from the opposition, the poor are often accused of using welfare for illegitimate purposes (drugs or gambling). However, Mary and her family used their welfare payments to acquire goods that allowed them to feel that their lives were improving.

Through these scenes and voices, I have put together an account that goes beyond fragmented knowledge and identifies money as a medium to create and express personal and collective life.

Community and pieces of money

Now, I want to highlight how I propose to use the moral sociology of money to rethink the connections between money and community.

Bruno Théret suggests that a community is recognized through a common unit of accounts and payment; a group forms around a quantitative way of calculating value[4]. In keeping with this idea, I propose considering pieces of money as accounting units and moral currency. I would like to suggest that communities not only use money as an accounting and payment unit but also recognize the plurality, competition and hierarchy of different types of money.

I want to show this dynamic at level of the families and at level of the social networks as political and religious groups.

The family's monetary dynamic.

The pieces of money form a unit that allows us to observe and understand the family universe. The people create and recreate the family social order in the monetary sphere, which carries solidarities and conflicts, reinforcing family projects or tearing them apart.

As we have seen Mary's household budget was comprised of heterogeneous pieces of money like *militant* money, money *earned* and *donated*. She managed the family finances, which were an arena for negotiating economic goods and social status. These multiple pieces had to be organized within the set of feelings and perspectives of *safeguarded* money. The power of *safeguarded* money was achieved through concessions, tensions and strategies for protecting it, such as savings.

The money *earned* by her children was a great source of arguments and resistance. When they didn't deliver, the tension in the household grew. When it reached Mary's hands, the money *earned* by her children was transformed into *safeguarded* money in a very specific form. The money *earned* by her children became savings when it was set aside in her daughter's house.

Through money *safeguarded* through savings, the sons' commitments and responsibilities in household finances were objectified and quantified. Ultimately, this was a quantitative indicator of the power of *safeguarded* money over the rest of the pieces that comprised Mary's household income.

The exploration of these heterogeneous *pieces* of money (their hierarchies, their conversions, their negotiations) allows us to understand how solidarity and

4 Bruno Theret, "El tripode de la moneda", 64–84.

conflict intermingle in Mary's family life. The contradictory nature of this monetary unit is based on an income earmarked for the economic survival of the household (Mary's money) and a supplementary income (that of her children). While the former positioned Mary as the breadwinner of the household, the second was used for activities associated with masculine sociability (alcohol, betting, going out). The money *earned* became *safeguarded* through negotiations and in spite of the resistance of certain members. This transformation ultimately served (in Mary's view) to inculcate her sons in the masculine responsibilities of providing for a wife and family.

The monetary competition between social groups.

Sonia was one of the coordinators of the parish church soup kitchen, a Paraguayan woman of around sixty, mother of two and grandmother of five. When Father Suárez cornered Luis Salcedo to demand salaries for his people, Sonia informed him that she did not wish to be on salary. My initial understanding of this refusal was based on Sonia's explanation of the matter. Because of her involvement in existing political networks, she said, she still had some old friends in other wings of the party and thus was already getting a political salary. That was why she didn't want another one. However, more than a personal decision, her refusal revealed a social order based on a monetary hierarchy.

When Sonia refused to take the money, she constructed a sacrificial microrite. *Sacrificed* money indicates the virtues of transcendence above material goods. Sonia avoids *militant* money by affirming that she is part of a social order that will acknowledge her for her *sacrificed* money. By relinquishing this money, she is building her hierarchy within the religious group lead by Father Suárez. This shows that *sacrificed* money works as a critical piece in Father Suárez's group. Sonia and the others always introduced themselves as volunteers, making it clear that they were not being paid for their work. The sociability of *sacrificed* money involves strict regulation. Sonia would say to me, "We're here to serve. This is not a job." The church's collaborators incorporated this perspective and their efforts categorized as "volunteer work."

These disputes between types of money represent conflicts between sources of moral capital. While it was perfectly all right for Sonia to receive a political payment from an old friend from her days in the Peronist party, accepting *militant* money in the religious group would have reduced her prestige. Sonia was not against *militant* money as an abstract concept; she was against it only within the social universe whose set of feelings and perspectives were associated with another piece of the money puzzle. Accepting that money would have meant acknowledging how the balance of power between these two types of money had been altered, or abandoning the distinction that the religious group provided her and jeopardizing her position in the neighborhood.

While circulating, *militant* and *sacrificed* money carry a series of overlapping social and monetary orders. Each piece is indecipherable outside of the monetary hierarchy and at the same time it projects a social hierarchy. Between the two, there is intense competition for the many objects and people involved. The widespread nature of *militant* money met with resistance, because accepting it meant eclipsing the social and monetary hierarchy of the *sacrificed* money.

I would like to emphasize that the use of *sacrificed* money is not exclusive to the religious world. In this regard, the presence of *sacrificed* money in the world of politics, or of *militant* money in the world of religion, speaks of the moral economies that regulate the tension and subordination between the people in these spaces.

Conclusion

The moral sociology of money contradicts the interpretation of money as an "acid" that dissolves community ties. The dynamics of money and their hierarchies, tensions and conversions affect a person's status and power in specific social orders. Beyond the analytical opposition between community and society, the moral sociology of money invites us to understand social life as a continual space where moral hierarchies are created and maintained through money. As Jane Guyer showed in her work *Marginal Gains*[5], in my telling the monetary hierarchies (the ranking of the pieces of money) and the moral hierarchies (people's statuses) appear as connected realities.

From this perspective, it is not conducive to consider money as an independent variable of a process one wishes to explain: its presence alone does not explain its role in social life. If we approach it as an isolated fact, we tend to view it homogenously, as if it always produced the same effects regardless of the context. In contrast, if we consider that its meanings depend on a hierarchy, we are obliged to reconstruct the connections and differences in its use as *pieces*.

The reconstruction of the multiple pieces of money and its dynamics gives a much more comprehensive and less distorted image of social life than the unilateral narratives of the moral value of money.

5 Jane GUYER, *Marginal Gains*.

Bibliography

Guyer, Jane, *Marginal Gains. Monetary Transactions in Atlantic Africa* (Lewis Henry Morgan Lectures, 1997). Chicago, The University of Chicago Press, 2004.
Mills, Wright, *Sociological Imagination*. Oxford, Cambridge University Press, 2000.
Theret, Bruno "El tripode de la moneda: deuda, soberanía y confianza" In *El laberinto de la moneda y las finanzas. La vida social de la economía*, edited by Ariel Wilkis and Alexandre Roig. Buenos Aires, Biblos, 2015, 64–84.
Wilkis, Ariel, *Las sospechas del dinero. Moral y economía en el mundo popular* (Latinoamericana, 13). Buenos Aires, Paidos, 2013.
Zelizer, Viviana, *The Social Meaning of Money: Pin Money, Paychecks, Poor Relief, and Other Currencies*. Nueva Jersey, Princeton University Press, 1994.

Historical Structures

Dardo Scavino

Community and Individual Autonomy: Genealogy of a Challenge

I

In order to deal with the challenge that the community presents to individual autonomy, we will start by recalling a familiar text from the fifth century AD: *The City of God* by St Augustine. The Latin title is actually *De Civitate Dei*, which not only means *About God's City* but also *About God's State*. It might be useful at this stage to remember that the exact title of the book is: *De Civitate Dei contra paganos*. For some obscure reason, the reference to pagans is often neglected and gets deleted from the title, perhaps because the translation of this term is difficult. Everyone knows that a pagan is an "infidel" or a "polytheist", but this meaning was barely known when Saint Augustine wrote his text. Furthermore, if this term managed to prevail it was due to the popularity of Augustine's treatise.

It was one of Augustine's teachers, Gaius Marius Victorinus, who used this term for the first time when referring to those who refused to adopt Jesus Christ's religion. He did so while he was writing a comment on the Letter to the Galatians. When referring to the Greek gentiles, Victorinus wrote: *apud Graecos, id est apud paganos*[1]. The term *pagus* not only referred to one's country or region, the pagan community, as we often recall the tough resistance of the rural regions of the Empire to the missionaries of the new religion. A *Pagus* was also one's folk or tribe, that is, an ethnical group or population sharing the same ancestral customs. As a matter of fact, such expressions as *mi paisano* in Spanish or *mon paysan* in French are still used to refer to a member of one's community. Hence, *pagus* was the opposite of *civitas* because a *paganus*, as opposed to a *civis*, was a person who abode by the customary rules of their community. But this did not necessarily mean that the *paganus* and the *civis* were two different people. It was rather two aspects of the same individual: the member of a community, on the one hand, who complied with the customs and traditions of a community; and the citizen, on the other hand, who abode by the laws of the State.

In this sense, *Pagus* was a usual translation of the Greek *ethnê*, the term used by Saint Paul when referring to Gentiles, that is, those who were neither Jews

1 Caius Marinus VICTORINUS, *In Epistolam Pauli ad Galatas*. In *Opera exegetica*, 105.

nor Christians. A *paganus* was an individual considered to be a member of an ethnic group. That is the reason why Christians never talked about a *Pagus Dei* but a *Civitas Dei*. Christianity brought the good news to the Gentiles, and that good news identified with no particular ethnic tradition. Christianity claimed to be universal.

Having said that, Gaius Marius Victorinus used the term *paganus* with this new meaning while he was commentating on a very precise passage of the Letter to the Galatians. Recalling the context in which this letter was written, between the years AD 47 and 48 Paul had converted the Galatians to Christianity. However, no sooner had the former left the region than some Jewish preachers succeeded in converting them to Abraham's faith. Once he found out what had happened, Paul sent them that letter in which he explained why Jesus's religion should not be mistaken with Judaism, even though Jesus was actually a Jew and the selfsame Messiah expected by the Hebrew people: with his coming this Messiah had abolished all traditions and taboos in force until then, thus liberating his followers from Moses' tutelage and from Judaism itself. Those who observed the rituals of Judaism were called Jews, but they ceased being Jews with the coming of the Messiah whom they themselves had been expecting. Paul had a similar idea regarding the Gentiles although they had not been anticipating a Redeemer.

In order for the Galatian churches to understand that conversion to Judaism meant, from a Christian perspective, a regression, the apostle resorted to a legal concept common to Greek and Roman law:

> My point is this: heirs, as long as they are minors, are no better than slaves, though they are the owners of all the property; but they remain under tutors and trustees until the date set by the father. So with us; while we were minors, we were enslaved to the elemental spirits of the world. But when the fullness of time had come, God sent his Son, born of a woman, born under the law, in order to redeem those who were under the law, so that we might receive adoption as children. (Gl. 4, 1–5)

As it is still the case today, no minor could dispose of their inheritance until they came of age. They depended for this on the decisions made by their tutors and trustees. But once the "fullness of time had come", they were freed from tutelage. In this way, evangelization resembled a minor's emancipation, and the latter was comparable to a slave's liberation, as they both used to live under the *mancipium* of a *pater familias*. Until the promised coming of Christ, Jews had lived, according to Paul, like children under the tutelage of their Father –of their Father and his Torah. However, the Son of God would have come to redeem them from that tutelage that had once been necessary, certainly, but then became outdated. They would be able to dispose of their due inheritance. The past tutelage had thus been replaced by an "adoption" (*huiothesía*) whose aim would no longer be to keep the

Jews as minors but, on the contrary, to emancipate them from all those childish duties and prohibitions. By getting circumcised, observing *Shabbat*, eating *kosher* or submitting themselves to a series of rituals and prohibitions dictated by tradition, the Galatians would never reach redemption. According to Paul, they would be regressing to the condition of minors which, while it lasted, did not differ much from slavery, even if that slave was paradoxically the "owner" of their inheritance. In the same way as Moses had freed the Hebrew people from slavery in Egypt, Christ had emancipated them from Mosaic harshness: the Messiah had come to abolish ritual obligations, replacing them with love for one's neighbor and for God and thus freeing the Jews from the dictates of tradition. Messianic time would be precisely the lapse between their adoption of the Christian faith and emancipation or redemption. That is the reason why, in his epistle to the Corinthians, this apostle once again says that in the end Christ would restore the kingdom of God, once he had "destroyed all dominion, authority and power" (1 Co 15, 24). Therefore, the coming of God's kingdom, what St Augustine would later call *Civitas Dei*, coincided with the emancipation of believers, that is, their coming of age.

When Caius Marius Victorinus referred to pagans, he was suggesting that Greek polytheists were in the same situation as the Jews in Paul's passage: they were still under the tutelage of tradition, and as long as they did not accept to convert, they were like children who refused to become emancipated or come of age. This meant that they preferred to carry on living and thinking under someone else's guidance, *alieni iuris*, instead of by themselves, that is, *sui iuris*. Christians became emancipated by abandoning *ethnê* or *pagus*, which explains a great deal the popularity of *xeniteía* or *peregrinatio* in ancient Christianity. One was supposed to retreat not only materially but above all spiritually from local habits: conversion meant precisely that.

The controversy started by Alasdair Macintyre about the importance of the *agraphoi nomoi* - of customs and customary law - in Aristotle's ethical thinking will not be dealt with here[2]. However, when the Greek philosopher established a distinction between those who governed themselves and those who were governed by others, he did not believe that respect for the *agraphoi nomoi* in a community transformed an individual into either a slave or a minor. Christianity, on the other hand, modified this position: pagans were not exactly slaves in their community but they lived under its tutelage, an in this respect they were in a similar situation to slaves or, more precisely, minors. It is clear that for Saint Paul and his followers, redemption, that is freedom from slavery, then appeared to be closely related to emancipation from communitarian tutelage or respect of the *agraphoi nomoi*.

---

2 Alasdair MacIntyre, *After Virtue*, 2007.

II

There is an episode at the origins of modernity in which both the Letter to the Galatians and *De Civitate Dei contra paganos* acquire vital importance for us all. When the Spanish conquered America they were confronted with the lack of a legal concept to regulate the new colonial domination. A jurist from Salamanca University, Francisco de Vitoria, resorted then to tutelage, a concept of private law, and explained that the relationship between the Spanish and the Indians resembled that between a tutor and a pupil[3]. Vitoria even invoked that very passage by Saint Paul in which he refers to tutelage. Indians were like children, he said, incapable of ruling themselves, and as minors they needed an adult, capable of ruling themselves, who would manage their property until they came of age in Christian education. Colonized peoples were thus associated with pagans and *alieni iuris* individuals according to Roman law. Colonizing peoples, on the other hand, were more like missionaries or Roman tutors, that is, *sui iuris* individuals, adult men, independent and rational.

This concept of colonial tutelage invented by Vitoria would keep its validity until the mid-twentieth century. Reproducing the Spanish jurist's position almost literally, article 22 of the 1919 Covenant of the League of Nations would continue to defend the "sacred mission of colonization" claiming that there still existed "peoples not yet able to stand by themselves under the strenuous conditions of the modern world", whose "well-being" and "development" were part of the "sacred trust of civilization"[4]. This article concluded that "the tutelage of such peoples should be entrusted to advanced nations who by reason of their resources, their experience or their geographical position can best undertake this responsibility, and who are willing to accept it"[5]. Even the 1945 United Nations Charter would still ask its members to assume "responsibilities for the administration of territories whose peoples have not yet attained a full measure of self-governance", recognizing "the principle that the interests of the inhabitants of these territories are paramount" and declaring as a sacred trust that these States were to promote the capacity of such populations "to develop self-government". They would take due account of "the political aspirations" of the peoples, and to assist them in the "progressive development of their free political institutions.[6]" It was not until 1960, after the Bandung Conference and under the pressure of national liberation movements in Asia and Africa, that the UN added an annex recognizing for

---

3 Francisco de VITORIA, *Relección de indios y del derecho de guerra*, 49.

4 *Covenant of the League of Nations*, http://avalon.law.yale.edu/20th_century/leagcov.asp

5 Ibid.

6 *United Nations Charter*, http://www.yale.edu/lawweb/avalon/un/unchart.htm

the first time the colonies' right to full self-determination on condition that these were territories geographically separate and distinct linguistically and culturally from the metropolitan state administering them. And only thirty-two years later another UN resolution required that its member countries respect the languages and traditions of their so-called "minorities".

However, the notion of this "sacred mission" of civilization, of Christian *civitas* or of European imperialism would not have persisted as late as the twentieth century if another Catholic priest, the French Jacques-Bénigne Bossuet, had not written his *Discourse on Universal History* in the late XVII century. Bossuet recalled there that Jesus Christ had put forward to his sons "new, more perfect and pure ideas of virtue" which would pull the peoples away from their barbarian habits and absurd taboos[7]. Universal History was therefore the long process chosen by Divine Providence to convert, and as a consequence, redeem the peoples of the world, that is, to build the *Civitas Dei* which was none other than the Great Universal Empire submitted to the one and only Emperor. This character would be a sort of representative, and equivalent, of God on Earth as he would put an end to the disputes between the peoples and bring the yearned-for long-lasting peace to all of humanity. Bossuet wrote that towards the end of history we would see the birth of "the kingdom of the Son of Man": "All peoples are subjects of this kingdom big and peaceful", he wrote, "the only one whose power will not pass over to another empire"[8]. For this French priest, if Europe had succeeded in prevailing upon the rest of the peoples, it was because God had chosen it to carry out his purpose.

Throughout four centuries the shift made by Francisco de Vitoria from private law to colonial law had massive repercussions upon western thought. For a start, it introduced the idea that peoples were like individuals with a childhood, youth, middle and even old age, and in this sense he associated history and progress. Savage peoples would no longer be compared to animals but to children, to the extent that a "primitive" mentality would end up being confused with childish thinking. Civilized peoples, on the other hand, would be perceived as those who had reached the so-called "age of reason". Primitive peoples would not be able to think by themselves and would always follow tradition and authority. Civilized peoples, instead, would be those who, as Kant put it, did not need anyone's supervision in thinking and were able to make decisions by themselves.

Furthermore, Vitoria's shift also transformed emancipation and self-determination into two main concepts of modern political thought. In the same way as a mature individual can rule themselves, an independent people is capable of dic-

7 Jacques-Bénigne Bossuet, *Discours sur l'Histoire Universelle*, 172.
8 Ibid.

tating its own laws. The difference between major and minor peoples coincides with the distinction between peoples with and without a State. Immature peoples would carry on living under the tutelage of their ancestral laws. Grown-up peoples would deliberate about their rules for living and would create the appropriate institutions for such purpose. An individual from a primitive nation is a *homo alieni iuris* as they do not decide for themselves which rules they want to live under: they follow the habits of their ancestors. On the other hand, an individual from a modern nation is a *homo sui iuris* because they can reflect upon the laws which rule their life and decide whether or not they wish to change them. Or better said in modern terms: minority corresponds with moral heteronomy whereas majority corresponds with autonomy.

"The history of a single world-historical nation", Hegel wrote in his *Philosophy of Right*, "contains the development of its principle from its latent embryonic stage until it blossoms into the self-conscious freedom of ethical life and presses in upon world history"[9]. A nation", he added, "does not begin by being a state. The transition from a family, a horde, a clan, a multitude, etc., to political conditions is the realization of the Idea in the form of that nation"[10]. If a nation does not achieve this formal realization, he continued, if it does not pass its own laws, if it does not achieve self-determination, other nations will never recognize its sovereignty or its independence and it will always be considered as a minor population prone to colonization. Hegel was thus repeating the interpretation that Vitoria had made of Saint Paul: nations without a State are ethnic groups composed of minor pagans, whereas nations with a State are civil societies made up of fully-grown citizens.

In order to explain the causes of this inequality between barbarian and civilized peoples when all humankind was supposed to be equally rational, Hegel resorted to the old Aristotelian distinction between potentiality and actuality: although everybody might have that potential rationality, only a few, as he saw it, would exert it effectively, whereas for everyone else it would appear as a more or less remote possibility. So as to illustrate such distinction, the German philosopher also evoked the concept of the Ages of Man: a "child only has capacities or the actual possibility of reason" but, as he does not put it into practice, "it is just the same as if he had no reason"[11]. Only as from the moment when "what man is at first implicitly becomes explicit", that is, only when rationality shifts from potentiality to actuality, or from what is possible to what is real, we can say that "man has actuality", as he starts living according to reason and no longer needs

9 Georg Wilhelm Friedrich HEGEL, *Philosophy of Rights*, 343.
10 Ibid., 345.
11 Georg Wilhelm Friedrich HEGEL, *Lectures on the history of philosophy*, 21.

someone else to guide him[12]. A man has actuality when he becomes a *homo sui iuris*. According to Hegel, the difference between child and man, or between a minor and an adult, would at the end of the day be the same as "between the Africans and the Asiatics on the one hand, and the Greeks, Romans, and moderns on the other"[13]. Furthermore, this difference is that "the latter know and it is explicit for them, that they are free, but the others are so without knowing that they are, and thus without existing as being free"[14]. Although they are rational and capable of rational actions, reason does not rule their lives yet.

For Hegel, the difference between a child and an adult resembled the gap between sense-perception and reason, and as a result, the evolution of humanity from primitive peoples, bound to the particularities of their senses and feelings - that is, to idols, figures and myths-, to civilized peoples, who deal with general things due to their use of understanding and reason and hence prefer argumentation and concepts. "Man's ends and objects", he said, are "abstract in general affairs", such as "in maintaining his family or performing his business duties", and therefore he "contributes to a great objective organic whole, whose progress he advances and directs", whereas "in the acts of a child only a childish and, indeed, momentary "I" prevails, as he is taken by sensory immediacy and attracted by a multiplicity of different and diverse stimuli[15]. When the youth are not acting "randomly", they are between childhood and adulthood as their "main aim" is their "subjective constitution" or education, a personal goal that transforms them nevertheless into individuals capable of assuming general aims.

For Hegel, the adult man is still an *oikonómos* and a *politês*, an *actor* and a *civis*, the administrator of the home and the city, the one who takes responsibility for public and private matters. The child, on the other hand, is a savage who has not yet managed to rise above his immediate desires. Hegel concludes that a people "is always rooted in history", at some stage of its evolution, progress or maturation. He goes on to explain that "in the same way as an individual is educated within a State" and "in the same way as they are raised as individuals to a general level", through this raising the child becomes a man. "In that same way also", he concludes, "the whole people is educated: its state of childhood or barbarism is thus exchanged for a rational state".

12 Ibid.
13 Ibid., 22
14 Ibid.
15 Ibid., 437.

III

According to Hegel, communities evolve in a similar way to individuals. At either pole of their evolution are savage peoples, without a State, and civilized peoples, with a State. An individual's relationships with both are not the same either. In order to understand this difference, it would be wise to leave Hegel aside for a moment and recall an article published a century later by Cairo magistrate René Maunier. Maunier denounced the supposed mistakes made by certain European journalists and intellectuals on the North African anti-colonialist rebellions. According to him, such nationalist revolutions were neither revolutionary nor nationalist because their nation and revolution, on the contrary, arose from the colonial tutelage they were rebelling against, from that civilizing process they were resisting through ignorance[16].

Tutelage was, to start with, a "revolution" as it had introduced that "profound and radical change" which might have "abrogated or altered" "the aboriginal inhabitants' conceptions and traditions". Such alteration of traditions arose from the institution of a "laic or written law" which ruled "everyone no matter their religion", and without which no modern nation could exist[17]. Hence, colonial tutelage introduced "two true revolutions in aboriginal life" given that "the inhabitants' right used to be religious and customary", traditional or ancestral. Due to the "abolition of traditions" and the subsequent modernization of these regions of the globe, colonial occupation for the first time introduced the idea of a nation. This was understood not only as a group of individuals submitted to the same laws, but also as individuals, that is, independent legal subjects, more precisely, now emancipated from their family, their tribe or their clan.

In tribal or family law individuality was "always submitted to the community" whereas colonial occupation would have fulfilled the mission of "emancipating" the individual, separating them from their "relatives and neighbors", hence embracing "two new feelings which are the true source of yearning for freedom: the individual's independence, the individual's vindication.[18]" According to Maunier, children had been submitted to the patriarchs' political authority until they would be able to replace them, and to their ancestors' moral authority for the rest of their lives. The reason why it was then possible to talk about "emancipation", was that "from now on every individual has become a legal subject" with "rights and duties as individuals". For this reason they were different from

16 René Maunier, "L'autonomisme aux colonies", 25.
17 Ibid., 26.
18 Ibid., 27.

"the father or the chief, who represented or symbolized their group"[19]. Hence, the law of the state freed individuals from patriarchal tutelage. This French magistrate added that even colonial censuses counted individuals and no longer homes or families. This was always resisted with indignation by the chiefs and patriarchs, similar to the leaders of the Jewish resistance to roman censuses. According to them, the true unity, the genuine social cell, was the family or the tribe.

In the same way as a young person became a *cives* when they came of age, for Maunier peoples reached adult age when they no longer were composed of disperse clans and tribes and became *civitates*, which has meant nation states since the nineteenth century. The national *civitas* ended up liberating the individuals from communitarian duties. According to him, that is why a *civitas* was not composed of a multitude of minorities but was a fragmentation of such minorities. This was a widespread point of view in those years in which European national states were formed. As far as Rome is concerned, German thinker Theodor Mommsen celebrated the "necessary transition from cantonal particularism, where the history of all nations begins, to national unity, where the revolution of its progress ends or must end", whereas his countryman Max Weber considered the passage from an ethnic nation to a civil nation – from *ethnos* to *polis* or from *pagus* to *civitas* – as the result of a "rationalization" process due to which national patriotism substituted atavistic tribal or family honour.

Maunier had no prejudices about presenting colonial intervention as apostolic "preaching" which aimed at the "conversion" of all pagans. "Between us french people, since the beginning we have tried to convert locals religiously and morally, politically and socially", propagating "our way of thinking, our feelings" through their instruction, french-frying the locals by teaching them our state of mind"[20]. For this magistrate, Bonaparte's expedition to Egypt had not only been "ethnographic" but also "democratic", since he had decided to "reform and educate the locals" as well as to "indoctrinate the Egyptians so as to transmit our customs to them". Napoleon might not have invaded Egypt only to get to know its *folk* but also to establish a *civitas*. This was thus "the missionary role of dominant nations": "between us", as concluded Maunier, "to colonize is to teach", and for that reason colons are "all missionaries of progress" who even in trivial café talks fulfill the function of "mentors", since they operate a "big change of mentality among the inhabitants" of the colonies. This gives the locals, through their example, "a taste for freedom" or a desire for emancipation of the individuals from the tutelage of retrograde tribes and patriarchal bondages.

---

19 Ibid.
20 Ibid.

Maunier believed that there was no conflict whatsoever between individual and State, because individuals were emancipated from family tutelage, and came of age when they became citizens, due to the substitution by the State of customary traditions for the civil code or the substitution of religious obligations to ancestral customs for citizen subjection to a laic law. Hegel had referred to this emancipation operated by the State in the paragraphs devoted to the emergence of civil society in his *Philosophy of Right*. In his opinion, the civil society or the association of citizens was not composed of an accumulation of families or local communities. It was made up of individuals emancipated from the tutelage of their families or their tribes, that is "private persons whose aim is their own interests" and act accordingly. However, the emancipation from family or tribal tutelage, the passage from minority to majority, from childhood to adulthood, or from being a child to being a citizen, would not be possible without the State mediation, in such a way that an individual's "selfish aim" would only come true thanks to the social aim of the State. In this aspect Hegel relied on the incalculable ambiguity of the German term *Bürger*: *oikonómos* and *politês*, *actor* and *cives*, the one that looks after their own private interests as well as general interest. Hobbes or Locke's "possessive individualism" had not existed, according to him, before the creation of Leviathan, but afterwards. Before the creation of the State, there were no selfish individuals but families and tribes; there was no absence of law or a natural state but customary law; there was no civil society but family or communal solidarity; there was no rational judgment but mythical narration. Hegel thought that the radical difference between the East and the West relied on this contrast.

The end of the 20th century debate between Communitarians and Individualists, or between Particularists and Universalists, would repeat thus the old Augustinian dichotomy between *Civitas Dei* and pagans such as was interpreted by modern colonialism. Partisans of individual moral autonomy such as John Rawls or Jürgen Habermas, would defend the nation state as opposed to ethnical minorities and universalism against community particularism. Communitarians such as Alasdair MacIntyre, Michael Sandel or Amitai Etzioni, would defend ethnical minorities, moral heteronomy and communitarian particularism against illusory individual autonomy and supposed state or imperial universalism. In other words, whereas the former continue to adhere to a secularized version of *Civitas Dei*, the latter vindicate pagans and minorities. For Communitarians we are still like children who live and think under someone else's guidance, and there is no way out of this moral heteronomy: an individual who does not abide by the moral dictates of their community becomes simply amoral. Only communities may aspire to a certain degree of autonomy, that is to say, independence of their values and private practices. Communitarians consider that state or hegemonic cultures ought to respect them. But it would be a mistake for them to relate ethics or morals with individual autonomy, that is, with their emancipation from communitarian tutelage as suggested by Paul's soteriology. For Universalists, on the other hand, to sacralize such particularisms and force individuals not to profane or criticize a minority's values would mean to condemn them to live under the community's tutelage, denying their autonomy and critical thought. Communitarianism would mean, for them, discarding the project of modernity and renouncing to the Kantian concept of man coming of age.

When defending minority cultures, Communitarians implicitly opposes the parallelism set forth by Saint Paul between an individual inserted in a traditional community and a minor under tutelage. That opposition to Saint Paul's tenacious metaphor suggests to what extent Communitarians is a symptom of the crisis of modern colonialism during the second half of the 20th century. Besides, the disagreement between Communitarians and Universalists carries on reproducing a contradiction inherent to modernity: the search for universal human values and the criticism to any universalization, considered as an imperialistic generalization of individual values or an ideological naturalization of historical practices.

Two recent attempts to solve these paradoxes stand out. Explicitly opposing Anglo-Saxon communitarians, Alain Badiou finds universality to be separate from imperialism in a remarkable interpretation of Saint Paul's texts[21]. On the other hand, Toni Negri and Michael Hardt think that universality is ineluctably associated to the project of a world empire, but this new empire is no longer one people's domination over others but the replacement of the different peoples with the planetary cooperation of the multitude[22]. In order to confirm their thesis they suggest a materialistic interpretation of Saint Augustine's *The City of God*. In this way, they both seek to solve the contradiction of morality by returning to its sources.

Translation: Cecilia Beaudoin

Bibliography

Badiou, Alain, *Saint Paul. La fondation de l'universalisme*. Paris, Presses universitaires de France, 1997.

Bossuet, Jacques-Bénigne, *Discours sur l'Histoire Universelle*. Paris, Garnier, 1873.

Hardt, Michael and Antonio Negri, *Empire*. New York, Harvard University Press, 2000.

---

21 Alain Badiou, *Saint Paul. La fondation de l'universalisme*.

22 Michael Hardt and Antonio Negri, *Empire*.

Hegel, Georg Wilhelm Friedrich, *Philosophy of Rights*. London, George Bell, 1896.
Hegel, Georg Wilhelm Friedrich, *Lectures on the history of philosophy*. London, Kegan Paul, 1892.
MacIntyre, Alasdair, *After Virtue. A Study in Moral Theory*. Indiana, University of Notre Dame Press, 2007.
Maunier, René, " L'autonomisme aux colonies. Motifs et objets", *Politique étrangère* 4 (1936), 23–31.
Victorinus, Caius Marius, *In Epistolam Pauli ad Galatas*. In *Opera exegetica*, edited by Franco Gori (Corpus scriptorum ecclesiasticorum Latinorum, 83/2). Vienna, Hoelder-Pichler-Tempsky, 1986.
Vitoria, Francisco de, *Relección de indios y del derecho de guerra*. Madrid, Espasa Calpe, 1928.

Horacio A. Gianneschi

The Contemporary "Divinization" of Individual Human Beings, or the Difficult Community

A Metaphysical View

> τοῦ λόγου δ' ἐόντος ξυνοῦ ζώουσιν οἱ πολλοὶ ὡς ἰδίαν ἔχοντες φρόνησιν (Heraclitus, DK 22 B 2; Sextus Empiricus, *Adversus Mathematicos* VII 133)
>
> πάντων χρημάτων μέτρον ἐστὶν ἄνθρωπος, τῶν μὲν ὄντων ὡς ἔστιν, τῶν δὲ οὐκ ὄντων ὡς οὐκ ἔστιν (Protagoras, DK 80 B 1; Sextus Empiricus, *Adversus Mathematicos* VII 60–61)
>
> ...la sagesse des nations... est fort triste (Jean-Paul Sartre, *L'existentialisme est un humanisme*, 24)

To a great extent, throughout its entire history, philosophy – which coincides with its own history – has understood itself as an attempt to become a universal inquiry, a universal discourse, even a universal body of knowledge, which universality should be understood in two dimensions: the dimension of 'extension', in accordance with which philosophy has aspired to become an inquiry into, a discourse on, a body of knowledge of, reality as a whole (in some cases, without further ado; in others, from specific perspectives), and the dimension of 'depth', in accordance with which philosophy has intended to inquire about, or to acquire knowledge of, the first principles and causes providing the foundations of reality as a whole and, precisely by reason of its universal nature, also forming the foundations of our knowledge thereof and our action thereon. In other words, we can say that, in large measure, philosophy has understood itself as "metaphysics" ("Philosophie ist Metaphysik"[1]), as a questioning which can fit the original expression τὰ μετὰ τὰ φυσικά,[2] at least in a sense which can be etymologically

1 Martin HEIDEGGER, "Das Ende der Philosophie und die Aufgabe des Denkens", in *Zur Sache des Denkens*, 61. Cf. also ID., *Nietzsche II*, 196: "...die *Meta-physik*... prägt... das Wesen der abendländischen Philosophie. *Deren Geschichte ist... Geschichte der Metaphysik*" (emphasized by Heidegger himself).

2 Greek literature makes no mention of μεταφυσική at all, with a late exception, as is the Commentary by SIMPLICIUS (6th century A.D.) on ARISTOTLE's *De Caelo* to make reference to the work we know as ARISTOTLE's *Metaphysics* (cf. SIMPLICIUS, *In Ar. de Caelo*, HEIBERG's edition, 503, 34). This is a problematic mention, though. As can be seen from the critical apparatus in HEIBERG's edition, whereas manuscript A, the oldest one, dating back to the 13th–14th centuries, bears the term Μεταφυσική, in manuscript F, dating back to the 14th century, we can read Μετὰ τὰ φυσικά instead. The fact that manuscript A is older than F explains why, in his 1894 edition, HEIBERG

defended (its original meaning, available documentation in hand, we know it, proves inaccessible to us[3]), to wit: by understanding μετά (+ the corresponding accusative) – as understood, e. g., by A. Schopenhauer – as "behind", τὰ μετά thus meaning "what is behind", as a 'foundation' (perhaps 'hidden')[4] of that

opted for Μεταφυσική – unlike S. KARSTEN (cf. SIMPLICIUS, *In Ar. de Caelo*, KARSTEN's edition, 225 b 19), who had printed Μετὰ τὰ φυσικά in his 1865 edition. However, it seems reasonable to think, in line with Luc BRISSON, that a scholar from the 13th–14th centuries might have been able to correct the traditional title Μετὰ τὰ φυσικά by replacing it with Μεταφυσική by using the Latin translation in just one word as a model: *metaphysica* – imposed as from the 12th century with Jacob of Venice – as there is no definite article in Latin. This gains importance if we consider – in addition to some linguistic difficulties related to the possibility of forming an adjective such as μεταφυσικός-ή-όν in the Greek language: the verb μεταφύομαι, existing in the Ionic dialect and of which testimony is provided by, e. g., Empedocles and Hippocrates, conveys only one concrete meaning, which is "to become from change" or "to grow or become after", whereas the noun μετάφυσις simply does not exist – that SIMPLICIUS employs the expression τὰ μετὰ τὰ φυσικά everywhere in his work in order to refer to the work by ARISTOTLE in question. Cf. Luc BRISSON, "Un si long anonymat", 41–43 (this author, surely by mistake, argues that S. KARSTEN published Μεταφυσική,when, actually – cf. loc. cit. –, he printed Μετὰ τὰ φυσικά).

3 The case of τὰ μετὰ τὰ φυσικά, or of the only word resulting as from the Latinization of such expression, is analogous to what Immanuel KANT argued with regard to the word "absolut" (cf. *Kritik der reinen Vernunft* A 324 = B 380 sq.), to wit: we are in front of an expression which, on the one hand, we cannot seemingly do without in philosophy but which, on the other, we cannot employ safely due to the multiplicity of meanings it encompasses following its long use and abuse throughout its history; we are certainly in front of one of those expressions which, in its original meaning, was appropriate to name something which no other expression available in the same language could exactly convey, but we have lost – seemingly irremediably – such original meaning, which has undoubtedly led to the variety of uses made of the expression. Very important researches were conducted by the middle of the 20th century, especially by Paul MORAUX (*Les listes anciennes des ouvrages d'Aristote*) and Hans REINER ("Die Entstehung des Namens Metaphysik", 210–237), on, among other documents, those by the ancient commentators of ARISTOTLE and the ancient catalogs of ARISTOTLE's works, but the results thereof were no more than reasonable guesswork as to the possible original meaning of the expression and as to the person who might have coined it: Aristo of Ceos? (cf. Paul MORAUX, op. cit., 314), Eudemus of Rhodes? (cf. Hans REINER, art. cit., 235–237). Even so, these researches demolished, in our opinion (cf., however, Ingemar DÜRING – among others, who even knew of such researches –, *Aristoteles*, 591 sq. with n. 38 and 286 sq. with n. 250b), the well-known legend about the merely library-related roots – by the hand of Andronicus of Rhodes, the famous editor of ARISTOTLE's works in the 1st century B.C – of the expression. This legend was first explicitly narrated in detail in 1788 by philosophy historian and philologist Johann G. BUHLE in his work entitled "Ueber die Aechtheit der Metaphysik des Aristoteles" (cf. 41 sq.) and since then it has been regarded as an allegedly scientific truth, reiterated over and over again in handbooks and even in the best encyclopedias of the world and among renowned Aristotelians (cf. Hans REINER, "Die Enstehung der Lehre", 84 sqq.).

4 Cf. Arthur SCHOPENHAUER, *Die Welt als Wille und Vorstellung II*, 212: "Unter *Metaphysik* verstehe ich jede angebliche Erkenntnis, welche über die Möglichkeit der Erfahrung, also über die Natur oder die gegebene Erscheinung der Dinge hinausgeht, um Aufschluß zu erteilen über das,

behind which it is: in this case, precisely τὰ φυσικά, the part of the expression which, based on an ancient meaning of φύσις, we can understand as "reality as a whole", as "'what is' as a whole" (τὸ ὄν, τὰ ὄντα).[5]

The philosophical question, the metaphysical question so understood, i. e., the question about reality as a whole, about the principles or ultimate foundations of reality, first appears 'explicitly' within the framework of what H.-G. Gadamer called the "solid ground (terreno solido)" of philosophy – from the textual viewpoint, of course –, that is, in Plato and Aristotle[6]: τί τὸ ὄν;[7] τί ὄντως ὄν;[8] τί πρώτως ὄν;[9]. This does not prevent us from going back in time and even talking, as H.-G. Gadamer did, about a "prehistory of metaphysics (Vorgeschichte der Metaphysik)".[10] Perhaps we can say that the philosophical, metaphysical inquiry, as understood herein, seems to have come as determined in its "Vorgeschichte", particularly from the – text known to us as – *Περὶ φύσεως* by Parmenides, as an inquiry that cannot be answered but always with modesty, humility (although we often forget this). Indeed, the so called "cosmological" or, even better, "cosmological-cosmogonical-theogonical" part of Parmenides's poem (cf.

---

wodurch jene in einem oder dem andern Sinne bedingt wäre; oder populär zu reden: über das, was hinter der Natur steckt und sie möglich macht."; ibid., 237: "In diesem Sinne also geht die Metaphysik über die Erscheinung, d. i. die Natur hinaus zu dem in oder hinter ihr Verborgenen (τὸ μετὰ τὸ φυσικόν), es jedoch immer nur als das in ihr Erscheinende, nicht aber unabhängig von aller Erscheinung betrachtend: sie bleibt daher immanent und wird nicht transzcendent."; ID., *Die Welt als Wille und Vorstellung I*, 578: "Ich sage daher, daß die Lösung des Rätsels der Welt aus dem Verständnis der Welt selbst hervorgehn muß; daß also die Aufgabe der Metaphysik nicht ist, die Erfahrung, in der die Welt dasteht, zu überfliegen, sondern sie von Grund aus zu verstehn, indem Erfahrung, äußere und innere, allerdings die Hauptquelle aller Erkenntnis ist; daß daher nur durch die gehörige und am rechten Punkt vollzogene Anknüpfung der äußern Erfahrung an die innere und dadurch zustande gebrachte Verbindung dieser zwei so heterogenen Erkenntnisquellen die Lösung des Rätsels der Welt möglich ist; wiewohl auch so nur innerhalb gewisser Schranken, die von unserer endlichen Natur unzertrennlich sind, mithin so, daß wir zum richtigen Verständnis der Welt selbst gelangen, ohne jedoch eine abgeschlossene und alle ferneren Probleme aufhebende Erklärung ihres Daseins zu erreichen."

5 Cf., e. g., ARISTOTLE, *Metaphysics* Γ 3, 1005 a 29–33.

6 Cf. chapter III of lessons on *L'inizio della filosofia occidentale* issued by Hans-Georg GADAMER in 1989 at Istituto Italiano per gli Studi Filosofici (IISF) in Naples and collected by Vittorio DE CESARE for its initial publication in 1993, a chapter entitled, precisely, "Il terreno solido: Platone e Aristotele" (cf. Hans-Georg GADAMER, op. cit., 41–50). Later, those same lessons, with some modifications by the same Gadamer, were published in German under the title *Der Anfang der Philosophie* and the chapter in question was entitled in the same way: "Fester Boden. Platon und Aristoteles" (cf. Hans-Georg GADAMER, op. cit., 43–53).

7 ARISTOTLE, *Metaphysics* Z 1, 1028 b 4.

8 Cf. PLATO, *Republic* VI 490 b; ID., *Sophist* 240 b.

9 Cf. ARISTOTLE, *Metaphysics* Z 1, 1028 a 30; Θ 1, 1045 b 27.

10 Cf. Hans-Georg GADAMER, "Zur Vorgeschichte der Metaphysik", 364–390.

DK 28, from B 8, v. 50, hereinafter), which is, so to speak, his attempt to provide an explanation for the entire φύσις, for the reality as a whole, is, in the poem, in the path of δόξα and not in the path of ἀλήθεια, where tautological truths reign. Philosophy, metaphysics, in its inmost beginnings, is already a "metaphysics of humility"[11], it is already φιλοσοφία (this word carrying its inherent modesty[12]), it is, upfront, a farewell to any possibility of thinking without assuming, a farewell to any alleged "philosophy without assumptions".[13] The issue is, already from the beginning, the philosophical awareness of the 'lack of absoluteness of the starting point', the philosophical awareness that metaphysics and, in general, each and every thought, is and will be inevitably suspended in a 'decision'.[14] It is by

11 Antonio Machado, *Juan de Mairena*, vol. 2, 111.

12 Cf. Friedrich Nietzsche, *Die fröhliche Wissenschaft*, § 351, 588: "Die *Bescheidenheit* war es, welche in Griechenland das Wort 'Philosoph' erfunden hat...".

13 The beginning of *Posterior Analytics* by Aristotle is no other thing, we believe, but the assumption of this initial, prehistoric trait of metaphysics (and of thinking in general): πᾶσα διδασκλία καὶ πᾶσα μάθησις διανοητικὴ ἐκ προϋπαρχούσης γίνεται γνώσεως (I 1, 71 a 1–2; cf., more broadly, the entire opening chapter: 71 a 1 – 71 b 8). Regarding this beginning of *Posterior Analytics*, see, in a similar direction, Elmar Treptow, *Der Zusammenhang*, 42, who also adds: "Nur Hegel würde verneinen, daß das eigentliche Wissen – nicht nur vorläufig, sondern endgültig – an ein vorgegebenes Faktum als seine Grundlage anknüpfen müßte und diesen Anknüpfungspunkt niemals selbst erreichen und begründen könnte" (ibid.).

14 Ultimately, what this is about – if it is assumed that all metaphysics, all ontology, is, through the word "be", a theory of reference – is but what W. V. O. Quine has developed as "the inscrutability of reference" (cf., e. g., Willard. V. O. Quine, *Ontological Relativity*, 27–68; for the expression, see 37 sqq.; in a similar direction, less developed, perhaps the Ludwig Wittgenstein's *Tractatus logico-philosophicus*: cf. 3.221: "Die Gegenstände kann ich nur *nennen*. Zeichen vertreten sie. Ich kann nur *von* ihnen sprechen, sie *aussprechen* kann ich nicht. Ein Satz kann nur sagen, *wie* Ding ist, nicht *was* es ist.", jointly with 3.262, 3.32–3.328 and 6.211 for the "use" as the answer to why a name designates what it designates). There was not and there will not be any intrinsic need for the choice, among different ways and senses of 'being', of one way or sense or another as the ὄντως ὄν or the πρώτως ὄν, for the decision of which is the first and fundamental way or sense (πρῶτον, πρότερον, ἀρχή). There was not and there will not be any intrinsic need, as noted by Pierre Aubenque ("Sens et function", 18), even for the choice of 'being' as a general title of the reference, that is, there was not and there will not be any need for metaphysics to have been, to be or to continue being ontology *stricto sensu* (think, for instance, when something was or is determined as a principle "beyond reality" or "beyond being"). These choices and decisions about what serves as the foundation, the starting point of each philosophy, of each metaphysics – and, in general, of each thought – escape, by definition, any empirical verification, for the essential reason that it is precisely these choices or decisions the ones that establish or institute the rules and criteria of any verification, rules and criteria which, therefore, cannot be applied, without moving in a vicious circle, to these decisions or choices that establish or institute them as such (cf. Pierre Aubenque, *Faut-il déconstruire?*, 83, and Willard. V. O. Quine, loc. cit., 53: "What makes ontological questions meaningless when taken absolutely is not universality but circularity"). In return, any metaphysics – strictly speaking, any non-dogmatic or critical metaphysics

reason of this precise and precious trait – which some do not find until Socrates[15]

– necessarily involves the immanent and unceasing reflection on its conditions of possibility, that is, on the conditions that make it possible to understand the reality as a whole as proposed by such metaphysics; it involves the immanent reflection on what within that understanding serves as the ultimate foundation, the starting point, that is, what that understanding assumes and takes as evident, knowing (and this wisdom is really ancient) that that assumption is not, nor can it be, absolute. This immanent reflection of any critical metaphysics inevitably leaves the door open to both its own mutation as well as to the possibility of its coexistence with other metaphysics and even – everything must be said – to questioning its own legitimacy and the legitimacy of any possible metaphysics. "So viel Misstrauen – we may say with Nietzsche –, so viel Philosophie" (Friedrich NIETZSCHE, *Die fröhliche Wissenschaft*, § 346, 580; cf. ibid. § 344, 574–577, and "Vorrede zur zweiten Ausgabe", § 3, 349–351; ID., *Nietzsche contra Wagner*, "Epilog.", § 1, 436 sq.; ID., *Menschliches, Allzumenschliches*, Erster Band, "Vorrede", § 1, 13–15). On "suspicion", on general "mistrust" in every thinking, and especially in the thinking of oneself, which reflects an "extreme skepticism" (very different from a "dogmatic skepticism"), that is, a skeptical position even – and this is stated for the sake of coherence – as regards skepticism, see the coherent and beautiful pages written by Antonio MACHADO (*Juan de Mairena*, vol. 1, 11, 31 sq., 66–68, 78, 80 sq., 104–106, 140–142, 150, 157, 166 sq., 168–172, 175–177, 180–184; *Juan de Mairena*, vol. 2, 29 sq., 43 sq., 49, 54–56, 60, 69, 72, 83 sq., 88–90, 99–100, 106–108, 138–141).

15 According to Pierre AUBENQUE, it is with SOCRATES – for whom, unlike for his senior disciple, dialectics was not the art of "questioning and answering (ἐρωτᾶν καὶ ἀποκρίνεσθαι)" (it is precisely PLATO, through his character SOCRATES, of course, who affirms this: cf., e. g., *Cratylus*, 390 c), but only of questioning (cf. XENOPHON, *Memorabilia* IV 4, 9–10 and, above all else, ARISTOTLE, *On Sophistical Refutations* 34, 183 b 7 sq.: ... Σωκράτης ἠρώτα ἀλλ' οὐκ ἀπεκρίνετο· ὡμολόγει γὰρ οὐκ εἰδέναι, jointly with 11, 172 a 23–24, 30–32) – that what we have called, in line with Antonio MACHADO, the "metaphysics of humility" begins. These are, in fact, the words used by the distinguished French professor to conclude his "Sens et function de l'aporie" (18 sq.): "Qu'il y ait une histoire de l'ontologie au sens large, qu'il y ait des 'ontologies alternatives', n'est pas un accident, mais la conséquence inévitable de l'indétermination, de l'inscrutabilité de son point de départ, qui obligent à des 'décisions' incommensurables entre elles et intraduisibles, si ce n'est dans les catégories d'une autre théorie. Qu'il n'y ait pas d'absoluité du point de départ, puisqu'il n'y en a pas d'évidence ni de démonstration possible sans cercle vicieux, que toute ontologie soit suspendue à une décision, à un 'commitment', à un engagement qui peut avoir de bonnes raisons pour lui, sinon logiques, du moins pragmatiques ou éthiques, c'est là, me semble-t-il, l'effet principal du questionnement socratique. En ce sens, c'est bien Socrate qui est le véritable initiateur de la métaphysique, mais d'une métaphysique aporétique et critique et, dans la mesure où elle inclut nécessairement la réflexion sur ses conditions de possibilité, 'transcendantale'". In this same direction as considering that the metaphysics of modesty began with SOCRATES, perhaps we should comprehend, as surely as many others, Arthur SCHOPENHAUER when he writes (*Die Welt als Wille und Vorstellung II*, 242): "Nichts kann jedoch der Auffassung auch nur des *Problems* der Metaphysik so fest entgegenstehen wie eine ihm vorhergängige, aufgedrungene und dem Geiste früh eingeimpfte Lösung desselben: denn der notwendige Ausgangspunkt zu allem echten Philosophieren ist die tiefe Empfindung des Sokratischen: 'Dies eine weiß ich, daß ich nicht weiß'". From a position like that of A. MACHADO, referred to above, in our n. 14, this Socratic saying "contains the arrogance of excessive wisdom as it forgot to add: *and yet of this very*

–, and by way of no other reason, that we believe it can be said, paraphrasing the famous expression in M. Heidegger's *Einführung in die Metaphysik*, that Parmenides's poem overrides entire libraries of philosophical literature.[16]

Each history era seems to have 'decided' to give an answer (at least one prevailing answer) – which characterizes the era in question as such – to the question about the first principles or foundations of reality (and of our knowledge of reality and our action thereon), i. e., to the philosophical question, to the metaphysical question. And the sequence of different answers to this question, the sequence of the various "mutations métaphysiques"[17], and the corresponding sequence of various different eras – or periods within each era – does not need

same I am not entirely sure." (Antonio MACHADO, *Juan de Mairena*, vol. 2, 139 sq.; as emphasized by MACHADO himself).

16 Cf. Martin HEIDEGGER, *Einführung in die Metaphysik*, 104: "Was wir vom Lehrgedicht des *Parmenides* noch besitzen, geht in ein dünnes Heft zusammen, das freilich ganze Bibliotheken philosophischer Literatur in der vermeintlichen Notwendigkeit ihrer Existenz widerlegt. Wer die Maßstäbe solchen denkenden Sagens kennt, muß als Heutiger alle Lust verlieren, Bücher zu schreiben."

17 For the expression, cf. Michel HOUELLEBECQ, *Les particules élémentaires*, 7 sq.: "En réalité, la vision du monde la plus couramment adoptée, à un moment donné, par les membres d'une société détermine son économie, sa politique et ses moeurs. Les mutations métaphysiques – c'est-à-dire les transformations radicales et globales de la vision du monde adoptée par le plus grand nombre – sont rares dans l'histoire de l'humanité. Par exemple, on peut citer l'apparition du christianisme. Dès lors qu'une mutation métaphysique s'est produite, elle se développe sans rencontrer de résistance jusqu'à ses conséquences ultimes. Elle balaie sans même y prêter attention les systèmes économiques et politiques, les jugements esthétiques, les hiérarchies sociales. Aucune force humaine ne peut interrompre son cours – aucune autre force que l'apparition d'une nouvelle mutation métaphysique". Much of what HOUELLEBECQ says here appears in the lessons issued by Pierre AUBENQUE within the framework of the *Chaire de métaphysique Étienne-Gilson* in 1997–1998, which were not published until a decade later. Cf. Pierre AUBENQUE, *Faut-il déconstruire?*, e. g., 20, 82 sq. and 86–88. AUBENQUE himself discusses there "l'histoire des mutations de cette compréhension [sc. of our comprehension of the sense of being] et des décisions qui les fondent", whose object is the object of study of metaphysics, is the metaphysics itself (cf. ibid., 82 sq.).

to be understood, in a Hegelian way, as a necessary sequence,[18] but perhaps, less

---

18 Cf. Georg W. F. HEGEL, *Einleitung in die Geschichte der Philosophie*, 92 (cf. ID., *Geschichte der Philosophie I*, 38): "...die Taten der denkenden Vernunft keine Abenteuer sind. Auch die Weltgeschichte ist nicht nur romantisch, sie ist nicht eine Sammlung von zufälligen Taten und Begebenheiten; nicht die Zufälligkeit herrscht in ihr vor. Die Begebenheiten in ihr sind nicht Fahrten von irrenden Rittern, Taten von Helden, die sich nutzlos herumschlagen, abmühen und aufopfern um einem zufälligen Gegenstand; und ihre Wirksamkeit ist nicht spurlos verschwunden. Sondern in ihren Begebenheiten ist ein notwendiger Zusammenhang. Dasselbe ist in der Geschichte der Philosophie der Fall. Es ist darin nicht von Einfällen, Meinungen usw. die Rede, welche Jeder nach der Besonderheit seines Geistes herausgefunden oder nach Willkür nicht ausgeklügelt hätte; sondern indem hier die reine Tätigkeit und die Notwendigkeit des Geistes zu betrachten ist, muß auch in der ganzen Bewegung des denkenden Geistes ein notwendiger und wesentlicher Zusammenhang sein. Mit diesem Glauben an den Weltgeist müssen wir an die Geschichte und besonders an die Geschichte der Philosophie herangehen."; ID., *Einleitung in die Geschichte der Philosophie*, 122 (cf. ID., *Geschichte der Philosophie I*, 54): "Wir nehmen die Gedanken geschichtlich auf, wie sie bei den einzelnen Individuen usw. erschienen sind; es ist eine Entwicklung in der Zeit, aber nach der inneren Notwendigkeit des Begriffs. Dies ist die einzig würdige Ansicht der Geschichte der Philosophie, oder es ist das wahrhafte Interesse der Geschichte der Philosophie, daß sie zeigt, daß es vernünftig zugegangen ist in der Welt auch nach dieser Seite. Dies hat von vornherein schon eine große Präsumtion für sich; die Geschichte der Philosophie ist die Entwicklung der denkenden Vernunft; im Werden derselben wird es daher wohl vernünftig zugegangen sein. ... – Daß es vernünftig zugegangen sei, diesen Glauben kann man mitbringen. Es ist der Glaube an die Vorsehung, nur in anderer Weise. Das Beste in der Welt ist das, was der Gedanke hervorgebracht hat. Daher ist es unpassend, wenn man glaubt, nur in der Natur, nicht auch im Geistigen, in der Geschichte usf. sei Vernunft. Wenn man einerseits meint, die Vorsehung habe die Welt regiert, und andererseits doch die Weltbegebenheiten im Gebiete des Geistes – und das sind die Philosophien – für Zufälligkeiten hält, so widerspricht diese Vorstellung der ersten; oder vielmehr es ist mit dem Glauben an die Vorsehung nicht Ernst; er ist nur ein leeres Gerede. Was aber geschehen ist, ist durch den Gedanken der Vorsehung geschehen."; ID., *Einleitung in die Geschichte der Philosophie*, 145 sq.: "Wir haben die einzelnen Philosophien zu betrachten als die Entwicklungsstufen *einer* Idee. Jede Philosophie stellt sich dar als eine notwendige Denkbestimmung der Idee. In der Folge der Philosophie findet keine Willkürlichkeit statt; die Ordnung, in der sie hervortreten, ist bestimmt durch die Notwendigkeit. Wie diese beschaffen ist, wird sich in der Ausführung der Geschichte der Philosophie selbst näher zeigen. Jedes Moment faßt das Ganze der Idee in einer einseitigen Form, hebt sich dieser Einseitigkeit wegen auf, und sich so als Letztes widerlegend, schließt es sich mit seiner entgegengesetzten Bestimmung, die ihm mangelte, zusammen und wird so tiefer und reicher. Das ist die Dialektik dieser Bestimmungen. Diese Bewegung schließt aber nicht mit dem Nichts, sondern die aufgehobenen Bestimmungen sind selbst affirmativer Natur. In diesem Sinne ist es, daß wir die Geschichte der Philosophie abzuhandeln haben." See also ID., *Einleitung in die Geschichte der Philosophie*, 7 (*Geschichte der Philosophie I*, 15); *Einleitung in die Geschichte der Philosophie*, 123, 124 sq., 125 sq. (cf. *Geschichte der Philosophie I*, 55 sq.). In the context of the conclusion of his lessons, cf. ID., *Geschichte der Philosophie III*, 461: "Hiermit ist diese Geschichte der Philosophie *beschlossen*. Ich wünsche, daß Sie daraus ersehen haben, daß die Geschichte der Philosophie nicht eine blinde Sammlung von Einfällen, ein zufälliger Fortgang ist. Ich habe vielmehr ihr notwendiges Hervorgehen auseinan-

pretentiously, in a Heideggarian way, as a sequence not merely casual.[19]

A schematic journey through the different eras into which the history of western philosophy is traditionally divided and their corresponding answers to the metaphysical question may adopt the relation between 'the knower and the known (knowable)', 'the percipient and the perceived (perceptible)' or 'the comprehender and the comprehended (comprehensible)', as a common thread – among other possible threads and with the limitations involved. Such relation seems to emerge as being relatively apt for this journey insofar as the foundation in question appears to have been, always prevailingly, on the side of the known or knowable, perceived or perceptible, comprehended or comprehensible (in Ancient Ages), or on the side of the knower, percipient or comprehender (with variations, in the Modern and Contemporary Ages), or on neither side of the relation, but on the side of a third party (God), who even grounds the relation itself and both sides thereof (in the Middle Ages). Naturally, the journey may only be a partial one, which necessarily implies a choice, by the one proposing the journey, of some moments or topics, even within each of the four traditionally distinguished historical eras.

As to this journey, here we are only interested in highlighting the direction taken by the answer to the metaphysical question from the beginning of the Modern era to the current times, to slightly dwell on the latter, aware of the difficulties involved in attempting to understand the times we are living in, since, as expressed by Romano Guardini in the "Vorbemerkung" of his *Das Ende der Neuzeit*, the true nature of an era only becomes fully visible once such era has faded away.[20] We are interested in this return to the beginning of the Modern era

der aufzuzeigen versucht, so daß die eine Philosophie schlechthin notwendig die vorhergehende voraussetzt. Das allgemeine Resultat der Geschichte der Philosophie ist: 1. daß zu aller Zeit nur *eine* Philosophie gewesen ist, deren gleichzeitige Differenzen die notwendigen Seiten des *einen* Prinzips ausmachen; 2. daß die Folge der philosophischen Systeme keine zufällige, sondern die notwendige Stufenfolge der Entwicklung dieser Wissenschaft darstellt; 3. daß die letzte Philosophie einer Zeit das Resultat dieser Entwicklung und die Wahrheit in der höchsten Gestalt ist, die sich das Selbstbewußtsein des Geistes über sich gibt. Die letzte Philosophie enthält daher die vorhergehenden, faßt alle Stufen in sich, ist Produkt und Resultat aller vorhergehenden."

19 Cf. Martin Heidegger, e. g., "Zeit und Sein", in *Zur Sache des Denkens*, 9: "Die Folge der Epochen im Geschick von Sein ist weder zufällig, noch läßt sie sich als notwendig errechnen."; Id., *Was ist das – die Philosophie*?, in *Identität und Differenz*, 18: "...eine freie Folge, weil auf keine Weise einsichtig gemacht werden kann, daß die einzelnen Philosophien und die Epochen der Philosophie im Sinne der Notwendigkeit eines dialektischen Prozesses auseinander hervorgehen."

20 Cf. Romano Guardini, *Das Ende der Neuzeit*, 9 sq.: "... eine Zeitgestalt aber erst dann ganz sichtbar wird, wenn sie sinkt." Cf., in addition, Georg W. F. Hegel, *Grundlinien der Philosophie des Rechts*, 27 sq.: "Um noch über das *Belehren*, wie die Welt sein soll, ein Wort zu sagen, so

since, having admitted Heidegger's characterization of such era as the era of the internalization of the ὑποκείμενον, of the internalization of the foundation, as the era in which "the Subiectität becomes Subjectivität"[21], or, simply, as the era of the "metaphysics of subjectivity (Metaphysik der Subjektivität)"[22] (within the framework herein proposed as a common thread, as the era in which the prevailing decision has been to place the foundation on the side of the knower, the percipient or the comprehender), the era we call "Contemporary" seems to still be a part, just another moment, of the Modernity so depicted. Just another moment, with all its corresponding peculiarities, of course, but in the same manner each of the different moments, each of the different types of founding-subjects coming on stage from the beginning of the Modern era are peculiar as well, and ranging from the Cartesian *cogito* as the *fundamentum inconcussum* or *fundamentum absolutum inconcussum veritatis,*[23] to Nietzsche's "Übermensch" as a "super-species (Über-

---

kommt dazu ohnehin die Philosophie immer zu spät. Als der *Gedanke* der Welt erscheint sie erst in der Zeit, nachdem die Wirklichkeit ihren Bildungsprozeß vollendet und sich fertig gemacht hat. ...die Eule der Minerva beginnt erst mit der einbrechenden Dämmerung ihren Flug."

21 Martin HEIDEGGER, *Nietzsche II*, 411. Here we can adopt HEIDEGGER's distinction between "Subiectität" and "Subjektivität". The first term is more general and is used to refer to the characterization applicable to all metaphysics, from the beginning thereof to the current times, to the extent it addresses the *subiectum*, the foundation – in a broad sense –, regardless of the foundation discussed, regardless of whether such foundation is a self (whichever the self); in HEIDEGGER's terms: "the term 'Subiectität' names the unitary history of being", or: "in its history as metaphysics, being is constantly 'Subiectität'." The second, "Subjektivität", is a specific form of "Subiectität", since it makes exclusive reference to the self (whichever the type) as founding-subject, that is, it exclusively refers to the *subiectum* highlighted since DESCARTES onwards. This way of providing a foundation, this form of "Subiectität", appears when the human representation, when the man, who become "der Repräsentant" of 'what is', exclusively reclaims for himself the name of "subject", in such way that *subiectum* and *ego*, the subject and the self, will become synonymous, that represented by such founding-subject thus becoming the *obiectum*, the thing that lies (*iectum*) versus (*ob*) the subject which provides its foundation. Cf. ibid., 410–413 and 395 sq.

22 Cf. Martin HEIDEGGER, e. g., *Nietzsche II*, 170, 171, 177, 206, 211, 218, 343, 347, 350; ID., "Brief über den Humanismus", in *Wegmarken*, 318; ID., "Zur Seinsfrage", in *Wegmarken*, 404; ID., *Metaphysik und Nihilismus*, 5, 134, 189, 241, 242, 243, 247. For further references regarding this expression in HEIDEGGER's work, see François JARAN and Christophe PERRIN, *The Heidegger Concordance*, vol. 2, 84.

23 These expressions, used by Martin HEIDEGGER in relation to the Cartesian *cogito* (cf., respectively and to refer only to two very well-known places of his work, *Sein und Zeit*, § 6, 24 and "Die Zeit des Weltbildes", in *Holzwege*, 106) – although not only in relation to the principle of philosophy of the philosopher from La Flèche –, are literally alien to DESCARTES himself. Cf., however, the "Secundae responsiones" of the *Meditationes*, where DESCARTES discusses the *fundamentum, cui omnis humana certitudo niti posse mihi videtur* (René DESCARTES, *Meditationes*, 144, 24–25); the "Meditatio segunda", which reads: ...*magna quoque speranda sunt, si vel minimum quid invenero quod certum sit et inconcussum* (ibid., 24, 11–13); the "Quatrième partie" of

Art)"[24] of man – in the opinion of whom, of course, there is no old or ancient God that is not dead[25] – who assumes himself as "will of power" (= "the most intimate essence of being")[26] and, therefore, as the unconditioned foundation of the imposition of values or – which for Nietzsche would be the same – as the unconditioned foundation of reality, of 'what is'; such range including – please, let us be extremely reductive and schematic here – the Kantian "transcendental subject", Hegel's "absolute subject", the "social class" of the "proletariat" as founding-subject destined to and apt for the transformation of reality in K. Marx's philosophy. It seems that the transition from the beginning of the Modern era up to these days can be conceived as an uninterrupted transition from the "metaphysics of the first person plural" (one founding-subject who was initially a self equal to we-all-of-us and, following his most significant expansion with Hegel's work, gradually restricted to we-most-of-us and, then, to we-some-of-us) to the "metaphysics of the first person singular" (a plurality of founding-subjects equivalent to the individual 'selves' individually presenting or trying to present themselves as the only founding-subject of reality and of the action thereon). In Heraclitus's terms, as a transition from a λόγος (reason, foundation) common (ξυνός) to us to a plurality of scattered and unrelated λόγοι ('reasons', foundations), to a plurality of ἴδιαι φρονήσεις, private wisdoms of each individual human being.[27]

Such *factum* which, from a metaphysical point of view, seems to depict the current era, still placing it within the "metaphysics of subjectivity", such placement (attempted placement) of every individual human being as such, and in each case, as the only and ultimate foundation of 'what is' and, therefore, of 'what is to be done' (ethically and politically speaking), may be called – in the light of Plato's peculiar interpretation of Protagoras's *homo mensura* thesis in *Theaetetus* –

---

the *Discours de la méthode*: "Remarquant que cette vérité: je pense, donc je suis, était si ferme et si assurée, que toutes les plus extravagantes suppositions des sceptiques n'étaient pas capables de l'ébranler, je jugeai que je pouvais la recevoir, sans scrupule, pour le premier principe de la philosophie, que je cherchais" (René DESCARTES, *Discours*, 32, 18–23); or the "Seconde Partie": "Jamais mon dessein ne s'est étendu plus avant que de tâcher à réformer mes propres pensées, et de bâtir dans un fonds qui est tout à moi" (ibid., 15, 4–6).

24 Cf. Friedrich NIETZSCHE, *Also sprach Zarathustra*, 98.

25 Cf. ibid., 102 (These are the words that close the "First Part" of the book): "'*Todt sind alle Götter: nun wollen wir, dass der Übermensch lebe.*' – dies sei einst am grossen Mittage unser letzter Wille! – Also sprach Zarathustra."; ibid., 357: "Gott starb: nun wollen *wir*, – dass der Übermensch lebe." (as always emphasized by NIETZSCHE).

26 Cf. Friedrich NIETZSCHE, *Nachgelassene Fragmente 1887–1889*, Nachlaß Frühjahr 1888, 14[80], 260 (= Friedrich NIETZSCHE, *Der Wille zur Macht*, § 693, 156): "Wenn das innerste Wesen des Seins Wille zur Macht ist, ...".

27 Cf. HERACLITUS, DK 22 B 2; SEXTUS EMPIRICUS, *Adversus Mathematicos* VII 133 (cit. above, as epigraph).

"divinization (or deification) of the individual human being as an individual". Let us briefly examine Plato's text in order to clarify the possibility of such designation.

Through Socrates, in 152 a, Plato introduces Protagoras's famous thesis for the first time in the dialogue. He does so with the following words, which are allegedly the words of the sophist from Abdera himself:

> For he [sc. Protagoras] says somewhere that man is the measure of all things, of the things which are, insofar as they are, and of the things which are not, insofar as they are not (φησὶ γάρ που πάντων χρημάτων μέτρον ἄνθρωπον εἶναι, τῶν μὲν ὄντων, ὡς ἔστι, τῶν δὲ μὴ ὄντων, ὡς οὐκ ἔστιν)[28]

A relatively attentive look at the development of the dialog shows that Plato understands the "man is the measure of all things" thesis not in the sense of "man" as species or human gender, but in the sense of all men and each one of them. In other words, he understands that "man" has a *suppositio distributa,* i. e., *supponit pro omnibus et singulis,* emphasized in each of the latter. We believe that this is explicitly evidenced by at least the following two quotes from the dialogue, where Plato reformulates Protagoras's thesis through the words of Socrates:

> ...each one of us is the measure of the things that are and those that are not (...μέτρον γὰρ ἕκαστον ἡμῶν εἶναι τῶν τε ὄντων καὶ μή)" (166 d)

> ...every man is the measure of all things... (...πάντ' ἄνδρα πάντων χρημάτων μέτρον εἶναι ...) (183 b–c).

What turns out to be very interesting – from the perspective aimed at comprehending what seems to be the predominant contemporary metaphysical position – are the two related consequences expounded below, which Plato, always through Socrates, extracts from Protagoras's thesis, having understood it as stated above. First of all (cf. 161 b – 162 a), "if each person is the measure of his own wisdom (...μέτρῳ ὄντι αὐτῷ ἑκάστῳ τῆς αὐτοῦ σοφίας)" (161 e) – here we find another reformulation of the thesis or a direct implication therefrom – there is no reason to consider, on the one hand, Protagoras as a wise man and teacher of the rest and, on the other hand, the others as ignorant people who must attend (and, above all, pay for) Protagoras's lessons. The consequence for the sophist as a teacher is, thus, fatal: nobody can be the teacher of another person since, in virtue of the hypothesis in question, no-one is wiser than another person. There-

28 Cf. Sextus Empiricus, *Adversus Mathematicos* VII 60–61 (DK 80 B 1), cit. above as epigraph. In *Lives of Eminent Philosopher* by Diogenes Laertius, IX 51 (DK 80 A 1), we find the exact fragment referred to by Sextus Empiricus (the absence of ἐστὶν is not relevant, of course): πάντων χρημάτων μέτρον ἄνθρωπος, τῶν μὲν ὄντων ὡς ἔστιν, τῶν δὲ οὐκ ὄντων ὡς οὐκ ἔστιν.

fore, it proves to be "complete nonsense" to examine and attempt to refute what an individual man believes that 'what is' is, if the opinions of each and every one are, in virtue of the hypothesis, correct.[29] Secondly (cf. 162 c), no man is inferior "in wisdom to any other men or even any of the gods (σοφίαν ὁτουοῦν ἀνθρώπων ἢ καὶ θεῶν)". As we can see, subject to Plato's interpretation, the implication of the *homo mensura* thesis in accordance with which "there is no other wisdom than each one's wisdom" – as expressed this time by A. Machado[30] – is associated with a sort of divinization of the individual human being, as his own wisdom does not acknowledge any other superior instance.

The individual human being seems to say now: ἐγώ εἰμι ὁ θεός, *ego sum deus* (*Psalm* 46:11; cf. *Deuteronomy* 32:39; *Ezekiel* 12:16, etc.), ἐγώ εἰμι ὁ θεός καὶ οὐκ ἔστιν ἔτι πλὴν ἐμοῦ, *ego sum Deus, et non est ultra deus, nec est similis mei* (*Isaiah* 46:9). Now the ancient God (or ancient Gods), consistently, depends on the deified individual human being, such dependence being not only gnoseological but also ontological, metaphysical. Apparently, only from this fundamental metaphysical position, from a deep understanding of the metaphysical position which contemporaneously seems to prevail, can a contemporary monk 'pray' in the following way (and can we, each one of us, understand his 'prayer'):

> Was wirst du tun, Gott, wenn ich sterbe?
> Ich bin dein Krug (wenn ich zerscherbe?)
> Ich bin dein Trank (wenn ich verderbe?)
> Bin dein Gewand und dein Gewerbe,
> mit mir verlierst du deinen Sinn.
>
> Nach mir hast du kein Haus, darin
> dich Worte, nah und warm, begrüßen.
> Es fällt von deinen müden Füßen
> die Samtsandale, die ich bin.
> Dein großer Mantel läßt dich los.
> Dein Blick, den ich mit meiner Wange
> warm, wie mit einem Pfühl, empfange,
> wird kommen, wird mich suchen, lange –
> und legt beim Sonnenuntergange
> sich fremden Steinen in den Schoß.
>
> Was wirst du tun, Gott? Ich bin bange.[31]

---

29 Cf. PLATO, *Cratylus*, 386 c–d, where it is concluded as a consequence of the thesis by Protagoras referred to here that it is impossible to state that some people are intelligent and other people are not.

30 Antonio MACHADO, *Juan de Mairena*, vol. 1, 171.

31 Rainer Maria RILKE, *Das Stunden-Buch*, "Erstes Buch: Das Buch vom mönchischen Leben (1899)", 22 sq. This first book was initially entitled *Die Gebete*, the same title RILKE gave to the

It does not seem that there are no more gods today. Rather, we, the individual human beings as individuals, have taken the place of the ancient gods.[32] It does not seem that we have freed ourselves of cults and altars. The altars remain as places where individual human beings worship themselves.[33] The emergence of a text like *La sculpture de soi* by M. Onfray nowadays – among so many texts, including many by this same author – may not be necessary, but it does not appear to be a mere casualty.[34]

---

whole project that finally included the three books which make up *Das Stunden-Buch*.

32 We can recall the words of “Zarathustra the atheist” in his conversation with the “old Pope”, already “retired and masterless”, in the section entitled “Retired” of the “Fourth Part” of *Also sprach Zarathustra* (324 sq.): “Es giebt auch in der Frömmigkeit guten Geschmack: *der* sprach endlich ‘Fort mit einem *solchen* Gotte! Lieber keinen Gott, lieber auf eigne Faust Schicksal machen, lieber Narr sein, lieber selber Gott sein!’” (Acknowledgement from the parties makes discoveries non-essential!). Cf. also the words that NIETZSCHE puts in the mouth of the Pope, clearly retired (ibid., 325): “‘Was höre ich! sprach hier der alte Papst mit gespitzten Ohren; oh Zarathustra, du bist frömmer als du glaubst, mit einem solchen Unglauben! Irgend ein Gott in dir bekehrte dich zu deiner Gottlosigkeit. Ist es nicht deine Frömmigkeit selber, die dich nicht mehr an einen Gott glauben lässt? Und deine übergrosse Redlichkeit wird dich auch noch jenseits von Gut und Böse wegführen! Siehe doch, was blieb dir aufgespart? Du hast Augen und Hand und Mund, die sind zum Segnen vorher bestimmt seit Ewigkeit. Man segnet nicht mit der Hand allein. In deiner Nähe, ob du schon der Gottloseste sein willst, wittere ich einen heimlichen Weih- und Wohlgeruch von langen Segnungen: mir wird wohl und wehe dabei. ...’” Cf. ID., *Die fröhliche Wissenschaft* – text in which, as from § 108, there appears the first explicit mention of the topic of *God’s death* in NIETZSCHE’s published work –, § 125, 480 sq.: “Der tolle Mensch”. NIETZSCHE puts in his mouth the following statements: “‘Wohin ist Gott? rief er, ich will es euch sagen! *Wir haben ihn getödtet*, – ihr und ich! Wir alle sind seine Mörder!... Gott ist todt! Gott bleibt todt! Und wir haben ihn getödtet! Wie trösten wir uns, die Mörder aller Mörder? Das Heiligste und Mächtigste, was die Welt bisher besass, es ist unter unseren Messern verblutet, – wer wischt dies Blut von uns ab? Mit welchem Wasser können wir uns reinigen? Welche Sühnfeiern, welche heiligen Spiele werden wir erfinden müssen? Ist nicht die Grösse dieser That zu gross für uns? Müssen wir nicht selber zu Göttern werden, um nur ihrer würdig zu erscheinen?...’” (emphasized by NIETZSCHE).

33 Cf. Ernesto SABATO, *La resistencia*, 62: “What have men placed in the place of God? They have not freed themselves of cults and altars. The altar is no longer the place of sacrifice and abnegation, but a place of welfare, of cult of the self...”

34 Michel ONFRAY, *La sculpture de soi*. Cf. ID., *Politique du rebelle*, 36 sq.: “cette ontologie... impose qu’on sache essentiel l’*individu*, sûrement pas le sujet, l’homme ou la personne. ...ce qui fait l’irréductibilité d’un être, c’est son individualité, et non sa subjectivité, son humanité o sa personnalité. C’est l’individu qui souffre, peine, a froid et faim, va mourir ou s’en tire, c’est lui, dans sa chair, donc dans son âme, qui subit les courps, sent progresser les parasites aussi bien que la faiblesse, la mort ou le pire. Toute nouvelle figure à inscrire sur le sable après la mort de l’homme passe par cette volonté délibérée de réaliser l’individu et rien d’autre.”; ibid., 158: “L’individu surgissant de ce mois de Mai constitue le pendant politique de la naissance du Je dans la philosophie de Descartes. Émancipé de toute attache scolastique et théologique, l’individu formulé

In an attempt to clarify the characterization of the *factum* in question a little, perhaps we can say that the contemporaneous individual does not seem to have put on any 'God's suit'. The "divinization of the individual human being" seems to inherit certain features from preexisting philosophical conceptions and, more precisely, as it could not be otherwise, from preexisting philosophical-theological conceptions.

On the one hand, e. g., it appears to be the heir to a specific theological position (monotheistic, of course), to wit, the so called "theological voluntarism", which privileges, among the divine attributes, the infinite freedom and omnipotence above all other attributes, the reason among them. In fact, this divinization of the individual human beings is not a divinization of what they have in common with the other individual human beings, but a divinization of what is most theirs, of what contributes the most to make them individuals, to wit, a divinization of the 'will', of their wants (we should take a deeper look into the direct relation between this dimension and the current absolute priority that πρᾶξις has over θεωρία).[35]

---

en Mai 68 se définit moins par son rapport au travail, à la famille et à la patrie, à la société et au Léviathan, que dans la relation entretenue avec lui-même. L'autonomie, au sens étymologique, c'est-à-dire la capacité à être pour soi sa propre fin, sa propre cause et propre raison, apparaît comme la quête essentielle de tout un chacun qui se trouva concerné par les événements de cette époque."; ibid., 232: "...le seul espoir, solipsiste, gît dans la possibilité d'une sculpture de soi."

35 The priority that πρᾶξις has over θεωρία has been practically a principle during all the Modern era. Suffice it to remember, not to bore anybody with the famous but, in this respect, belated MARX's XI Thesis on Feuerbach, the paradigmatic but not less famous translation that the Faust of GOETHE made, from "the sacred original" to his "beloved German" (cf. Johann W. von GOETHE, *Faust*, vv. 1222 sq.), of the first verse of *John's Gospel*: Ἐν ἀρχῇ ἦν ὁ λόγος. After three attempts that do not convince him: "Im Anfang war das Wort!" (v. 1224), "Im Anfang war der Sinn." (v. 1229), "Im Anfang war die Kraft!" (v. 1233), he finally says: "Im Anfang war die Tat!" (v. 1237). On the other hand, we can remember, not to expand on this and only refer to the pinnacle of Modernity, what HEGEL himself says about "der Geist der Welt", namely: "Sein Leben ist *Tat*" (cf. Georg W. F. HEGEL, *Einleitung in die Geschichte der Philosophie*, 22; HEGEL himself highlighted it – or his lesson editor –). So, if we want to come closer to the present: "il n'y a de réalité que dans l'action" (Jean-P. SARTRE, *L'existentialisme est un humanisme*, 51). Within this framework of the priority assigned to the πρᾶξις, the role of the will, in its different forms, could be no other thing but essential to the "metaphysics of subjectivity". Cf., in this respect, Martin HEIDEGGER, *Nietzsche II*, 411 sq.: "Für die neuzeitliche Geschichte der Metaphysik spricht aber der Name Subjektivität nur dann das volle Wesen des Seins aus, wenn nicht nur und nicht einmal vorwiegend an den Vorstellungscharakter des Seins gedacht wird, sondern wenn der appetitus und seine Entfaltungen als Grundzug des Seins offenkundig geworden sind. Sein ist seit dem vollen Beginn der neuzeitlichen Metaphysik *Wille*, d. h. *exigentia essentiae*. 'Der Wille' birgt vielfaches Wesen in sich. Er ist der Wille der Vernunft oder der Wille des Geistes, er ist der Wille der Liene oder der Wille zur Macht.". It is our opinion that, at the current moment of Modernity (the era we are still living in), it is the individual human being who assumes himself as will of power, not NIETZSCHE's

On the other hand, the current divinization of individual human beings seems to be an heir of "nominalism", while at the same time differing from this position: the current absolute privilege of individual human beings as individuals implied in the divinization in question is not only to the detriment of the universal "man" (*universales sunt nomina*), at which point it would concur with nominalism,[36] but also, and above all, to the detriment of the 'other' individual human beings, which is why it differs from said position. Jorge Luis Borges, one of those poets whose work usually reflects the determining aspects of an era, shows this concept in his poem "You (Tú)", in the first verses of which we can read:

> In the entire world, one man has been born, one man has died.
>
> To insist otherwise is nothing more than statistics, an impossible addition.
>
> No less impossible than adding the smell of rain to your dream of two nights ago.[37]

Borges had surely read Antonio Machado's *Juan de Mairena* in detail, as we will show below. One of the first pages of that work reads:

> *L'individualité enveloppe l'infini.* –The individual human being is everything. – And what is society then? A mere addition of individuals. (Prove the superfluity of both the addition and society.)
>
> As much as I think about it – Mairena said – I cannot find the way to add up individuals.
>
> /…/
>
> The soul of every human being – Mairena once quoted his teacher – could be a pure intimacy – lyrically speaking, a monad without windows or doors –; a melody sung to the self and heard by a self, deaf and indifferent to all other melodies – the same or differet melodies? – produced by the other souls. The conductor's baton would obviously be useless here. Rather, we would have to postulate Leibnitz's brilliant hypothesis of a pre-established harmony. And we would have to suppose a great ear interested in listening to a great symphony. And why not, for that matter, a great bedlam?[38]

Let us resume the poem "Tú":

"Übermensch", who, as such and as we have emphasized, is not an individual human being, but a type of man, a "super-species" of man.

36 Cf. Rainer Maria RILKE, *Die Aufzeichnungen des Malte Laurids Brigge*, 13: "Ist es möglich, daß man 'die Frauen' sagt, 'die Kinder', 'die Knaben' und nicht ahnt (bei aller Bildung nicht ahnt), daß diese Worte längst keine Mehrzahl mehr haben, sondern nur unzählige Einzahlen?".

37 Jorge L. BORGES, "Tú", 491. (We adopt Alastair REID's translation, in Jorge L. BORGES, *The Book of the Sand*, 115, but in this case with some modifications).

38 Antonio MACHADO, *Juan de Mairena*, vol. 1, 10.

> The man is Ulysses, Abel, Cain, the first to make constellations of the stars, to build the first pyramid, the man who contrived the hexagrams of the Book of Changes, the smith who engraved runes on the sword of Hengist, Einar Tamberskelver the archer, Luis de León, the bookseller who fathered Samuel Johnson, Voltaire's gardener, Darwin aboard the Beagle, a Jew in the death chamber, and, in time, you and I.
>
> One man alone has died at Troy, at Metaurus, at Hastings, at Austerlitz, at Trafalgar, at Gettysburg.
>
> One man alone has died in hospitals, in boats, in painful solitude, in the rooms of habit and of love.
>
> One man alone has looked on the enormity of dawn.
>
> One man alone has felt on his tongue the fresh quenching of water, the flavour of fruit and of flesh.[39]

Although human beings are legion, although individuals abound, there is no more than 'one individual' in each case. Although the planet is flooded by the crowd, there are no shared worlds, only intersecting worlds. Each individual human being is a world, his world. "Like the god of the monotheisms is odd by nature, each man is alone"[40]. The poet so finds out and, at the end, to clear up all doubts, expresses the irony embedded in the title of his poem, as no genuine 'you' appears therein for the individual 'selves' mentioned:

> I speak about the only one, the one, the one that is always alone.[41]

If each of the individual human beings is a world, the world for each of them neither surpasses them nor survives them. Borges brilliantly uses the poetic device of an individual who decides to commit suicide. What the suicidal person effectively knows, what he has understood, perceived, imagined, created, everything that is for him, unless he fails in his attempt, everything will be lost forever. We can read in "The Suicide (El suicida)":

> Not a single star will be left in the night.
> The night will not be left.
> I will die and, with me,
> the weight of the intolerable universe.
> I shall erase the pyramids, the medallions,

---

39 Jorge L. Borges, "Tú", 491. (Alastair Reid's translation, op. cit., 115).
40 Cf. Julián Serna Arango, *Ontologías alternativas*, 58 sq. We have closely followed this author's interpretation of the three texts by J. L. Borges that we address in this paper. Cf. ibid., esp. 57–63.
41 Jorge L. Borges, "Tú", 491. (Alastair Reid's translation, op. cit., 115).

> the continents and faces.
> I shall erase the accumulated past.
> I shall make dust of history, dust of dust.
> Now I am looking on the final sunset.
> I am hearing the last bird.
> I bequeath nothingness to no one.[42]

The individual is shown here as the lord and master of the world, of 'his own unique' world. A world that is at his disposal even until the very end. The last verse, "I bequeath nothingness to no one", which works as testament (not of he who runs, not of he who abandons, but of he who destroys), goes beyond all of this, maximizing – or rather, explicitly stating – the coherence: If he achieves his aim, not only will there be nothing left to bequeath, there will be 'no-one' to bequeath anything to either. This is because, in line with the "metaphysics of the first person singular", for this individual and his world, there has never been an 'other' outside of himself.[43]

---

42 Jorge L. BORGES, "El suicida", 86. (Alastair REID's translation, op. cit., 149).

43 Ernesto SABATO's entire first novel does not seem to reflect something different. It is enough here just to remember the novel's epigraph: "...in any case, there was a single tunnel, dark and solitary: mine" (Ernesto SABATO, *El túnel*, 59). Cf. Antonio MACHADO, *Juan de Mairena*, vol. 1, 180–183: "The problem with the love thy neighbor concept – Mairena says to his students – ... sharply poses a different one...: that of the real existence of our neighbor, our fellow-man... Because if our neighbor does not exist, we could hardly love him. I ingenuously tell you that this is a serious matter. Think about it. ... solipsism may or may not respond to an absolute reality, may or may not be true; but there's nothing absurd about it. It is the inevitable and perfectly logical conclusion of all subjectivism taken to the extreme. ... It is clear that any philosophical position – sensualist or rationalist – that calls into question the real existence of the external world turns *eo ipso* into a problem, the problem of our neighbor. Only a pragmatic, profoundly illogical, thought is able to affirm the existence of our neighbor with the same degree of certainty as the existence of the self and at the same time acknowledge that this neighbor seems to us to be encompassed within the outside world – a mere creation of our spirit –, without any feature that reveals its heterogeneity. In other words: if nothing is in itself more than myself, what is there to do but declare the absolute unreality of our neighbor? My thought erases and expels you from existence – from an existence in itself – in the company of those same benches on which you sit. ...In short, behind the borders of myself begins the kingdom of nothingness. ...The self is able to love itself with an absolute love of an infinite radius. And the love given to the neighbor, to the other self that is nothing in itself, to the self represented in the absolute self, is only to be professed paying lip service to him. We have arrived at this conclusion *d'enfants terribles* – and what else are we? – of logic. And now notice that 'love your neighbor as yourself, and even more, if it is necessary', that is the true Christian precept, implies an altruistic faith, a belief in an absolute reality, in the existence of the other self in itself. ... how would I dare, within this Christian faith, debase my neighbor so profoundly and substantially to the point that his existence is taken away and turned into a mere representation, a mere ghost of myself?... – And into a bad-shadow ghost

If 'what is' is 'what is for each individual', then 'what should be done' is 'what each individual believes should be done'. Borges makes these two dimensions appear to be intertwined in his short story about the Cretan minotaur, who was killed by the hands of Theseus. Starting with the title and the epigraph of the story, Borges, as in his two previous poems, seems to have intended to depict the determining aspects of our era. As epigraph, the Argentine writer uses a phrase from Apollodorus – we could not expect any different –, the only text from the classical Antiquity in which the son of Pasiphaë and the bull is called by his proper name, "Asterion". In all other sources, he is called by his generic name: "minotaur". This seems to announce something that could be wrongly understood as a mere matter of literary style. Just like each one of us, Asterion looks at the world from his own perspective. "The House of Asterion (La casa de Asterión)" aims to recover that perspective. The story's narrative style is not insignificant: unlike every author that preceded him, Borges writes the story – could it be any other way? – in 'first person'. The narrator of the story is Asterion himself, who, like each one of us, is unique. "The fact is that I am unique"[44], he declares with precision. The only other, the only 'you', is an other made up by Asterion himself, an other that is not an other but rather himself: "But of all the games, I prefer the one about the other Asterion. I pretend that he comes to visit me and that I show him my house".[45] Asterion will say that his house "is the same size as the world; or rather, it is the world"[46] and that it is unique: "a house like no other on the face of the earth. (There are those who declare there is a similar one in Egypt, but they lie.)".[47] This is because his house is a work of his own, of himself, who is unique: "Perhaps I have created the stars and the sun and this enormous house, but I no longer remember".[48] As in the poem "El suicida", the individual Asterion has himself – and as a result, his own world – at his disposal until the very end, even in the very end. Indeed, an astonished Theseus – the individual redeemer Asterion had been waiting for and who in the story only speaks after killing that who had had the word before (there is no lack of metaphysical implications) –

– the quietest in the class dared to add. ... Into a bad-shadow ghost who is capable of paying me back in the same coin. I mean that I should think about him as a ghost of myself which is at the same time able to turn me into a ghost of himself. – Very good, Mr. García – Mairena exclaimed –; you have provided a definition of the other self that is a little Gideonic, but accurate, within the *solus ipse*: a bad-shadow ghost, that is extremely disturbing."

44 Jorge L. Borges, "La casa de Asterión", 569 (we adopt Donald A. Yates and James E. Irby's translation, in Jorge L. Borges, *Labyrinths*, 139).

45 Ibid., 570. (Donald A. Yates and James E. Irby's translation, op. cit., 139).

46 Ibid. (Donald A. Yates and James E. Irby's translation, op. cit., 139).

47 Ibid., 569. (Donald A. Yates and James E. Irby's translation, op. cit., 138).

48 Ibid., 570. (Donald A. Yates and James E. Irby's translation, op. cit., 140).

tells Ariadne at the very end of "La casa de Asterión": "The minotaur scarcely defended himself".[49] One dimension not explicitly stated in the two Borges's poems referred to above appears here in this story. Asterion proves, through his own careful meditation about his house, which is his world, that "everything is repeated many times, fourteen times" – which in the story is equivalent to infinite times. However, as Asterion says, uncovering the relevance of what this implies, "two things in the world seem to be only once: above, the intricate sun; below, Asterion".[50] It should escape no one that, with these words, Borges's intention is to parody these extremely famous opening words to the "Conclusion" of Kant's *Kritik der praktischen Vernunft:* "Two things fill the mind with ever new and increasing admiration and awe, the oftener and the more steadily we reflect on them: the starry heavens above me and the moral law within me".[51] Just as for Kant 'the' world is his world, while for Asterion 'his' world is the world, for Kant there is a moral law that he and everyone must obey, while for Asterion there is no moral law other than that which he decides is moral 'law'.

In order to go a bit further with this rough sketch of the characterization of the *factum* of the divinization of the individual human being, it is important to specify – although this seems to be something relatively evident – that not all individuals-gods are Zeus. There are gods and there are gods, so to speak. In order to better determine the differences of effectiveness between divinized individual human beings, it might be useful to consider the Hegelian distinction between "Realität" and "Wirklichkeit" – but let us leave that topic for another time. In addition to the foregoing, another issue to elaborate on would be the relationship between the divinization of human individuals with the attempts (carried out precisely by individuals) to divinize certain cultures when, as it has been well pointed out, "culture is said in the plural form".[52] The attempt in *The End of History and the Last Man*, by Francis Fukuyama, is, or was, both famous and paradigmatic. Perhaps that poem by Mario Benedetti, "Fukuyama says (Lo dice Fukuyama)", is enough to demystify that concrete and futile attempt of the former member of the Policy Planning Staff of the U.S. Department of State and his followers, although it could also be enough to demystify others that have appeared

---

49 Ibid. (Donald A. YATES and James E. IRBY's translation, op. cit., 140).

50 Ibid. (Donald A. YATES and James E. IRBY's translation, op. cit., 139 sq.).

51 Cf. Immanuel KANT, *Kritik der praktischen Vernunft*, 205: "Zwei Dinge erfüllen das Gemüt mit immer neuer und zunehmender Bewunderung und Ehrfurcht, je öfter und anhaltender sich das Nachdenken damit beschäftigt: *der bestirnte Himmel über mir und das moralische Gesetz in mir.*" (The emphasis belongs to the Kantian text itself).

52 Patxi LANCEROS, "Antropología hermenéutica", 50. Cf. the whole article, 45–57.

and are likely to appear. Perhaps the last five verses are enough, and not just because (in the original version) they rhyme:

> has history ended?
> will this be the end of its wandering path?
> will this world end up
> lethargic and motionless?
> or will this be the start of the second volume?[53]

Among the direct implications of the divinization of the individual, which are closely linked to each other – and to some of which we have had to, at least, allude –, we might mention:

- What we could call "absolute relativism": 'What everyone says is' is 'what is', and 'what everyone says should be done' is 'what should be done' (ethically and politically). This is always under a pretension of absolute validity because what each individual says, according to each individual, does not depend on any other instance. In this context, the word, the λόγος, which is 'common' to all of us, seems to have lost its *status*, its intrinsic value: now it lacks the validity of foundation, of justification (to the others) and, as a consequence, its appearance on stage is not even required (because there is no 'other' to whom we must justify our acts or our speech). "Every interpretation is valid" – Sabato wrote a few years ago – "and words serve more for absolving ourselves of our acts than for being held accountable for them".[54]
- Hardly could there be an "ethic" today (no matter how limited the universality it intends to achieve may be), hardly could a "philosophical discipline" be thought of as "ethics", if the foundation is currently placed where it effectively seems to be. Every individual, just like Asterion in Borges's story, seems to have his own moral 'law'. To say it in words similar to those of Nietzsche: Every individual seems to speak his own language of good and evil; his neighbor does not understand it or does not want to understand it. Over every individual seems to be suspended one table of values: it is the table of 'his' victories, the voice of 'his' will of power.[55]

---

53 Mario BENEDETTI, "Lo dice Fukuyama", 87 sq.: "la historia ¿habrá acabado? / ¿será el fin de su paso vagabundo? / ¿quedará aletargado / e inmóvil este mundo? / ¿o será que empezó el tomo segundo?"

54 Ernesto SABATO, *La resistencia*, 51.

55 Cf. Friedrich NIETZSCHE, *Also sprach Zarathustra*, 61: "jedes Volk spricht seine Zunge des Guten und Bösen: die versteht der Nachbar nicht."; ibid., 74: "Eine Tafel der Güter hängt über jedem Volke. Siehe, es ist seiner Überwindungen Tafel; siehe, es ist die Stimme seines Willens zur Macht." It should be noted, among other differences, that where NIETZSCHE said "every Volk" we say "every individual".

- The so called "political philosophy" (the philosophical thinking of the political community) seems to face serious difficulties if it currently intends to turn into something other than just a party speech. In no way do we want to deflate or become supporters of deflating political activity, quite the contrary in fact: despite everything, we, each one of us, must engage in politics, although those who intend to enter politics 'without' each one of us and, naturally, 'against' each one of us suggest otherwise. However, seemingly, in order not to downgrade it, we must engage in politics in a way that is open, free from, e. g., any 'philosophical' or 'religious' disguises.[56] Connected with the difficulties faced by a purported political philosophy is the following implication, which lies at its bottom.
- In this current era of divinization of the human individual as an individual – or *l'ere de l'individu*, if we want to call it using A. Renaut's book title[57] –, not only the 'actual' existence but also the 'possibility' of the existence of a 'genuine' human community (irrespective of its dimension) – that is to say, not a mere "addition of individuals", not a mere gathering of individuals, not just a group of people getting together, or anything whose unity annihilates the diversity of those who constitute such unity – appear, at the very least, difficult. It seems to be essential to speak of "community" rather than of "identity" when a genuine human community is involved. Perhaps we could refer to Aristotle, especially to Aristotle the metaphysician,[58] and differentiate something that is 'common and identical' – or 'unique' – in the gathered things from something 'common which cannot be identical or unique' in the gathered things.[59] Just

---

56 Cf. Antonio MACHADO, *Juan de Mairena*, vol. 1, 75: "Politics, gentlemen – Mairena goes on – is a highly important activity... I will never tell you to be *apolitical*, but, ultimately, feel disdain for bad politics, that practiced by social climbers and opportunists for no purpose other than making a profit and securing positions for their relatives. You must *engage in politics*, although those who intend to enter such activity without you and, naturally, against you tell you otherwise. I only dare to suggest that you become involved in politics openly; in the worst case scenario, wear a political mask, without any other disguises, as is the case of literature, philosophy or religion. Otherwise, you will be contributing to the corruption of excellent activities, at least, such as politics, and to making politics muddy in such a way that we may never understand each other."

57 Cf. Alain RENAUT, *L'ere de l'individu*.

58 Let us resort to this usage here, although we are aware that among the fundamental differences between PLATO and ARISTOTLE is the latter's lack of integration of practical and theoretical philosophy, which in PLATO is present in the form of a "universal ontological dialectics". We agree on this issue, though not in every detail, with Hans KRÄMER (*Platone e i fondamenti della metafisica*, esp. 233–235), although we do not agree on others.

59 Cf. *Metaphysics* Γ 2, 1005 a 9, in which ARISTOTLE denies in respect of τὸ ὄν (and also in respect of τὸ ἕν) that it is a καθόλου καὶ ταὐτὸ ἐπὶ πάντων (2, 1005 a 9), i. e., a universal which is identical, univocal, generic, and predicable as such of, or present as such in, all things.

as, for Aristotle, 'being' can only be common to everything (κοινὸν… πᾶσι[60]) because it is intrinsically, immediately (εὐθύς) diverse,[61] as it carries the differences at its core, which 'are' and constitute the plurality and the diversity of reality as a whole,[62] no human community, no community of those who have λόγος, is homogeneous; no human community is a community composed of homogeneous, 'identical' interests. A genuine human community is intrinsically, constituently and originally diverse, heterogeneous. There is no human community which can be sustainable over time, which can survive through time, if it fails to insist on guaranteeing such diversity, such plurality, which originally constitutes it as such. Originally diverse, the human community (irrespective of its dimension), just like 'being', does not have an 'essence'.[63] It is different from a generic or specific community, which is also something common, but identical and unique in the plurality gathered. There is nothing which 'identifies' a human community, which remains 'identical' in it, even though this does not mean that a community cannot be recognized as such. Not even the goals or projects thereof, as community projects, are 'identical' (they are neither 'identical' in the eye of each and every monad composing such community nor do they remain 'identical' over time as they arise from diverse and changing forces) as they are objectives which result from the confluence of different forces. They are 'common' goals and projects, and recognizable as such, though 'not identical', i. e., the diversity of the directions

60 *Metaphysics* Γ 2, 1004 b 20: κοινὸν δὲ πᾶσι τὸ ὄν ἐστιν. Cf. also 3, 1005 a 27–28.

61 Cf. *Metaphysics* Γ 2, 1004 a 4–5: ὑπάρχει… εὐθὺς γένη ἔχοντα τὸ ὄν καὶ τὸ ἕν. Here (cf. 1004 a 2–9) the γένη into which "being" and "one" are said to be immediately divided are the classes or types of the first category, i. e., the classes or types of οὐσίαι, namely, the immobile or prime οὐσία (cf. τῆν πρώτην οὐσίαν, 3, 1005 a 35) and the mobile or sensible οὐσία. Cf. *Metaphysics* H 6, 1045 a 36 – b 7, the passage which reads that each category genus is immediately a being or a one: εὐθὺς ὅπερ ἕν τί [εἶναί] ἐστιν ἕκαστον, ὥσπερ καὶ ὅπερ ὄν τι… (1045 a 36 – b 1).

62 This is the fundamental reason behind ARISTOTLE's main argument (*Metaphysics* B 3, 998 b 22–27, complemented by *Topics* VI 6, 144 a 31 – b 3) in favor of the nuclear thesis of his metaphysics – with which he probably summed up his opposition to the platonic-academic conception of being –, to quote it in his *locus clasicissimus* (*Posterior Analytics* II 7, 92 b 14): οὐ… γένος τὸ ὄν (cf. *Metaphysics* B 3, 998 b 22; H 6, 1045 b 5–6; I 2, 1053 b 22–24; K 1, 1059 b 33–34). In *Topics*, ARISTOTLE had already argued that being (like one) can neither constitute a *genus* nor constitute or be a part of a definition (ὅρος) or a property (ἴδιον); it can neither be an accident (συμβεβηκός) nor constitute or be a part of a difference (διαφορά) or species (εἶδος) of a genus. With regard to any mentions of ARISTOTLE the metaphysician, let us refer to an extensive work of ours published in two issues: Horacio A. GIANNESCHI, "Aristóteles o el ente (I)", 93–122, and "Aristóteles o el ente (II)", 79–122.

63 Cf. *Posterior Analytics* II 7, 92 b 13–14: …τὸ δ' εἶναι οὐκ οὐσία οὐδενί. Also *Metaphysics* Z 16, 1040 b 18–19: …φανερὸν ὅτι οὔτε τὸ ἓν οὔτε τὸ ὂν ἐνδέχεται οὐσίαν εἶναι τῶν πραγμάτων…

coming from the different forces is at their core, such diversity deriving from the fact that, in good hermeneutics, every reading (of reality) is always a position in itself (towards it),[64] and there is no interpretation (of reality) without assumptions, there is no interpretation which is not grounded, at its core, on a 'decision' concerning the starting point of the interpretation in question. For the rest, just as Aristotle argued that no being or type of being can arrogate to itself the being, no member of a human community, none of the possessors of λόγος, can arrogate to themselves the community in question. Such community may bear the mark of one or more of its members, which mark may be recognized in it, but the best mark (because it constitutes and maintains the community as such) can only be that which always seeks to preserve the 'diversity', without this entailing the oblivion of what is 'common', that is to say, without such preservation entailing the 'dispersion' of the several individual forces.

However, the difficulties involved in the formation of a genuine human community, irrespective of its nature, lie in the fact that the community is precisely one of the 'othernesses', to the detriment of which the core of the prevailing metaphysical position in this era (i. e., the absolute privilege of the individual human being as an individual) seems to be consolidating itself. We have stated that such absolute privilege exists to the detriment of a divine, transcendent Otherness. The individual's position as the ultimate and unique foundation seems to imply the internalization of such Otherness; in other words, the individual arrogates to himself this divine Otherness, which precisely led us to speak about the "divinization" of the individual. We have sufficiently insisted that this divinization is also to the detriment of the other individuals. And, if this is so, the divinization in question would thus be detrimental for the human community (irrespective of its nature and dimension) as well, which equates to an otherness for the individual since it would necessarily be a 'community' – which is not him – with 'others' – who are not him.[65]

64 Cf. Martin Heidegger, "Logos (Heraklit, Fragment 50) (1951)", in *Vorträge und Aufsätze*, 216: "Jedes Lesen ist schon Legen."

65 These three dimensions of 'otherness' to the detriment of which the 'individual' is privileged can be clearly seen already, whether one agrees with all its content or not, in the following text by A. Machado: "An atheistic communism – my teacher said – would always be a very superficial social phenomenon. Atheism is an essentially individualistic belief: it is that of a man who takes his own existence as the evidence on which he bases a realm of nothingness outside of the limits of his ego. This man either does not believe in God, or he believes himself to be God, which amounts to the same thing. This man also does not believe in his fellowman, in the absolute reality of his neighbor. In neither case has he seen any proof of the other; he has no intuition of

Having said all this, if we still feel like returning to Kant's three philosophical questions, to wit: what can I know?, what ought I to do?, what may I hope?,[66] today it might prove difficult to find metaphysically consistent answers to them (i. e., consistent with what seems to prevail as a foundation nowadays) which do not offer a 'saddening' outlook.[67] The thing is that having to lay one's hopes exclusively on the individual as such produces, in general and from a viewpoint intended to be philosophical, 'sadness' rather than any other effect. Philosophy itself, even – and above all else – understood as metaphysics (if we have not considered yet – although we should not discard it – that nowadays philosophy "is rhetorical thinking, neither objective nor descriptive but, rather, persuasive"[68], which serves no other purpose but to help maintain a conversation, far from any assertions of ultimate principles and incontrovertible truths[69]), 'saddens', or, so as to be coherent – at this point let me move on, for the first time, from the first person plural, which I only used out of modesty, to the first person singular –, it 'saddens me'.[70] That is perhaps the reason why the beginning of the saying attributed to the philosopher of the Garden persistently comes into my mind:

---

otherness, without which one cannot pass from the I to the you. One of the profound teachings of all superior religions is that it is the excessive love of oneself that separates man from God. That he is separated from his fellowman is also implicit in this affirmation. Still, there are vital moments in history when man only believes in his own existence, when he attributes being only to himself. In these moments it is very difficult to accept the existence of God, and also the existence, in an ontological sense, of the watchman in the street. With this *self-made-man*..., with this auto-sufficient monad, you can never speak of communion, or of community, nor of communism. With what, or with whom, could this man commune?" (*Juan de Mairena*, vol. 1, 151 sq.; emphasis by MACHADO).

66 Immanuel KANT, *Kritik der reinen Vernunft* A 804 sq. = B 832 sq.

67 Cf. Antonio MACHADO, *Juan de Mairena*, vol. 2, 29: "... I do not think that culture, let alone wisdom, must necessarily be joyful and a playing matter"; Jean-P. SARTRE, *L'existentialisme est un humanisme*, 24, text cit. above as epigraph: "...la sagesse des nations... est fort triste". Cf. also Gilles DELEUZE, *Nietzsche et la philosophie*, 120: "La philosophie sert à *attrister*. Une philosophie qui n'attriste personne... n'est une philosophie."; Friedrich NIETZSCHE, *Unzeitgemässe Betrachtungen III*, 8, 426 sq.: "und... wäre zu sagen, was Diogenes, als man einen Philosophen lobte, seinerseits einwendete: 'Was hat er denn Grosses aufzuweisen, da er so lange Philosophie treibt und noch Niemanden *betrübt* hat?' Ja, so sollte es auf der Grabschrift der Universitätsphilosophie heissen: 'sie hat Niemanden betrübt'."

68 Gianni VATTIMO, in Teresa OÑATE and Santiago B. OLMO, 151.

69 Richard RORTY, *Philosophy and the Mirror of Nature*, esp. 373, 378, 394.

70 I cannot achieve, at least fully and constantly – perhaps I only do so in some lucid intervals – what MACHADO's *Juan de Mairena* – SARTRE will do so as well but in his own way –, by maximizing the coherence of his extreme skepticism, achieves: being the owner of a sad way of thinking without that making him feel sad. "The background of my thought is sad, but I'm not a sad man, and I don't believe I sadden anyone else. In other words, the fact that I do not adhere to my own thinking saves me from its evil spell, or, rather, deeper than my own thinking is my faith in its

Γελᾶν ἅμα δεῖ καὶ φιλοσοφεῖν..., "We must laugh and philosophize at the same time..." (*Vatican Saying* 41). Every time I am in the presence of this text, I become aware of the fact that, much as I try to fully put it into practice – and I do say this with as much seriousness as I am capable of –, only at times, during the rare moments which are perhaps among those which save me, I achieve the former, probably not as well as I wish, but just the former.

This 'aporetic' atmosphere, without solutions, in which there is no way out – or, if we wish to use common terms, this 'pessimism', which appears to emanate from these pages –, clearly breaks with the ones which now seem to emerge as utopian and millenarist fantasies from the two previous centuries. Those "lendemains qui chantent", the immediate preparation of which was announced Gabriel Péri before gloriously dying in 1941[71], never came. We do not see "the solved puzzle of history (das aufgelöste Rätsel der Geschichte)" announced by young Marx in 1844 with regard to a near future, to wit: "the true solution of the conflict between man and nature, and between man and man; the true solution of the antagonism between existence and essence, between objectification and self-affirmation, between freedom and necessity, between the individual and the species".[72] We do not glimpse "the big change (den großen Umschwung)" which Engels also as a young man in that same year already announced for that century within the framework of the "universal progress of humanity", to wit, "human beings' reconciliation with nature and with themselves".[73] Instead, it is conceivable to think that the owl of Minerva, even though it has not turned into – it could not – a bird of ill omen yet, is far from presenting a positive balance upon flying off at twilight.

Now, it would be absolutely naive to blame this 'sunset' on metaphysics, on philosophy, which is in no way its root, but its symptom. It would be perhaps naive, perhaps no longer possible, for us, the individuals, to lay our hopes on possible innovations in the (common) domain of θεωρία, which, however, gives us the chance to remain awake (and that is something). The door left open is

inanity, the fountain of Youth in which my heart is continually bathing" (*Juan de Mairena*, vol. 2, 140; see, more broadly, 138–140, and also 43 sq.).

71 "Je crois toujours, en cette nuit, que mon cher Paul-Vaillant Couturier avait raison de dire que 'le communisme est la jeunesse du monde' et 'qu'il prépare des lendemains qui chantent'. Je vais préparer tout à l'heure des 'lendemains qui chantent'" ("Lettre d'adieu de Gabriel Péri", in Gabriel PÉRI, *Les lendemains qui chantent*, 59). These are known to have been the last words written by Gabriel PÉRI before he was executed by firearm on December 15, 1941 by the Nazis occupying France. The same words, as is widely known as well, served as a title for his autobiography, published posthumously in 1947.

72 Karl MARX, *Ökonomisch-philosophische Manuskripte*, 536.

73 Friedrich ENGELS, "Umrisse", 505.

that of the domain of πρᾶξις, the domain of action ("only he who acts learns (nur der Thäter lernt)"[74]), in which we have the opportunity, and maybe the duty (the eventual fulfillment of which, however, does not carry, *a priori*, any promises of reward), to be as lucid as possible.[75] Perhaps it is only within this domain – pressed by the inevitable nature of certain decisions, even though a foundation thereof cannot be entirely provided, and in view of the fact that refusing to make a decision is a decision in itself as well – that we can cling to the hope, the paradoxical hope, of that very ancient, obscure wisdom:

> ἐὰν μὴ ἔλπηται, ἀνέλπιστον οὐκ ἐξευρήσει, ἀνεξερεύνητον ἐὸν καὶ ἄπορον[76]

Let me translate it as follows:

> If we do not hope, we will not find that which cannot be hoped for, which is unfindable and to which there is no access.

We are perhaps left with the not so paradoxical hope of that who is capable of understanding that, ultimately,

> The courage of living
> feeds on all of its defeats.[77]

Bibliography

Aristotle, *Metaphysics. A Revised Text with Introduction and Commentary*, 2 vols., edited by William D. Ross. Oxford, Clarendon Press, 1924.

Aristotle, *Prior and Posterior Analytics. A Revised Text with Introduction and Commentary*, edited by William D. Ross. London, Oxford University Press, ²1965.

Aristotle, *Topica et sophistici elenchi. Recensvit brevique adnotatione critica instrvxit*, edited by William D. Ross. Oxford, Oxford University Press, ⁹1991.

Aubenque, Pierre, "Sens et function de l'aporie socratique", *Philosophie antique* 3 (2003), 5–20.

Aubenque, Pierre, *Faut-il déconstruire la métaphysique?* Paris, Presses Universitaires de France, 2009.

Benedetti, Mario, "Lo dice Fukuyama". In Mario Benedetti, *Las soledades de Babel*. Buenos Aires, Seix Barral, 1993, 87 sq.

---

74 Friedrich Nietzsche, *Also sprach Zarathustra*, 331, through Zarathustra.

75 Cf. Pierre Aubenque, *Faut-il déconstruire?*, 3 sq.

76 Heraclitus, DK 22 B 18; Clement of Alexandria, *Stromata* II 17, 4.

77 Carlos R. Ruta, *Brizna perdida*, 58: "El coraje de vivir / se nutre de todas sus derrotas".

Borges, Jorge L., “La casa de Asterión”. In Jorge L. Borges, El *Aleph* (1949). In Jorge L. Borges, *Obras Completas*, vol. 1, edited by Carlos V. Frías. Buenos Aires, Emecé, 1993, 569 sq.
Borges, Jorge L., “Tú”. In Jorge L. Borges, *El oro de los tigres* (1972). In Jorge L. Borges, *Obras Completas*, vol. 2, edited by Carlos V. Frías. Buenos Aires, Emecé, 1993, 491.
Borges, Jorge L., “El suicida”. In Jorge L. Borges, *La rosa profunda* (1975). In Jorge L. Borges, *Obras completas*, vol. 3, edited by Carlos V. Frías. Buenos Aires, Emecé, 1994, 86.
Borges, Jorge L., *Labyrinths. Selected Stories and Others Writings*, edited by Donald A. Yates and James E. Irby. New York, New Directions Publishing, 1964 (34th printing, s. d.).
Borges, Jorge L., *The Book of Sand. The Book of Sand – The Gold of the Tigers (Selected Later Poems)*, translated by Norman Th. di Giovanni and Alastair Reid. Harmondsworth, Penguin Books, 1979 (reprinted 1981).
Brisson, Luc, “Un si long anonymat”. In *La métaphysique. Son histoire, sa critique, ses enjeux*, edited by Jean-Marc Narbonne and Luc Langlois. Paris/Québec, Vrin/Les Presses de l’Université Laval, 1999, 37–60.
Buhle, Johann G., “Ueber die Aechtheit der Metaphysik des Aristoteles”, *Bibliothek der alten Literatur und Kunst* 4 (1788), 1–42.
Clement of Alexandria, *Stromata, Buch I-VI*. In *Clemens Alexandrinus*, vol. 2, edited by Otto Stählin. Leipzig, J. C. Hinrichs’sche Buchhandlung, 1906.
Deleuze, Gilles, *Nietzsche et la philosophie*. Paris, Press Universitaires de France, 1962.
Descartes, René, *Discours de la méthode*. In *Oeuvres de Descartes*, vol. 6, edited by Charles Adam and Paul Tannery. Paris, L. Cerf, 1902, 1–78.
Descartes, René, *Meditationes de prima philosophia*. In *Oeuvres de Descartes*, vol. 7, edited by Charles Adam and Paul Tannery. Paris, L. Cerf, 1904.
Diels, Hermann A. and Walther Kranz, eds., *Die Fragmente der Vorsokratiker. Griechisch und Deutsch*, 3 vols. Berlin, Weidmann, 61951–1952.
Diogenes Laertius, *Lives of Eminent Philosopher*, edited by Tiziano Dorandi. Cambridge, Cambridge University Press, 2013.
Düring, Ingemar, *Aristoteles. Darstellung und Interpretation seines Denkens*. Heidelberg, Carl Winter, 1966.
Engels, Friedrich, “Umrisse zu einer Kritik der Nationalökonomie”. In Karl Marx and Friedrich Engels, *Werke*, vol. 1. Berlin, Dietz Verlag, 1956, 499–524.
Epicurus, *Opere. Introduzione, testo critico, traduzione e note*, edited by Graziano Arrighetti. Torino, Einaudi, 1960.
Fukuyama, Francis, *The End of the History and the Last Man*. New York, The Free Press, 1992.
Gadamer, Hans-Georg, “Zur Vorgeschichte der Metaphysik”. In *Um die Begriffswelt der Vorsokratiker*, edited by Hans-Georg Gadamer. Darmstadt, Wissenschaftliche Buchgesellschaft, 31989, 364–390.
Gadamer, Hans-Georg, *L’inizio della filosofia occidentale. Lezioni raccolte da Vittorio De Cesare*. Milano, Istituto Italiano per gli Studi Filosofici/A. Guerini e associati, 22014.
Gadamer, Hans-Georg, *Der Anfang der Philosophie*. Stuttgart, Reclam, 1996.
Gianneschi, Horacio A., “Aristóteles o el ente no solamente no es un género (I)”, *Ordia Prima. Revista de Estudios Clásicos* 10 (2011), 93–122.
Gianneschi, Horacio A., “Aristóteles o el ente no solamente no es un género (II)”, *Ordia Prima. Revista de Estudios Clásicos* 11/12 (2012–2013), 79–122.
Goethe, Johann W., *Faust. Erster Teil*, edited by Karl J. Schröer. Leipzig, O. R. Reisland, 51907.
Guardini, Romano, *Das Ende der Neuzeit. Ein Versuch zur Orientierung*. Basel, Hess Verlag, 1950.

Hegel, Georg W. F., *Einleitung in die Geschichte der Philosophie*, edited by Johannes Hoffmeister. Hamburg, F. Meiner, [3]1966.

Hegel, Georg W. F., *Grundlinien der Philosophie des Rechts oder Naturrecht und Staatswissenschaft im Grundrisse. Mit Hegels eigenhändigen Notizen und den mündlichen Zusätzen*. In Georg W. F. Hegel, *Werke in zwanzig Bänden* (Auf der Grundlage der *Werke* von 1832–1845 neu edierte Ausgabe. Redaktion E. Moldenhauer und K. M. Michel), vol. 7. Frankfurt a. M., Suhrkamp Verlag, [2]1989 (reprinted 1970).

Hegel, Georg W. F., *Vorlesungen über die Geschichte der Philosophie I*. In Georg W. F. Hegel, *Werke in zwanzig Bänden* (Auf der Grundlage der *Werke* von 1832–1845 neu edierte Ausgabe. Redaktion E. Moldenhauer und K. M. Michel), vol. 18. Frankfurt a. M., Suhrkamp Verlag, 1986 (reprinted 1971).

Hegel, Georg W. F., *Vorlesungen über die Geschichte der Philosophie III*. In Georg W. F. Hegel, *Werke in zwanzig Bänden* (Auf der Grundlage der *Werke* von 1832–1845 neu edierte Ausgabe. Redaktion E. Moldenhauer und K. M. Michel), vol. 20. Frankfurt a. M., Suhrkamp Verlag, 1986 (reprinted 1971).

Heidegger, Martin, *Sein und Zeit*. Tübingen, Niemeyer, [11]1967.

Heidegger, Martin, *Einführung in die Metaphysik (Sommersemester 1935)*, edited by Petra Jaeger. Frakfurt a. M., Klostermann, 1983.

Heidegger, Martin, *Vorträge und Aufsätze (1936–1953)*, edited by Friedrich-Wilhelm von Herrmann. Frankfurt a. M., Klostermann, 2000.

Heidegger, Martin, *Nietzsche II (1939–1946)*, edited by Brigitte Schillbach. Frankfurt a. M., Klostermann, 1997.

Heidegger, Martin, *Zur Sache des Denkens (1962–1964)*. Tübingen, Max Niemeyer Verlag, 1969.

Heidegger, Martin, *Wegmarken (1919–1961)*, edited by Friedrich-Wilhelm von Herrmann. Frankfurt a. M., Klostermann, 1976.

Heidegger, Martin, *Holzwege (1935–1946)*, edited by Friedrich-Wilhelm von Herrmann. Frankfurt a. M., Klostermann, 1977.

Heidegger, Martin, *Metaphysik und Nihilismus. 1. Die Überwindung der Metaphysik. 2. Das Wesen des Nihilismus*, edited by Hans-Joachim Friedrich. Frankfurt a. M., Klostermann, 1999.

Heidegger, Martin, *Identität und Differenz*, edited by Friedrich-Wilhelm von Herrmann. Frankfurt a. M., Klostermann, 2006.

Houellebecq, Michel, *Les particules élémentaires*. Paris, Flammarion, 1998.

Jaran, François and Christophe Perrin, *The Heidegger Concordance*, vol. 2. London/New York, Bloomsbury, 2013.

Kant, Immanuel, *Kritik der reinen Vernunft*, edited by Raymund Schmidt. Hamburg, Felix Meiner, 1956.

Kant, Immanuel, *Kritik der praktischen Vernunft*, edited by Karl Vorländer. Leipzig, Felix Meiner, [8]1922.

Krämer, Hans, *Platone e i fondamenti della metafisica. Saggio sulla teoria dei principi e sulle doctrine non scritte con una raccolta dei documenti fondamentali in edizione bilingue e bibliografia*. Milano, Vita e Pensiero, [6]2001.

Lanceros, Patxi, "Antropología hermenéutica". In *Diccionario interdisciplinar de Hermenéutica*, edited by Andrés Ortiz Osés and Patxi Lanceros. Bilbao, Universidad de Deusto, [3]2001, 45–57.

Machado, Antonio, *Juan de Mairena – Sentencias, donaires, apuntes y recuerdos de un profesor apócrifo*, 2 vols. Buenos Aires, Losada, [6]1969.

Marx, Karl, *Ökonomisch-philosophische Manuskripte aus dem Jahre 1844*. In Karl Marx and Friedrich Engels, *Werke*, Ergänzungsband: *Schriften – Manuskripte – Briefe bis 1844, Erster Teil*. Berlin, Dietz Verlag, 1968, 465–588.
Moraux, Paul, *Les listes anciennes des ouvrages d'Aristote*. Louvain, Éditions Universitaires de Louvain, 1951.
Nietzsche, Friedrich, *Der Wille zur Macht. Versuch einer Umwerthung aller Werthe, Drittes und Viertes Buch*. In Friedrich Nietzsche, *Nachgelassene Werke*, vol. 16. Leipzig, A. Kröner Verlag, [2]1922.
Nietzsche, Friedrich, *Unzeitgemässe Betrachtungen III: Schopenhauer als Erzieher*. In Friedrich Nietzsche, *Sämtliche Werke. Kritische Studienausgabe (KSA)*, vol. 1, edited by Giorgio Colli and Mazzino Montinari. Berlin/New York, De Gruyter, [2]1988, 335–428.
Nietzsche, Friedrich, *Menschliches, Allzumenschliches. Ein Buch für freie Geister I-II*. In Friedrich Nietzsche, *Sämtliche Werke. Kritische Studienausgabe (KSA)*, vol. 2, edited by Giorgio Colli and Mazzino Montinari. Berlin/New York, De Gruyter, [2]1988.
Nietzsche, Friedrich, *Die fröhliche Wissenschaft ("la gaya scienza")*. In Friedrich Nietzsche, *Sämtliche Werke. Kritische Studienausgabe (KSA)*, vol. 3, edited by Giorgio Colli and Mazzino Montinari. Berlin/New York, De Gruyter, [2]1988, 343–651.
Nietzsche, Friedrich, *Also sprach Zarathustra. Ein Buch für Alle und Keine I-IV*. In Friedrich Nietzsche, *Sämtliche Werke. Kritische Studienausgabe (KSA)*, vol. 4, edited by Giorgio Colli and Mazzino Montinari. Berlin/New York, De Gruyter, [2]1988.
Nietzsche, Friedrich, *Nietzsche contra Wagner*. In Friedrich Nietzsche, *Sämtliche Werke. Kritische Studienausgabe (KSA)*, vol. 6, edited by Giorgio Colli and Mazzino Montinari. Berlin/New York, De Gruyter, [2]1988, 412–445.
Nietzsche, Friedrich, *Nachgelassene Fragmente 1887–1889*. In Friedrich Nietzsche, *Sämtliche Werke. Kritische Studienausgabe (KSA)*, vol. 13, edited by Giorgio Colli and Mazzino Montinari. Berlin/New York, De Gruyter, [2]1988.
Oñate, Teresa and Santiago B. Olmo, "Entrevista a Gianni Vattimo", *Suplementos Anthropos 10* (1988), 147–155.
Onfray, Michel, *La sculpture de soi. La morale esthétique*. Paris, Grasset & Fasquelle, [11]2014.
Onfray, Michel, *Politique du rebelle. Traité de résistance et d'insoumission*. Paris, Grasset & Fasquelle, [13]2013.
Péri, Gabriel, *Les lendemains qui chantent. Autobiographie présentée par Aragon*. Paris, Éditions sociales, 1947.
Plato, *Cratylus*. In *Platonis opera recognovit breviqve adnotatione critica instrvxit*, vol. 1: *Tetralogias I-II continens*, edited by John Burnet. Oxford, Oxford University Press, 1900.
Plato, *Theaetetus*. In *Platonis opera recognovit breviqve adnotatione critica instrvxit*, vol. 1: *Tetralogias I-II continens*, edited by John Burnet. Oxford, Oxford University Press, 1900.
Plato, *Sophist*. In *Platonis opera recognovit breviqve adnotatione critica instrvxit*, vol. 1: *Tetralogias I-II continens*, edited by John Burnet. Oxford, Oxford University Press, 1900.
Plato, *Republic*. In *Platonis opera recognovit breviqve adnotatione critica instrvxit*, vol. 4: *Tetralogiam VIII continens*, edited by John Burnet. Oxford, Oxford University Press, 1902.
Quine, Willard. V. O., *Ontological Relativity and Others Essays*. New York, Columbia University Press, 1969.
Reiner, Hans, "Die Entstehung und ursprüngliche Bedeutung des Namens Metaphysik", *Zeitschrift für philosophische Forschung* 8 (1954), 210–237.
Reiner, Hans, "Die Enstehung der Lehre vom bibliothekarischen Ursprung des Namens Metaphysik", *Zeitschrift für philosophische Forschung* 9 (1955), 77–99.

Renaut, Alain, *L'ere de l'individu. Contribution à une histoire de la subjectivité*. Paris, Gallimard, 1989.
Rilke, Rainer M., *Das Stunden-Buch enthaltend die drei Bücher. Vom moenchischen Leben – Von der Pilgerschaft – Von der Armuth und vom Tode*. Leipzig, Insel-Verlag, [2]1907.
Rilke, Rainer M., *Die Aufzeichnungen des Malte Laurids Brigge*, edited by Karl-Maria Guth. Berlin, Hofenberg, 2013.
Rorty, Richard, *Philosophy and the Mirror of Nature*. Princeton, Princeton University Press, 1979.
Ruta, Carlos R., *Brizna perdida*. Sevilla, Ediciones de la Isla de Siltolá, 2014.
Sabato, Ernesto, *El túnel*, edited by Ángel Leiva. Madrid, Cátedra, [21]1994.
Sabato, Ernesto, *La resistencia*. Buenos Aires, Seix Barral, 2000.
Sartre, Jean-Paul, *L'existentialisme est un humanisme. Présentation et notes par Arlette Elkaïm-Sartre*. Paris, Gallimard, 1996.
Serna Arango, Julián, *Ontologías alternativas. Aperturas al mundo desde el giro lingüístico*. Barcelona, Anthropos, 2007.
Sextus Emiricus, *Adversus dogmaticos libros quinque (Adv. Mathem. VII-XI) continens*. In *Sexti Empirici opera recensuit*, vol. 2, edited by Hermann Mutschmann. Leipzig, Teubner, 1914.
Schopenhauer, Arthur, *Die Welt als Wille und Vorstellung I*. In Arthur Schopenhauer, *Sämtliche Werke*, vol. I, edited by Wolfgang F. von Löhneysen. Stuttgart, Cotta-Insel, [2]1968 (reprinted, Darmstadt, Wissenschaftliche Buchgesellschaft, 2004).
Schopenhauer, Arthur, *Die Welt als Wille und Vorstellung II*. In Arthur Schopenhauer, *Sämtliche Werke*, vol. II, edited by Wolfgang F. von Löhneysen. Stuttgart, Cotta-Insel, [2]1968 (reprinted, Darmstadt, Wissenschaftliche Buchgesellschaft, 2004).
Simplicius, *Commentarius in IV libros Aristotelis de caelo*, edited by Simon Karsten. Utrecht, Kemink et filium, 1865.
Simplicius, *In Aristotelis de caelo commentaria*, edited by Johann L. Heiberg. Berlin, G. Reimer, 1894.
Treptow, Elmar, *Der Zusammenhang zwischen der Metaphysik und der Zweiten Analytik des Aristoteles*. München, Anton Pustet, 1966.
Wittgenstein, Ludwig, *Tractatus logico-philosophicus. The German Text of Ludwig Wittgenstein's Logisch-philosophische Abhandlung, with a New Translation by D. F. Pears & B. F. McGuinness, and with the Introduction by Bertrand Russell*. London, Routledge & Keagan Paul, [2]1963.
Xenophon, *Memorabilia (Ἀπομνημονευμάτων)*. In *Xenophontis opera omnia. Recognovit brevique adnotatione critica instrvxit*, vol. 2: *Commentarii, Oeconomicvs, Convivivm, Apologia Socratis*, edited by Edgar C. Marchant. Oxford, Clarendon Press, 1900.

Bernd Schneidmüller

Enduring Coherence and Distance

Monastic and Princely Communities in Medieval Europe[1]

Community was both a basic concept of human relations and a challenge of life which restricted privacy as well as individuality. The stimulating ideas of Gert Melville and Carlos Ruta provided the inspiration for this article.[2] The constraint of living in communities was a challenge for all individuals. Having just arrived coming from the lively capital of Buenos Aires at our secluded conference hotel in the Delta del Tigre I am truly able to compare different types of social communities.

For a European medievalist approaching the ideas and realities of community means a challenge as well. We are being trained to understand life in premodern periods as people living a life largely shaped by personal ties, a life that lacks any institutional framing for the most part.[3] Historians usually consider the early Middle Ages as a period marked by face-to-face communities established by physical relationships between individuals. The German term for this is 'Personenverband' (a group of people, of individuals) or even more sophisticated: 'Personenverbandsstaat'.[4] This term distinguishes the basic principle of medieval societies from later patterns of institutionalization which tied all subjects to political or administrative structures. Thus, the term "communities of individuals" defines the common cornerstone of social life in the Middle Ages. In fact, all individuals were involved and embedded in various relationships, namely kinship, confraternity, vicinity, equality, or subjection.[5]

Community obviously was a basic principle of medieval life. Hence, historical research has focused on the impacts of communities, their varying ranges, their shaping, and their alterity when compared to the same phenomena in the modern world. Even though there was a remarkable medieval discourse on com-

---

1 Extended version of my paper given at the International Conference "Community as Challenge of Life", Universidad Nacional de San Martín in Buenos Aires (Argentina), November 12, 2015.

2 See the introduction of this volume.

3 There is no article "Gemeinschaft" in the famous encyclopedia "Geschichtliche Grundbegriffe". See Otto BRUNNER et al., eds., *Geschichtliche Grundbegriffe*, vol. 6, articles "Staat und Souveränität", 1–154; "Stand, Klasse", 155–284.

4 Theodor MAYER, *Mittelalterliche Studien*; Karl BOSL, *Frühformen der Gesellschaft*; Hans K. SCHULZE, *Grundbegriffe*, vols. 1–2.

5 Gert MELVILLE and Martial STAUB, eds., *Enzyklopädie des Mittelalters*, vol. 1, 1–175.

munity which is worth studying[6], modern historians rather prefer considering the proportions of society and community or the symbolic meanings of community.[7]

In his well-known book on 'The Symbolic Construction of Community' Anthony P. Cohen introduced models of symbolizing the boundary, the communities of meaning, and the correlations between community and identity. His main argument is that communities existed as a mental construct. They come into being by the use of symbols, rituals, and by staging human gregariousness.[8] I will follow these ideas Cohen outlined by focusing on the variability and iconology of monastic and princely communities in medieval Europe (12th–14th c.).[9]

Let me elaborate this by pointing out four important principles:

1) Bishops and clerics defined medieval society as a God-given coexistence of major groups or *ordines* (such as clerics, nobles, and peasants).[10] Nevertheless, proximity was an enduring challenge for all individuals forced to live and act in groups.
2) Monastic rules elaborated on communal life (*vita communis*) of monks or nuns as a prefiguration of celestial paradise. The monastic communities of abbots and monks embodied Christ and his disciples in the present world. The need for and the actual life of communities were always based on divine transcendence.[11] The New Testament outlined the mandatory guidelines of charity and propinquity set out by Christ and his disciples, by the unity in mind and life founded on the Holy Spirit, and by the confraternity of the holy Apostles sharing everyday life as well as their possessions.
3) Attendance at the king's court was one of the most important feudal duties for all princes and noblemen in medieval politics. Assemblies organized personal government as well as the public performance of decision-making and

6 Hans-Jürgen DERDA, *Vita Communis*; Richard CORRADINI et al., eds., *The Construction of Communities*.

7 Mirko BREITENSTEIN, *Das Noviziat*; Jörg SONNTAG, *Klosterleben*; Anne MÜLLER and Karen STÖBER, eds., *Self-Representation*; Klaus SCHREINER, *Gemeinsam leben*; Mirko BREITENSTEIN et al., eds., *Rules and Observance*; Gert MELVILLE et al., eds., *Die Klöster der Franziskaner*; Gert MELVILLE, *The World of Medieval Monasticism*.

8 Anthony P. COHEN, *The Symbolic Construction of Community*, 39–118.

9 Bernd SCHNEIDMÜLLER, *Grenzerfahrung und monarchische Ordnung*, 171–187.

10 Georges DUBY, *Les trois ordres*; Otto Gerhard OEXLE, "Deutungsschemata der sozialen Wirklichkeit".

11 Gert MELVILLE, "Regeln – *Consuetudines*-Texte – Statuten"; Christina LUTTER, "Social Groups"; Gert MELVILLE et al., eds., *Innovationen durch Deuten und Gestalten*; Gert MELVILLE et al., eds., *Gerechtigkeit*; Gert MELVILLE, *The World of Medieval Monasticism*; Christina LUTTER, "*Vita communis*". For the Vienna project on 'Visions of community' see: http://www.univie.ac.at/viscom/ (April 3, 2016).

jurisdiction. In face-to-face communities it was unusual to voice opposition in a deliberate or open discussion.[12] Staying together and not dissolving already expressed consensus. Leaving the community without permission or not even attending the assembly signified open dissent. Therefore, proximity offered an opportunity to participate. But simultaneously, it compelled one to postpone certain individual actions as well as self-fulfillment.

4) Communities of individuals or bodies symbolized what we would later come to call institutions. In the 14th century the mere assemblies of the ruler and the princes represented the empire. Thus, the Golden Bull agreed upon by Emperor Charles IV and the prince electors in 1356 shaped the Holy Roman Empire by rituals of embodiment, i.e. honor services, common meals, or processions.[13]

I will outline these short assumptions by pointing out some examples, combining written texts with the iconography of different types of communities. My objective is not to discover teleological developments from premodern to modern times but rather to demonstrate the alterity of medieval designs of society.[14]

1 Society and community

The editors of this volume exhorted us not to mix the terms society and community. This reasonable claim is a result of a very complex process in modern sociological discussions. Starting with the famous book by Ferdinand Tönnies on 'Gemeinschaft und Gesellschaft', first published in 1887, the theoretical concepts of society and community have been described by employing a number of diverse typologies.[15] Today we are far from Tönnies' romantic views who had claimed a chronological precedence of community in the history of humankind. These ideas arose in the later 19th century to criticize modernity and to emphasize the ideals of a cozy past. His concept told a story of decline. After a precious medieval period of community characterized by organic human relations and intimate ways of living together modern society was instead constituted by individualism, by rationalism, as well as by the collapse of contiguousness, solidarity, and virtues. These assertions were

12 Gert MELVILLE, ed., *Institutionen und Geschichte*; Jörg PELTZER et al., eds., *Politische Versammlungen und ihre Rituale*; Julia DÜCKER, *Reichsversammlungen im Spätmittelalter*.

13 Bernd SCHNEIDMÜLLER, "Inszenierungen und Rituale"; Sandra WEFERLING, *Spätmittelalterliche Vorstellungen*.

14 Klaus RIDDER and Steffen PATZOLD, eds., *Die Aktualität der Vormoderne*; Otto Gerhard OEXLE, *Geschichtswissenschaft im Zeichen des Historismus*.

15 Ferdinand TÖNNIES, *Gemeinschaft und Gesellschaft*.

not based on an interpretation of historical sources. In fact, they served as arguments that were only superficially gilded by the use of alleged historical empirical data. However, the idea of using a distant past as a background has accompanied modern discussions in cultural studies as well as in history.[16]

Fig. 1: Christ and the three estates. Woodcut. LICHTENBERGER, *Prognosticatio*, Mainz, Jacob Meydenbach, 1492 [https://commons.wikimedia.org/wiki/File:Three-estates-prognosticatio-lichtenberger-mainz-1492.jpg]

16 Otto Gerhard OEXLE, "Das Mittelalter und das Unbehagen an der Moderne"; Nele SCHNEIDEREIT, *Die Dialektik von Gemeinschaft und Gesellschaft*.

However, fundamental medieval sources did not claim that there was a contrast between society and community. Instead, they portrayed society as an accumulation of communities. Soon after the year 1000 Bishop Adalbero of Laon and Bishop Gerard of Cambrai described the existence of humankind as an interaction of three estates, namely of praying men, peasants, and warriors. Gerard pointed out that this threefold partition originated from the creation of humans by God himself. The unknown chronicler of the Bishops of Cambrai paraphrased Gerard's doctrine in the following words:

> He taught that humankind was divided from the very beginning in a threefold way, in praying men, in peasants, and in warriors. He clearly outlined that these three parts supported one another on the right and on the left hand side.
>
> (*Genus humanum ab initio trifarium divisum esse monstravit, in oratoribus, agricultoribus, pugnatoribus; horum singulos alterutrum dextra laevaque foveri evidens documentum dedit*).[17]

Adalbero presented the unique house of God on earth as a mutual entanglement of different parts:

> The threefold house of God at once appeared as a unity. One part prays, the second fights, the third labors. But these parts are joined together without any disjunction. The efforts of two parts stand for the office of one. They give comfort to each other by mutual change, because this threefold combination is in fact unitary. As long as this law shines the world rests in peace.
>
> (*Triplex ergo Dei domus est quae creditur una.*
Nunc orant, alii pugnant aliique laborant.
Quae tria sunt simul et scissuram non patiuntur:
Alternis uicibus cunctis solamina prebent.
Est igitur simplex talis conexio triplex.
Dum lex preualit tunc mundus pace quieuit.)[18]

According to clerics of the central Middle Ages dominion or rulership came into existence because of the fall of Adam and Eve.[19] Emperor Frederick II or Emperor Charles IV claimed their domination as a God-given necessity to restrict the natural peccability of humans.[20] These theological doctrines affirmed the hierar-

17 *Gesta episcoporum Cameracensium*, see III 52, 485.
18 ADALBERO OF LAON, *Carmen ad Rotbertum regem* 22, lines 295–301. See Georges DUBY, *Les trois ordres*; Otto Gerhard OEXLE, "Deutungsschemata der sozialen Wirklichkeit".
19 Wolfgang STÜRNER, *Peccatum und Potestas*; Bernhard TÖPFER, *Urzustand und Sündenfall*.
20 Emperor Frederick II. in the preamble (Prooemium) of his constitutions for Sicily 1231: *Die Konstitutionen Friedrichs II.*, 145–148; see Wolfgang STÜRNER, "Rerum necessitas und divina

chy as well as the invariability of society which consisted of different parts being integrated by functional relations. The collocation of these three parts created the community of all humans. Concurrently, praying men, warriors, and peasants constituted distinct communities of their own.[21]

Combining ancient Roman ethnography with biblical traditions of the Old Testament late medieval authors presented the history of humankind as a process from individuality towards community. At the end of the Middle Ages the Dominican Annius of Viterbo († 1502) noticed in his so-called 'Pseudo-Berosus' that humans started to live together after the Noachian Deluge:

> Mankind was greatly multiplied, and necessity compelled the acquisition of new dwelling-places. Then father Janus encouraged the chiefs to seek new dwelling-places and live in a social community of men, and build cities. And so he designated three parts of the world, Asia, Africa, and Europe, as he had seen them before the Flood.
>
> (*Multiplicatum est in immensum genus humanum: et ad comparandas nouas sedes necessitas compellabat. Tum Ianus pater adhortatus est homines principes ad quaerendas nouas sedes et communem coetum inter homines agendum: et aedificandas urbes. Designauit itaque illas tres partes orbis Asiam Aphricam et Europam: ut ante diluuium uiderat*).[22]

In this perspective social complexity appeared as an historical advancement in humankind.

2 Monastic communities as prefiguration of paradise

The principal aim of Latin monasticism was not only to prepare for Judgment Day but to come to a reflection of community in celestial eternity in the present world.[23] Monks and nuns designed their cohabitation based on examples from

provisio". – Charles IV as King in the abortive constitutions for his kingdom of Bohemia 1355: *Maiestas Carolina. Der Kodifikationsentwurf Karls IV. für das Königreich Böhmen von 1355*, edited by Bernd-Ulrich HERGEMÖLLER (Veröffentlichungen des Collegium Carolinum, 74). München, Oldenbourg, 1995, 32.

21 Alfred HAVERKAMP, "Leben in Gemeinschaften"; Otto Gerhard OEXLE and Andrea von HÜLSEN-ESCH, eds., *Die Repräsentation der Gruppen*; Alfred HAVERKAMP, "Neue Formen von Bindung und Ausgrenzung".

22 Ronald E. ASHER, *National Myths*, 200 sq. See Thomas LEHR, *Was nach der Sintflut wirklich geschah.*

23 Gerd ALTHOFF et al., eds., *Person und Gemeinschaft*; Arnold ANGENENDT, *Geschichte der Religiosität*; Klaus SCHREINER, *Gemeinsam leben*; Gert MELVILLE, *Frommer Eifer und methodischer*

the New Testament. The Rule of Saint Benedict served as a guideline for monastic life. This famous text defined important rules and gave mandatory advice for the course of life as well as for the daily practice in monasteries. But looking closer we realize that it contained only marginal references to a theory of community. I quote some sentences concerning the abbot's calling on the brethren for counsel (chapter 3) and on the sense of stability in the community (chapter 4). Because monastic life was based on the principle of obedience the humble advice of the brethren was valuable but was merely a prerequisite for the abbot's final independent decision:

> Whenever any important business has to be done in the monastery, let the Abbot call together the whole community and state the matter to be acted upon. Then, having heard the brethren's advice, let him turn the matter over in his own mind and do what he shall judge to be most expedient. The reason we have said that all should be called for counsel is that the Lord often reveals to the younger what is best.
>
> Let the brethren give their advice with all the deference required by humility, and not presume stubbornly to defend their opinions; but let the decision rather depend on the Abbot's judgment, and all submit to whatever he shall decide for their welfare.
>
> However, just as it is proper for the disciples to obey their master, so also it is his function to dispose all things with prudence and justice.

Chapter 4 concludes with a request:

> Now the workshop in which we shall diligently execute all these tasks is the enclosure of the monastery and stability in the community.[24]

These rules highlighted the importance of humility and obedience. The authors of Saint Benedict's Rule obviously knew about deviance, opposition, or individuality. But they did not discuss or reflect upon the challenges of community in their regime of rules because they did not accept a broader variability of decision making. Since they based the principles of humility and obedience on God's mandatory order, no individual devoting his life to God was allowed to dissent or to diverge from the instructions given. Individuality therefore symbolized an error

---

Betrieb; Gert Melville and Carlos Ruta, eds., *Thinking the Body*; Gert Melville, *The World of Medieval Monasticism*.

24 *Saint Benedict's Rule for Monasteries*, translated from the Latin by Leonard J. Doyle OSB, of Saint John's Abbey (© Copyright 1948, 2001, by the Order of Saint Benedict, Collegeville, MN 56321): http://www.osb.org/rb/text/toc.html (April 3, 2016).

in a coherent system which understood a monastery as the house of God (*domus Dei*) on earth.[25]

Fig. 2: St Benedict as instructor of his monks, 1419. Miniature. Heidelberg, Universitätsbibliothek, Cpg 144, fol. 334v [http://digi.ub.uni-heidelberg.de/diglit/cpg144/0692?sid=be2358e590c03cd12471e7a4fdb719ab]

However, our medieval sources are full of stories concerning misguided monks and nuns. Their inability to bear the restrictions of their confined communities demonstrated the limits of a religious dream in reality.[26] On the one hand the change from secular existence to monastic life marked a considerable change in an individual's life and an irreversible conversion to God.[27] In theory the congregation had already won the harbor of salvation and the pleasures of paradise. On the other hand, these lofty ideas could only form a stronghold against the temptations and the demons all around. In our project at the Heidelberg Academy of Sciences and Humanities Julia Burkhardt and Verena Schenk zu Schweinsberg study

---

25 Mirko BREITENSTEIN et al., eds. *Rules and Observance*; Christina LUTTER, "Geistliche Gemeinschaften in der Welt"; Mirko BREITENSTEIN et al., eds., *Identität und Gemeinschaft.*
26 Steffen PATZOLD, *Konflikte im Kloster.*
27 Mirko BREITENSTEIN, *Das Noviziat.*

significant monastic texts of the 13th century describing the ideals of monastic communities as well as the permanent terrible attacks of demons against monks who only wanted to obtain sanctity. The 'Liber revelationum' of Richalm, Abbot of Schöntal, a Cistercian monastery in the diocese of Würzburg, reported in detail how legions of demons tempted poor Richalm.[28]

To overcome all these seductions prudent authors like the Cistercian Caesarius of Heisterbach[29] or the Dominican Thomas of Cantimpré compiled large collections of examples to support pastoral care as well as monastic education.[30] We can expect Julia Burkhardt's new edition of Thomas of Cantimpré's *Bonum universale de apibus* in the next few years. This important text compares monastic congregations to another particular community, a swarm of bees, in order to demonstrate that both – staying in a community as well as rulership and subjection – are common parts of life. In fact, Thomas of Cantimpré did not outline a comprehensive theory of community. He rather compiled various examples to present an ideal of monastic cooperation, defining the exemplary responsibilities of prelates and the overall duties of subjects. Many chapters start with only short references to the king of the bees and the great number of working bees as his servants.

In spite of the comprehensive title attributed to the work of Thomas (*Bonum commune*) in fact he did not evolve a profound theory of community. Instead he gave an allegorical explanation connecting examples with some ideas of tripartite hierarchies (*diuisio triplex*). In a very brief chapter between Book I (concerning the prelates) and Book II (concerning the subordinates) Thomas presented the ecclesiastical domain as an ensemble of bishops, priests, and clerics ruled by the pope. Aside from the church, the secular sphere integrates princes, warriors, and common men governed by the emperor. According to the bees being separated into three parts and unified under their king every monastic community consists of office-holders (*officiati*), claustral brothers (*claustrales*), and lay brothers (*conuersi*) all of them subject to one spiritual head (*praelatus*).[31]

---

28 Richalm of Schöntal, *Liber revelationum*, edited by Paul Gerhard SCHMIDT (Monumenta Germaniae Historica. Quellen zur Geistesgeschichte des Mittelalters, 24). Hannover, Hahn, 2009. See Paul Gerhard SCHMIDT, "Von der Allgegenwart der Dämonen".

29 Caesarius of Heisterbach, *Dialogus miraculorum. Dialog über die Wunder*, 5 vols., with an introduction by Horst SCHNEIDER, translated by Nikolaus NÖSGES and Horst SCHNEIDER (Fontes christiani, 86/1–5). Turnhout, Brepols, 2009.

30 Markus SCHÜRER, *Das Exemplum*.

31 For now: Thomas Cantipratanus, *Bonum universale de apibus*, edited by Georgius COLVENERIUS. Douai 1627, 107 sq. See Julia DÜCKER, "Vorstellungen von Gemeinschaft und sozialer Ordnung"; Julia BURKHARDT, "Die Welt der Mendikanten als Bienenschwarm und Vorstellung"; EAD., "Predigerbrüder im Bienenstock des Herrn".

In medieval monasticism the influence of communities furthermore varied in different stages, on the one hand being restricted to only one community as was the case with many Benedictine monasteries or – on the other hand – integrating a wider union of cloisters like the later monastic orders of the Cistercians, the Franciscans, or the Dominicans.[32] The iconography of medieval monastic communities often only presents an ensemble of outstanding charismatic leaders[33] together with ordinary monks or the distinctions between different monastic orders in their very specific vestments. The selected colors – such as black, grey, or white – proclaimed different types of self-assurance. In any case, spirituality and liturgy were the crucial foundations of all monastic communities.

3 Princely attendance as constraint or privilege

Historians have often described medieval society as a top-down hierarchy totally oriented towards a ruler. In fact, the system of feudal tenure evoked the idea of the lord's supremacy and the vassal's subjection.[34] But modern revisions of this traditional doctrine[35] have given rise to another reconstruction of medieval society established by entangled rulership as well as by the power of consensus.[36] Being present at the king's court and being a member of the king's army – these were the major duties of a vassal. In return, the ruler was expected to react favorably to advice and to be a prudent mediator. Medieval assemblies offered the best platform to perform the connectivity of a political community. The rituals of political decision-making and of its staging were a mirror of these ideals combining the magnanimity of a wise ruler with the affection and consensus of his self-reliant princes. Differing from our modern parliaments medieval courts or assemblies were an arena of demonstrating agreement, not of deliberate discussions or open controversy.[37]

---

32 Gert MELVILLE, *The World of Medieval Monasticism*; Gert MELVILLE and Anne MÜLLER, eds., *Mittelalterliche Orden und Klöster im Vergleich*; Jörg SONNTAG, *Klosterleben*; Anne MÜLLER and Karen STÖBER, eds., *Self-Representation*; Mirko BREITENSTEIN et al., eds., *Innovation in Klöstern und Orden*; Gert MELVILLE et al., eds., *Innovationen durch Deuten und Gestalten*.

33 Franz J. FELTEN et al., eds., *Institution und Charisma*.

34 François Louis GANSHOF, *Feudalism*.

35 Susan REYNOLDS, *Fiefs and Vassals*; Susan REYNOLDS, *Kingdoms and Communities*; Steffen PATZOLD, *Das Lehnswesen*.

36 Bernd SCHNEIDMÜLLER, "Rule by Consensus". On the practice of entangles rulership see ID., "Verantwortung aus Breite und Tiefe".

37 Jörg FEUCHTER and Johannes HELMRATH, eds., *Politische Redekultur*; Julia DÜCKER, *Reichsversammlungen im Spätmittelalter*; Jörg FEUCHTER and Johannes HELMRATH, eds., *Parlamentarische Kulturen*; Julia BURKHARDT, "Procedure, rules and meaning of political assemblies".

Fig. 3: King Henry II enthroned. Miniature on parchment, 1002–1014. Regensburg Sacramentary, München, Bayerische Staatsbibliothek, Clm 4456, fol. 11v [https://commons.wikimedia.org/wiki/File:Sacramentary_of_king_Henry_II_-_throne.jpg]

The famous conflict between Emperor Frederick I Barbarossa and Duke Henry the Lion is a good example one can use to understand medieval realities of proximity and distance. Henry had been present at the emperor's court for many years when he was still in Barbarossa's grace. But after falling out with his ruler Henry naturally stayed away from court and rejected all invitations to meet with Frederick or with the princes of the Empire. In public trial the emperor himself accused Henry of neglecting his feudal duties and of his offending the imperial majesty.

The princes as judges in court condemned their peer because of unbearable arrogance (in Latin: *contumacia*) and confiscated his fiefdoms. This example shows the paradoxes as well as the challenges of community. For Henry the Lion it had been impossible to furthermore accept being associated with his peers in the usual way. Therefore his peers used the community they formed to oust him as an outlaw from his former rank.[38]

The political elites of the central Middle Ages interacted with each other on various levels of coherence. The ruler together with his close entourage formed an exclusive court constantly moving through different parts of the Empire. This core group assembled with an additional larger community in solemn diets, usually four times a year. These assemblies formed the political body of the Empire in pre-institutional times – practically the Empire itself.[39] The rituals of interaction and performance represented the long-lasting idea of community in occasional temporality.

This specific type of political framing defined by a changing interaction of emperors and princes shaped the oligarchic outline of the Holy Roman Empire. In the long run from the 10th to the 15th century political theory as well as political iconography reflected the modified designs of entangled rulership.[40] The hierarchies in political communities became smaller and smaller, and gave rise to the obvious equality of Emperor and prince-electors.[41]

4 Performing political community

In the last part of this article I would like to focus on some medieval configurations presenting the joint community of monarchs and their subjects. We can observe a distinct change in the portrayals of rulers and their settings in political iconography from the 10th to the 15th century. This can be explained by the

---

38 Joachim EHLERS, *Heinrich der Löwe*; Knut GÖRICH, *Friedrich Barbarossa*; Bernd SCHNEIDMÜLLER, *Die Welfen*, 224–239.

39 Benjamin ARNOLD, *Princes and Territories*; Peter MORAW, ed., *Deutscher Königshof*; Gabriele ANNAS, *Hoftag*.

40 Henning OTTMANN, *Geschichte des politischen Denkens*; Bernd SCHNEIDMÜLLER and Stefan WEINFURTER, eds., *Heilig – Römisch – Deutsch*; Jürgen MIETHKE, *Politiktheorie im Mittelalter*.

41 Ernst SCHUBERT, *König und Reich*; Peter MORAW, *Von offener Verfassung*.

growing influence of consensus and participation, at first exercised by princes and later on by non-princely nobles and citizens. The early medieval concept of depicting a ruler as a "Christ of the Lord" in celestial transcendence ceased in the 11th, in the 12th century at the latest. Around the year 1000 illuminators of liturgical manuscripts portrayed their rulers such as Emperor Otto III or Emperor Henry II together with Christ, with the holy Apostles, or with saints.[42]

Fig. 4: Emperor with seven prince-electors. The Golden Bull. First illustrated incunable, Straßburg 1485.

These arrangements completely changed from the 13th century onwards.[43] First of all, the Holy Roman Empire presented itself as an ensemble comprising the Emperor and the prince-electors. The Golden Bull of Emperor Charles IV issued in 1356 finally established this group consisting of the Emperor and the seven prince-electors as the pivotal part of the Empire. In contrast to modern constitutions the Golden Bull orders the community of only these eight men: one

42 Ludger KÖRNTGEN, *Königsherrschaft und Gottes Gnade*; Wolfgang Eric WAGNER, *Die liturgische Gegenwart*; Eliza GARRISON, *Ottonian Imperial Art*.
43 Bernd SCHNEIDMÜLLER, *Grenzerfahrung und monarchische Ordnung*, 171–187.

emperor acting together with seven prince-electors. In this fundamental document, which served as the basic foundation of the Empire until its end in 1806, the entire authority of the Empire is only vested in a small elite. Furthermore, the many detailed chapters of the Golden Bull completely left out all other important groups in the hierarchy of the Empire as well as millions of subjects.[44]

The political iconography in the western European kingdoms such as France or England also presented a king and his distinguished advisers together. Differences are obvious. Miniatures of the French king sitting in a 'Lit de justice'[45] or of the coronation of Charles V of France with the assistance of his peerage[46] portrayed a wider range of social groups – such as kings, nobles, even counselors, citizens, or scholars – to form a complex entity representing the realm. The English *Modus tenendi parliamentum* of the 14th century integrated the royal administration into the political framework of monarchy, nobles, and clergy.[47] While the Empire appeared as quite an oligarchic corpus of the Emperor with seven princes, the western kingdoms were imagined as a broader ensemble of political participants.

Conclusion

At first glance, communities seem to be a simple matter in the Middle Ages. But a closer look reveals inconsistencies in two respects.

(1) Staying in a community was an intense challenge for all individuals – even long before the modern period of individualization. Thus, making these communities stable was a challenging task for all groups.

(2) Monastic and princely communities in medieval Europe were based on different foundations and provoked different kinds of challenges. For a monk or a nun daily life signified prefiguration of paradise work. This involved to rigorously subordinate to a spiritual order. Enduring community in political assemblies implied only temporary integration for an individual prince or nobleman. Hence, the frequency and quality of relinquishing one's respective individuality varied considerably.

44 Axel GOTTHARD, *Säulen des Reiches*; Matthias PUHLE and Claus-Peter HASSE, eds., *Heiliges Römisches Reich Deutscher Nation*; Ulrike HOHENSEE et al., eds., *Die Goldene Bulle*.

45 Elizabeth A. R. BROWN and Richard C. FAMIGLIETTI, *The Lit de Justice*.

46 Carra Ferguson O'MEARA, *Monarchy and Consent*.

47 Text of the *Modus tenendi parliamentum*: Nicholas PRONAY and John TAYLOR, *Parliamentary Texts*, 11–114. See W. C. WEBER, "The Purpose of the English *Modus Tenendi Parliamentum*".

But let us not disregard the main purpose of community in the mind of the medieval contemporary: Fulfilling God's promise on earth was still more valuable than ruling over a mere empire on earth.

Bibliography

Adalbero of Laon, *Carmen ad Rotbertum regem. Poème au roi Robert*, edited by Claude Carozzi (Les classiques de l'histoire de France au Moyen Age, 32). Paris, Société d'édition 'Les belles lettres', 1979.

Althoff, Gerd, Dieter Geuenich, Otto Gerhard Oexle, and Joachim Wollasch, eds., *Person und Gemeinschaft im Mittelalter. Festschrift für Karl Schmid zum fünfundsechzigsten Geburtstag*. Sigmaringen, Thorbecke, 1988.

Anderson, Benedict, *Imagined Communities. Reflections on the Origin and Spread of Nationalism*. Revised edition. London/New York, Verso, 2006.

Angenendt, Arnold, *Geschichte der Religiosität im Mittelalter*. Darmstadt, Wissenschaftliche Buchgesellschaft, ⁴2009.

Annas, Gabriele, *Hoftag – Gemeiner Tag – Reichstag. Studien zur strukturellen Entwicklung deutscher Reichsversammlungen des späten Mittelalters (1349–1471)* (Schriftenreihe der Historischen Kommission bei der Bayerischen Akademie der Wissenschaften, 68). 2 vols., Göttingen, Vandenhoeck & Ruprecht, 2004.

Arnold, Benjamin, *Princes and Territories in Medieval Germany*. Cambridge, Cambridge University Press, 1991.

Asher, Ronald E., *National Myths in Renaissance France. Francus, Samothes and the Druids*. Edinburgh, Edinburgh University Press, 1993.

Bosl, Karl, *Frühformen der Gesellschaft im mittelalterlichen Europa. Ausgewählte Beiträge zu einer Strukturanalyse der mittelalterlichen Welt*. München/Wien, Oldenbourg, 1964.

Breitenstein, Mirko, *Das Noviziat im hohen Mittelalter. Zur Organisation des Eintrittes bei den Cluniazensern, Cisterziensern und Franziskanern* (Vita regularis. Abhandlungen, 38). Berlin, LIT, 2008.

Breitenstein, Mirko, Julia Burkhardt, Stefan Burkhardt, and Jens Röhrkasten, eds., *Rules and Observance. Devising Forms of Communal Life* (Vita regularis. Abhandlungen, 60). Berlin, LIT, 2014.

Breitenstein, Mirko, Julia Burkhardt, Stefan Burkhardt, and Jörg Sonntag, eds., *Identität und Gemeinschaft. Vier Zugänge zu Eigengeschichten und Selbstbildern institutioneller Ordnung* (Vita regularis. Abhandlungen, 67). Berlin, LIT, 2015.

Breitenstein, Mirko, Stefan Burkhardt, and Julia Dücker, eds., *Innovation in Klöstern und Orden des Hohen Mittelalters. Aspekte und Pragmatik eines Begriffs* (Vita regularis. Abhandlungen, 48). Berlin, LIT, 2012.

Brown, Elizabeth A. R. and Richard C. Famiglietti, *The Lit de Justice: Semantics, Ceremonial, and the Parlement of Paris* (Beihefte der Francia, 31). Sigmaringen, Thorbecke, 1994.

Brunner, Otto, Werner Conze, and Reinhart Koselleck, eds., *Geschichtliche Grundbegriffe*, vol. 6. Stuttgart, Klett-Cotta, 1990.

Burkhardt, Julia, "Die Welt der Mendikanten als Bienenschwarm und Vorstellung. Zum Ideal religiöser Gemeinschaften bei Thomas von Cantimpré". In *Die Klöster der Franziskaner im Mittelalter. Räume, Nutzungen, Symbolik*, edited by Gert Melville, Leonie Silberer, and Bernd Schmies (Vita regularis. Abhandlungen, 63). Berlin, LIT, 2015, 73–88.

Burkhardt, Julia, "Predigerbrüder im Bienenstock des Herrn. Dominikanische Identitäten im 'Bienenbuch' des Thomas von Cantimpré". In *Die deutschen Dominikaner und Dominikanerinnen 1221–1515*, edited by Sabine von Heusinger, Walter Senner, Elias Füllenbach, and Klaus-Bernward Springer (Quellen und Forschungen zur Geschichte des Dominikanerordens. Neue Folge). Berlin/Boston, De Gruyter, 2016 [forthcoming].

Burkhardt, Julia, "Procedure, rules and meaning of political assemblies in Late Medieval Central Europe", *Parliaments, Estates and Representations* 35 (2015), 153–170 [published online: http://dx.doi.org/10.1080/02606755.2015.1023666].

Cohen, Anthony P., *The Symbolic Construction of Community*. E-Reprint. London/New York, Routledge, 2001.

Corradini, Richard, Max Diesenberger, and Helmut Reimitz, eds., *The Construction of Communities in the Early Middle Ages. Texts, Resources and Artefacts* (The Transformation of the Roman World, 12). Leiden/Boston, Brill, 2003.

Derda, Hans-Jürgen, *Vita Communis. Studien zur Geschichte einer Lebensform in Mittelalter und Neuzeit*. Köln/Weimar/Wien, Böhlau, 1992.

Duby, Georges, *Les trois ordres ou l'imaginaire du féodalisme* (Bibliothèque des histoires). Paris, Gallimard, 1978.

Dücker, Julia, "Vorstellungen von Gemeinschaft und sozialer Ordnung. Zum Innovativen in dominikanischen Schriften des 13. Jahrhunderts". In *Innovation in Klöstern und Orden des Hohen Mittelalters. Aspekte und Pragmatik eines Begriffs*, edited by Mirko Breitenstein, Stefan Burkhardt, and Julia Dücker (Vita regularis. Abhandlungen, 48). Berlin, LIT, 2012, 197–214.

Dücker, Julia, *Reichsversammlungen im Spätmittelalter. Politische Willensbildung in Polen, Ungarn und Deutschland* (Mittelalter-Forschungen, 37). Ostfildern, Thorbecke, 2011.

Ehlers, Joachim, *Heinrich der Löwe. Eine Biographie*. München, Siedler, 2008.

Felten, Franz J., Annette Kehnel, and Stefan Weinfurter, eds., *Institution und Charisma. Festschrift für Gert Melville zum 65. Geburtstag*. Köln/Weimar/Wien, Böhlau, 2009.

Feuchter, Jörg and Johannes Helmrath, eds., *Politische Redekultur in der Vormoderne. Die Oratorik europäischer Parlamente in Spätmittelalter und Früher Neuzeit* (Eigene und fremde Welten. Repräsentationen sozialer Ordnung im Vergleich, 9). Frankfurt/New York, Campus, 2008.

Feuchter, Jörg and Johannes Helmrath, eds., *Parlamentarische Kulturen vom Mittelalter bis in die Moderne. Reden – Räume – Bilder* (Beiträge zur Geschichte des Parlamentarismus und der politischen Parteien, 164). Düsseldorf, Droste, 2013.

Ganshof, François Louis, *Feudalism*. New York, Harper & Row, [3]1964.

Garrison, Eliza, *Ottonian Imperial Art and Portraiture. The Artistic Patronage of Otto III and Henry II*. Farnham, Ashgate, 2012.

Gesta episcoporum Cameracensium, edited by Ludwig C. Bethmann (Monumenta Germaniae Historica. Scriptores, 7). Hannover, Hahn, 1846, 393–525.

Görich, Knut, *Friedrich Barbarossa. Eine Biographie*. München, Beck, 2011.

Gotthard, Axel, *Säulen des Reiches. Die Kurfürsten im neuzeitlichen Reichsverband* (Historische Studien, 457/1–2). 2 vols., Husum, Matthiesen, 1999.

Haverkamp, Alfred, "Leben in Gemeinschaften: alte und neue Formen im 12. Jahrhundert". In *Aufbruch – Wandel – Erneuerung. Beiträge zur "Renaissance" des 12. Jahrhunderts*, edited by Georg Wieland. Stuttgart/Bad Cannstadt, Frommann-Holzboog, 1995, 11–44.

Haverkamp, Alfred, "Neue Formen von Bindung und Ausgrenzung. Konzepte und Gestaltungen von Gemeinschaften an der Wende zum 12. Jahrhundert". In *Salisches Kaisertum und neues Europa. Die Zeit Heinrichs IV. und Heinrichs V.*, edited by Bernd Schneidmüller and Stefan Weinfurter. Darmstadt, Wissenschaftliche Buchgesellschaft, 2007, 85–122.

Hofmann, Hasso, *Repräsentation. Studien zur Wort- und Begriffsgeschichte von der Antike bis ins 19. Jahrhundert* (Schriften zur Verfassungsgeschichte, 22). Berlin, Duncker & Humblot, [4]2003.

Hohensee, Ulrike, Mathias Lawo, Michael Lindner, Michael Menzel, and Olaf B. Rader, eds., *Die Goldene Bulle. Politik – Wahrnehmung – Rezeption* (Berlin-Brandenburgische Akademie der Wissenschaften. Berichte und Abhandlungen. Sonderband, 12). 2 vols., Berlin, Akademie, 2009.

Körntgen, Ludger, *Königsherrschaft und Gottes Gnade. Zu Kontext und Funktion sakraler Vorstellungen in Historiographie und Bildzeugnissen der ottonisch-frühsalischen Zeit* (Orbis mediaevalis. Vorstellungswelten des Mittelalters, 2). Berlin, Akademie, 2001.

Die Konstitutionen Friedrichs II. für das Königreich Sizilien, edited by Wolfgang Stürner (Monumenta Germaniae Historica. Constitutiones et acta publica imperatorum et regum, 2, Supplementum). Hannover, Hahn, 1996.

Lehr, Thomas, *Was nach der Sintflut wirklich geschah. Die 'Antiquitates' des Annius von Viterbo und ihre Rezeption in Deutschland im 16. Jahrhundert*. Frankfurt a. M., Lang, 2012.

Lutter, Christina, "Geistliche Gemeinschaften in der Welt. Kommentar zur Sektion Individuum und Gemeinschaft – Innen und Außen". In *Innovationen durch Deuten und Gestalten. Klöster im Mittelalter zwischen Jenseits und Welt*, edited by Gert Melville, Bernd Schneidmüller, and Stefan Weinfurter (Klöster als Innovationslabore. Studien und Texte, 1). Regensburg, Schnell & Steiner, 2014, 145–160.

Lutter, Christina, "Social Groups, Personal Relations, and the Making of Communities in Medieval *vita monastica*". In *Making Sense as a Cultural Practice. Historical Perspectives*, edited by Jörg Rogge (Mainzer Historische Kulturwissenschaften, 18). Bielefeld, Transcript, 2013, 45–61.

Lutter, Christina, "*Vita communis* in Central European Monastic Landscapes". In *Meanings of Community across Medieval Eurasia. Comparative Approaches*, edited by Eirik Hovden, Christina Lutter, and Walter Pohl. Leiden/Boston, Brill, 2016, 362–387.

Mayer, Theodor, *Mittelalterliche Studien. Gesammelte Aufsätze*. Lindau/Konstanz, Thorbecke, 1959.

Melville, Gert and Anne Müller, eds., *Mittelalterliche Orden und Klöster im Vergleich. Methodische Ansätze und Perspektiven* (Vita regularis. Abhandlungen, 34). Berlin, LIT, 2007.

Melville, Gert and Carlos Ruta, eds., *Life Configurations* (Challenges of Life. Essays on Philosophical and Cultural Anthropology, 1). Berlin/Boston, De Gruyter, 2014.

Melville, Gert and Carlos Ruta, eds., *Thinking the Body as a Basis, Provocation, and Burdon of Life. Studies in Intercultural and Historical Contexts* (Challenges of Life. Essays on Philosophical and Cultural Anthropology, 2). Berlin/Boston, De Gruyter, 2015.

Melville, Gert and Martial Staub, eds., *Enzyklopädie des Mittelalters*. 2 vols., Darmstadt, Wissenschaftliche Buchgesellschaft, [2]2013.

Melville, Gert and Peter von Moos, eds., *Das Öffentliche und Private in der Vormoderne* (Norm und Struktur. Studien zum Wandel in Mittelalter und früher Neuzeit, 10). Köln/Weimar/Wien, Böhlau, 1998.

Melville, Gert, "Im Spannungsfeld von religiösem Eifer und methodischem Betrieb. Zur Innovationskraft der mittelalterlichen Klöster", *Denkströme. Journal der Sächsischen Akademie der Wissenschaften zu Leipzig* 7 (2011), 72–92.

Melville, Gert, "Regeln – *Consuetudines*-Texte – Statuten. Positionen für eine Typologie des normativen Schrifttums religiöser Gemeinschaften im Mittelalter". In *Regulae – Consuetudines – Statuta. Studi sulle fonti normative degli ordini religiosi nei secoli centrali del Medioevo*, edited by Cristina Andenna and Gert Melville (Vita regularis. Abhandlungen, 25). Münster, LIT, 2005, 5–38.

Melville, Gert, Bernd Schneidmüller, and Stefan Weinfurter, eds., *Innovationen durch Deuten und Gestalten. Klöster im Mittelalter zwischen Jenseits und Welt* (Klöster als Innovationslabore. Studien und Texte, 1). Regensburg, Schnell & Steiner, 2014.

Melville, Gert, *The World of Medieval Monasticism. Its History and Forms of Life*, translated by James D. Mixson (Cistercian Studies Series, 263). Collegeville, Liturgical Press, 2016.

Melville, Gert, ed., *Institutionen und Geschichte. Theoretische Befunde und mittelalterliche Aspekte* (Norm und Struktur. Studien zum Wandel in Mittelalter und früher Neuzeit, 1). Köln/Weimar/Wien, Böhlau, 1992.

Melville, Gert, *Frommer Eifer und methodischer Betrieb. Beiträge zum mittelalterlichen Mönchtum*, edited by Cristina Andenna and Mirko Breitenstein. Köln/Weimar/Wien, Böhlau, 2014.

Melville, Gert, Gregor Vogt-Spira, and Mirko Breitenstein, eds., *Gerechtigkeit* (Europäische Grundbegriffe im Wandel. Verlangen nach Vollkommenheit, 1). Köln/Weimar/Wien, Böhlau, 2014.

Melville, Gert, Leonie Silberer, and Bernd Schmies, eds., *Die Klöster der Franziskaner im Mittelalter. Räume, Nutzungen, Symbolik* (Vita regularis. Abhandlungen, 63). Berlin, LIT, 2015.

Miethke, Jürgen, *Politiktheorie im Mittelalter. Von Thomas von Aquin bis Wilhelm von Ockham* (UTB, 3059). Tübingen, Mohr Siebeck, 2008.

Moraw, Peter, ed., *Deutscher Königshof, Hoftag und Reichstag im späteren Mittelalter* (Vorträge und Forschungen, 48). Stuttgart, Thorbecke, 2002.

Moraw, Peter, *Von offener Verfassung zu gestalteter Verdichtung. Das Reich im späten Mittelalter 1250–1490*. Frankfurt a. M. /Berlin, Propyläen, 1989.

Müller, Anne and Karen Stöber, eds., *Self-Representation of Medieval Religious Communities. The British Isles in Context* (Vita regularis. Abhandlungen, 40). Berlin, LIT, 2009.

O'Meara, Carra Ferguson, *Monarchy and Consent. The Coronation Book of Charles V of France. British Library MS Cotton Tiberius B. VIII*. London/Turnhout, Harvey Miller, 2001.

Oexle, Otto Gerhard and Andrea von Hülsen-Esch, eds., *Die Repräsentation der Gruppen. Texte – Bilder – Objekte* (Veröffentlichungen des Max-Planck-Instituts für Geschichte, 141). Göttingen, Vandenhoeck & Ruprecht, 1998.

Oexle, Otto Gerhard, "Das Mittelalter und das Unbehagen an der Moderne. Mittelalterbeschwörungen in der Weimarer Republik und danach". In *Spannungen und Widersprüche. Gedenkschrift für František Graus*, edited by Susanna Burghartz, Hans-Jörg Gilomen, Guy P. Marchal, Rainer C. Schwinges, and Katharina Simon-Muscheid. Sigmaringen, Thorbecke, 1992, 125–153.

Oexle, Otto Gerhard, "Deutungsschemata der sozialen Wirklichkeit im frühen und hohen Mittelalter. Ein Beitrag zur Geschichte des Wissens". In *Mentalitäten im Mittelalter. Methodische und inhaltliche Probleme*, edited by František Graus (Vorträge und Forschungen, 35). Sigmaringen, Thorbecke, 1987, 65–117.

Oexle, Otto Gerhard, "Die Moderne und ihr Mittelalter – eine folgenreiche Problemgeschichte". In *Mittelalter und Moderne. Entdeckung und Rekonstruktion der mittelalterlichen Welt. Kongreßakten des 6. Symposiums des Mediävistenverbandes in Bayreuth 1995*, edited by Peter Segl. Sigmaringen, Thorbecke, 1997, 307–364.

Oexle, Otto Gerhard, *Geschichtswissenschaft im Zeichen des Historismus. Studien zu Problemgeschichten der Moderne* (Kritische Studien zur Geschichtswissenschaft, 116). Göttingen, Vandenhoeck & Ruprecht, 1996.

Ottmann, Henning, *Geschichte des politischen Denkens*, vol. 2/2: *Das Mittelalter*. Stuttgart/Weimar, Metzler, 2005.

Patzold, Steffen, *Das Lehnswesen* (C. H. Beck Wissen, 2745). München, Beck, 2012.

Patzold, Steffen, *Konflikte im Kloster. Studien zu Auseinandersetzungen in monastischen Gemeinschaften des ottonisch-salischen Reichs* (Historische Studien, 463). Husum, Matthiesen, 2000.

Peltzer, Jörg, Gerald Schwedler, and Paul Töbelmann, eds., *Politische Versammlungen und ihre Rituale. Repräsentationsformen und Entscheidungsprozesse des Reichs und der Kirche im späten Mittelalter* (Mittelalter-Forschungen, 27). Ostfildern, Thorbecke, 2009.

Pronay, Nicholas and John Taylor, *Parliamentary Texts of the Later Middle Ages*. Oxford, Clarendon Press, 1980.

Puhle, Matthias and Claus-Peter Hasse, eds., *Heiliges Römisches Reich Deutscher Nation 962 bis 1806. Von Otto dem Großen bis zum Ausgang des Mittelalters*. 2 vols., Dresden, Sandstein, 2006.

Reynolds, Susan, *Fiefs and Vassals. The Medieval Evidence Reinterpreted*. Oxford, Oxford University Press, 1994.

Reynolds, Susan, *Kingdoms and Communities in Western Europe, 900–1300*. Oxford, Clarendon Press, [2]1997.

Ridder, Klaus and Steffen Patzold, eds., *Die Aktualität der Vormoderne. Epochenentwürfe zwischen Alterität und Kontinuität* (Europa im Mittelalter. Abhandlungen und Beiträge zur historischen Komparatistik, 23). Berlin, Akademie, 2013.

Schmidt, Paul Gerhard, "Von der Allgegenwart der Dämonen. Die Lebensängste des Zisterziensers Richalm von Schöntal", *Literaturwissenschaftliches Jahrbuch. Neue Folge* 36 (1995), 339–346.

Schneidereit, Nele, *Die Dialektik von Gemeinschaft und Gesellschaft. Grundbegriffe einer kritischen Sozialphilosophie* (Politische Ideen, 22). Berlin, Akademie, 2010.

Schneidmüller, Bernd and Stefan Weinfurter, eds., *Heilig – Römisch – Deutsch. Das Reich im mittelalterlichen Europa*. Dresden, Sandstein, 2006.

Schneidmüller, Bernd, "Inszenierungen und Rituale des spätmittelalterlichen Reichs. Die Goldene Bulle von 1356 in westeuropäischen Vergleichen". In *Die Goldene Bulle. Politik – Wahrnehmung – Rezeption*, vol. 1, edited by Ulrike Hohensee, Mathias Lawo, Michael Lindner, Michael Menzel, and Olaf B. Rader (Berlin-Brandenburgische Akademie der Wissenschaften. Berichte und Abhandlungen. Sonderband, 12). Berlin, Akademie, 2009, 261–297.

Schneidmüller, Bernd, "Rule by Consensus. Forms and Concepts of Political Order in the European Middle Ages", *The Medieval History Journal* 16 (2013), 449–471.

Schneidmüller, Bernd, "Verantwortung aus Breite und Tiefe. Verschränkte Herrschaft im 13. Jahrhundert". In *König, Reich und Fürsten im Mittelalter. Festschrift für Karl-Heinz Spieß*, edited by Oliver Auge. Stuttgart [forthcoming].

Schneidmüller, Bernd, *Die Welfen. Herrschaft und Erinnerung (819–1252)* (Urban-Taschenbücher, 465). Stuttgart, Kohlhammer, [2]2014.
Schneidmüller, Bernd, *Grenzerfahrung und monarchische Ordnung. Europa 1200–1500* (C. H. Beck Geschichte Europas). München, Beck, 2011.
Schreiner, Klaus, *Gemeinsam leben. Spiritualität, Lebens- und Verfassungsformen klösterlicher Gemeinschaften in Kirche und Gesellschaft des Mittelalters*, edited by Gert Melville in collaboration with Mirko Breitenstein (Vita regularis. Abhandlungen, 53). Berlin, LIT, 2013.
Schubert, Ernst, *König und Reich. Studien zur spätmittelalterlichen deutschen Verfassungsgeschichte* (Veröffentlichungen des Max-Planck-Instituts für Geschichte, 63). Göttingen, Vandenhoeck & Ruprecht, 1979.
Schulze, Hans K., *Grundstrukturen der Verfassung im Mittelalter*, vol. 1: *Stammesverband, Gefolgschaft, Lehnswesen, Grundherrschaft* (Urban-Taschenbücher, 371). Stuttgart, Kohlhammer, [4]2004. – Vol. 2: *Familie, Sippe und Geschlecht, Haus und Hof, Dorf und Markt, Burg, Pfalz und Königshof, Stadt* (Urban-Taschenbücher, 372). Stuttgart/Berlin/Köln/Mainz, Kohlhammer, 1986.
Schürer, Markus, *Das Exemplum oder die erzählte Institution. Studien zum Beispielgebrauch bei den Dominikanern und Franziskanern des 13. Jahrhunderts* (Vita regularis. Abhandlungen, 23). Berlin, LIT, 2005.
Sonntag, Jörg, *Klosterleben im Spiegel des Zeichenhaften. Symbolisches Denken und Handeln hochmittelalterlicher Mönche zwischen Dauer und Wandel, Regel und Gewohnheit* (Vita regularis. Abhandlungen, 35). Berlin, LIT, 2008.
Stürner, Wolfgang, "Rerum necessitas und divina provisio. Zur Interpretation des Prooemiums der Konstitutionen von Melfi (1231)", *Deutsches Archiv für Erforschung des Mittelalters* 39 (1983), 467–554.
Stürner, Wolfgang, *Peccatum und Potestas. Der Sündenfall und die Entstehung der herrscherlichen Gewalt im mittelalterlichen Staatsdenken* (Beiträge zur Geschichte und Quellenkunde des Mittelalters, 15). Sigmaringen, Thorbecke, 1987.
Tönnies, Ferdinand, *Gemeinschaft und Gesellschaft. Abhandlung des Communismus und des Socialismus als empirischer Culturformen*. Leipzig, Fues, 1887.
Töpfer, Bernhard, *Urzustand und Sündenfall in der mittelalterlichen Gesellschafts- und Staatstheorie* (Monographien zur Geschichte des Mittelalters, 45). Stuttgart, Hiersemann, 1999.
Wagner, Wolfgang Eric, *Die liturgische Gegenwart des abwesenden Königs. Gebetsverbrüderung und Herrscherbild im frühen Mittelalter* (Brill's Series on the Early Middle Ages, 19). Leiden/Boston, Brill, 2010.
Weber, W. C., "The Purpose of the English *Modus Tenendi Parliamentum*", *Parliamentary History* 17 (1998), 149–177.
Weferling, Sandra, *Spätmittelalterliche Vorstellungen vom Wandel politischer Ordnung. Französische Ständeversammlungen in der Geschichtsschreibung des 14. und 15. Jahrhunderts* (Heidelberger Abhandlungen zur mittleren und neueren Geschichte, 20). Heidelberg, Winter, 2014.

Gert Melville

"Singularitas" and Community

About a Relationship of Contradiction and Complement in Medieval Convents

The monastic form of life strives for total institutionalisation more stringently than any other – and in most cases, it reaches that goal to a higher degree than any other form of life, too. In this assessment, I essentially agree with Erving Goffman, who treats the monastery as a prototype of "total institution"[1]. The chief concern of the monastic form of life is an absolute correspondence between spiritual guiding principles and a communal life whose norms demand the voluntary and complete commitment of the individual. To be a member of a religious convent means the radical implementation of indisputable basic religious and moral values into everyday practical behaviour patterns, and the absolute submission to a community that exercises precise, detailed, and, above all, indispensable control over life in all spiritual and physical spheres, in terms of daily routine, of housing conditions, of clothing, food, etc.[2] Anything prescribed by such rules in written or oral form had to be adhered to almost unconditionally. Although instructions given in monastic everyday life were not allowed to interfere with the Rule or the will of God, they could be rigorous and almost impossible to bear. Still, their validity was not to be doubted. It was insisted time and again that obedience was each individual member's "guide to the virtues". A refusal of obedience meant rebellion and sin because the disobedient person would impertinently consider himself superior to Christ, who had been obedient to His father until His death.

Nevertheless, this characterization and, thus, Goffman's hypothesis, have to be subjected to a considerably stronger differentiation – namely, with regard to the following historical structures: the assertiveness of monastic validity claims and, hence, also the willingness to submit to the demands of a community lay in their transcendental anchorage.

This circumstance actually seemed to give the monastic form of live no room to move at all. It appeared to be as unalterable as God Himself. Even so, precisely this transcendental anchorage constituted the crucial point of a dichotomy that exhibited a fundamental and yet also fertile tension. That is to say, the transcendent anchorage turned a monastic life into a life in transition and a monas-

---

1 Erving GOFFMANN, *Asylums*.

2 Cf. Gert MELVILLE, *The World of Medieval Monasticism*, 332–349.

tic community into a temporary transit station of the individual between Earth and Heaven. This, however, meant that the weighting between community and individual could be perceived in different ways: was it necessary to strengthen the community whose institutionality guaranteed the realisation of such a transit station and that simply took the individual along in the process?[3] Or was it required to put the individual first and to see the community in the role of a supporting institution because, after all, the chief concern was gaining the individual's personal salvation?[4] Although, at first glance, one would rather expect the former version as being more appropriate to the Middle Ages, that epoch definitely knew and cultivated the latter version as well. For not least because of the tremors of the Investiture Controversy and the great Church Reform, there was a general drive for the internalisation of belief from the second half of the 11th century onwards, which now led to an individual search for God in an increased way. Reform-oriented monasteries were the vanguard in this search; they provided the patterns.

From that time onwards, monasteries allowed a form of community that permitted the self-responsibility of the individual and allowed him to pursue perfection in his own soul. One was perfectly aware that written as well as oral instructions that were meant to lead the members of a monastery to the right views and to sustain those views in them would not succeed if they were treated like juridical precepts, rules or statutes that could be complied with by applying exterior behaviour practices. These instructions could only achieve success if every individual they addressed actually 'internalised' the normative contents. With empathy, an anonymous author of the 12th century got to the heart of this matter: "Of what use are these writings, all that has been read and understood if you do not read and understand yourself? Therefore, apply yourself to inner perusal so that you read, investigate and recognise yourself."[5]

It was, above all, remarks like this one that induced researchers to speak of that time as the time when the individual or at least the 'self', the 'I' as one's un-exchangeable own was discovered. This was about the scope of the view into the interior of the self, into the *interior domus* ("the inner house") of the individual.[6] It was about the assessment of the meaning of sentences like "Dare to

3 About the importance of community cf. Caroline W. BYNUM, "The Spirituality of Regular Canons".
4 Cf. *Das Eigene und das Ganze*. About this relationship of tension see Gert MELVILLE, "The Innovational Power of Monastic Life".
5 ANONYMOUS, *Meditationes piissimae de cognitione humanae conditiones*, col. 508.
6 Cf. Gerhard BAUER, *Claustrum animae*; Ineke VAN 'T SPIJKER, *Fictions of the Inner Life*; Mirko BREITENSTEIN, "Der Traktat vom 'Inneren Haus'".

know yourself" or "Return into yourself".[7] They had old roots in patristics, yet their renewed topicality was clarified not least by contemporary works such as, first and foremost, Abelard's *Scito te ipsum* ("Know about yourself").[8] Such sentences explicitly asserted not only the ability to recognise one's own 'I' but also the ability to correct and control one's personal actions – an assertion that could definitely still be enhanced if it was contrasted with the lower capacity of external assessment: "To wit, no-one can know more about who you are than you do as you are conscious of yourself." For it was possible to hide one's exterior actions from others but not from oneself, as was said in, for instance, the treatise *De interiori domo*[9].

The method of gaining corresponding access to oneself was provided by a technique of cognition that had also been redeveloped again: the *conscientia*, the conscience.[10] In this, the issue was not only one's solipsistic self but also an individual relationship to God that was unique in each case. The conscience was perceived as a path that helped guide one towards God via one's own [self]. Accordingly, it was stated time and again that the conscience was the inevitable companion for oneself because nobody could flee from themselves; the conscience forced oneself to be one's own accuser and judge; in dealing with one's own conscience one placed oneself before oneself and passed judgement about oneself; one was bound by one's own law if one followed one's own conscience. However, this investigative journey to one's own inside, to one's 'own law' inescapably stood under the sign of God as the superordinate observer and examiner, the only entity that "hearts and thoughts lay open to". It was stated that if one's own inside was oriented in such a way, it would win its actual goal. For after its cleansing through the conscience, the respective *interior domus* of the individual religious person can be the dwelling of God and, thus, is able to reach the highest goal of an individual existence.[11] Therefore, this was a question of finding one's individual 'I' as a basic prerequisite for being able to transcend this 'I' to God from its earthly bondage. This postulate provided members of a monastic community with a high degree of autonomy and, ultimately, assigned the community predominantly with the role of a supporting, protecting and encouraging shell – which was definitely in the interest of the community. "To wit, if they are

---

7 ANONYMOUS, *Meditationes piissimae de cognitione humanae conditiones*, col. 494.

8 ABELARD, *Scito te ipsum*.

9 ANONYMOUS, *De interiore domo*, col. 533sq..

10 Cf. Marie-Dominique CHENU, *L'éveil de la conscience*; Ermenegildo BERTOLA, *Il problema della coscienza monastica*; Gert MELVILLE, "Der Mönch als Rebell", 172–181; Mirko BREITENSTEIN, "Die Verfügbarkeit der Transzendenz", 37–56.

11 Cf. Mirko BREITENSTEIN, "Der Traktat vom 'Inneren Haus'".

inside the fortress of the monastery walls, the inner and the outer man escape the attacks of the old foe and the fickle vicissitudes of temporal matters"[12], as Hugh of Folieto put it.

Besides those monks and nuns whose love of God induced them to strive for individual progress in the search for their salvation within the framework of a community of kindred spirits, and to submit to the dictate of this community for the same reason,[13] there also were others, however, who withdrew into themselves as well while causing the community great problems. These monks and nuns denied the community its actual task: namely – as has already been said – to serve as a transit station of the individual between Earth and Heaven. This was because their concern was not the individuality with and within God; they themselves were their sole reference point that everything revolved around. Such behaviour was called *singularitas* and was classified as a vice that threatened to disrupt the community or at least inflicted severe troubles upon it.

Singularitas is a medieval term that is hard to translate into a modern language by means of a single word. Originally, in a Christian tradition that hearkened back to Hieromymus, it denoted the solitary life of a monk in a very general way,[14] but soon its semantic field was broadened and differentiated. On the one hand, it was an ontological term that also played a special role in the works of many medieval authors up to Nicolaus Cusanus in explaining the trinity of God, the specificity of human individuality and the uniqueness of each and every existing person.[15] On the other hand, it also served to mark exactly that self-centred behaviour. Here, only the latter meaning is of interest. It oscillated, however, between self-complacency, egoism, self-will and isolation. In particular the Cistercian, Bernard of Clairvaux, one of the most reflective and constructive thinkers among the monks of that time, attempted a description by defining *singularitas* as the fourth of twelve degrees of pride (*superbia*) – after boasting and before arrogance – and characterising the person indulging in it with the following words, among others:

> He who places himself above others deems it disgraceful if he does not do anything with which he can present himself as superior to the others. Thus, what is demanded by the general observance or the examples of the superiors does not suffice to him. Yet he does

12 HUGO DE FOLIETO, *De claustro animae*, col. 1020.

13 By the example of the Canons Regular, the importance of mutual support and instruction in the convents is shown in: Caroline W. BYNUM, "The Spirituality of Regular Canons".

14 Cf. Paul ANTIN, "Á la source de *singularitas*".

15 Cf. Gerda VON BREDOW, "Participatio singularitatis"; Thomas LEINKAUF, "Die Bestimmung des Einzelseienden"; Wolfgang HÜBNER, "Der theologisch-philosophische Konservatismus", 173 sq.; Christian STRAUB, "Singularität des Individuums?"; Helmut MEINHARDT, "Singularität, Spontaneität, Subjektivität"; Christophe ERISMANN, "*Singularitas*".

> not seek to be better but to seem better. His desire is not to lead a better life but to appear superior in order to be able to say: "I am not like the other people." (Lk 18:11) [....] He is more afraid of the injury his honour might suffer than of the anguish of hunger. If he sees somebody who is gaunter and paler, he thinks himself low and cannot find peace of mind. And as he is unable to see his own face in the way it presents himself to the eyes of others, he regards the hands and arms, which he can see, he strokes the ribs, feels shoulders and loins in order to judge the paleness and colour of his visage according to how thin or less thin he finds the limbs of his body. In short, he is strict in everything that concerns his person but negligent in the collective exercises.[16]

Singularitas secedes from the community and searches for a special space, as Bernard points out elsewhere: "Where self-interest reigns, there is *singularitas*; but where there is *singularitas*, there is a nook; but where there is a nook, without a doubt there is filth and rust."[17] It had to be feared that such behaviour could fragment a community into separate monads that were not connected by a common reference to God anymore and that were not able to provide support in the fight against sin anymore, either. And, thus, Bernard comes to the final exhortation to each confrere in one of his other works:

> "These are people who isolate themselves, who are of a worldly mind and do not possess the spirit" (Jud 19). [...] It really is an absolutely wicked and pestilent evil when the hardening of an individual causes unrest in all others, becomes fuel for conflict between all others and the reason for annoyances. "Finally listen to the words of the prophet who speaks about the vineyard of the Lord: "One wild animal alone", he says, "has eaten it bare." (Ps 79:14) For this reason I beseech you, my brethren, flee all hypocrisy and the nooks of self-will, flee the unrest and the spirit of carelessness, flee stubbornness and the very bad vice of *singularitas*.[18]

Bernard, the former knight, compares the conventual community (*congregatio*) of the religious with an *acies ordinata*, an "orderly battle-line", in which "the many in the monastery fight in the same way"[19] and, thus, withstand the attacks of evil because – as is pointedly worded in a correct interpretation – "like-minded confreres stand by the individual's side; battle-tested monks can warn of the enemy's ruses"[20]. Only someone who indulged in the self-serving vice of *singularitas* was in danger there; and since he created a gap in the battle-line, he endangered the others as well. On this topic, Bernard wrote a parable whose allegoric language

16 Bernard of Clairvaux, De *gradibus humilitatis et superbia tractatus*, 49.
17 Id., *Epistola XI*, 55.
18 Id., *Sermo ad domenicam VI post Pentecosten, I*, 208.
19 Id., In circumcisione, sermo 3, 6, 287.
20 Michaela Diers, *Elitäre Frömmigkeit*, 85.

provides it with fascinating force of expression. I would like to present it to you in condensed form.[21]

David's army of virtues, which was well-ordered and led by a firm hand, stood against the chaotic but nonetheless dangerous Babylonian troops of vice and the evil spirits under the leadership of Nebuchadnezzar. Then a novice who had only just been furnished with weapons stepped forward from David's host, disregarding all discipline of the army camp and the protection from his comrades, as he was full of presumption and eager to make a name for himself (*ad faciendum sibi nomen*) in the fight against the enemy. He mounted his horse, which had become strong and splendid from the drink of the world. It was in vain that David warned him: "Woe to him who is alone; when he falls, he will have no-one to lift him up". With impudence he urged on his horse with the slashes of fasting and the spurs of vigils, driving it at an especially terrifying enemy, namely, Lust. All of David's army reasoned with him to desist from this single combat. He did not listen; and soon – without really sensing it – he found himself exposed to two female fighters the enemy army had sent to meet him, pride (*superbia*) and thirst for glory (*vana gloria*). Moreover, Lust feigned flight, which drove the novice even deeper into enemy territory. Soon his horse was exhausted and did not obey its master any longer. Wrath and Envy as further warriors of Babylon and, finally, Lust herself assaulted him, who now was defenceless, and threw him into the dungeon of despair. David mourned his loss and undertook to save him at once. He sent a member of his retinue who was particularly experienced in such matters: Fear (*timor*). Obedience (*oboedientia*) stood by her side. Fear found the prisoner, tore him out of his dungeon and handed him and his horse to Obedience. With iron force Obedience tamed the horse, which had become stubborn and rebellious, and placed it at the disposal of its master again. On his way back home into David's camp, the novice was given a first dwelling (*mansio*) with Love (*pietas*), so that his mind, scared with fright, would be uplifted again. He then found a second dwelling with Knowledge (*scientia*) in order to learn how to handle fear and love. The third dwelling was with Strength (*fortitudo*) in order to be able to endure the way of his return. The fourth dwelling was the Advice (*consilium*), so that he would act upon advice in future. Insight (*intellectus*) provided the fifth dwelling, so that he would not only act upon advice of humans but would also recognize through himself what was the will of the Lord. Finally, the sixth domicile permitted him to attain Wisdom (*sapientia*), so that he might relish the Lord's goods.

---

21 On the following, cf. BERNARD OF CLAIRVAUX, *Parabolae 3*, 274–276. Cf. Gert MELVILLE, "Einleitende Aspekte", XVIII-XXI.

The message of this story is as simple as it is suggestive. In the foreground, it was about a fight between virtues and vices[22] – each represented by the two armies. An inexperienced novice's feeling of strength induces him to commence a fight against the most dangerous of all vices, which is lust. Spiritual exercises like asceticism and vigils stimulate him. However, supposed initial successes quickly fill him with *superbia* and *vana gloria*. Thus blinded and deceived, he does not sense at all how his body – his horse – is wrested from him and how strongly lust already dominates him. Furthermore, despair is all he is left with. Yet fear pulls him up again, causes him to become obedient and then to devote himself to the virtues of *pietas*, *fortitudo*, *intellectus* and so forth, which lead him to the acceptance of the divine blessings.

At first glance, this story appears to be a biographical outline of the very individual experience of a crisis of body and soul. The psychological point it made was that desiring success against the vice already entailed defeat and that only a compelling force beyond virtues and vices – to wit, naked fear – could lead to physical and mental re-stabilisation. Yet still, paraenetically speaking, this was about much more – namely, about a demonstration showing that the individual monk absolutely required complete embedding in the monastic community in order to achieve success against vice.[23] This story can now also be retold saying that the novice had become guilty of *singularitas*; he wanted to make an individual 'name' for his 'I', as it were. Through this mentally self-excluding act he achieved that he indeed engaged in single combat outside of his community. He paid no heed to warnings from the disciplined and orderly battle-line of the monastic community and – alone as he was – suffered defeat. Salvation, however, he only found in the apparatus of the community. That is to say, the leadership of said community exhibited care for the novice, instilled fear into him, forced him to obey them again, opened his mind to advice from others, imparted virtues to him in depth and, thus, led him to an inner realisation of the monastic values.

Upon bringing together the different aspects of that parable, the monastery appears to be the place where, firstly, the repertoire of virtues required for the salvation of the soul and for control of the body was available reliably, orderly, and instantly, and – secondly – into which all had to integrate themselves with complete relinquishment of their self-will if their chief concern was to be saved from sinfulness and to be successful in the fight against vices. 'Woe to him who is alone' is the counter-argument offered to the individual whose *singularitas* prevents a total inclusion into the *acies ordinata*, because only the absolute unity

---

22 On these structures, cf. Thomas FÜSER, *Mönche im Konflikt*.

23 On these aspects and their affective implications, cf. Caroline W. BYNUM, "The Cistercian Conception", 77–81.

of the monastic community enables one to find others who will lift one up again after the fall.

Therefore, the monastery in particular gives reason to expect the prevention of personal sinfulness and the attainment of personal salvation in a threefold way: through an order that strictly and permanently permeates all areas of life, through the presence of a supporting community, and through the omnipresent availability of the spiritual means. Yet it is exactly these three dimensions that make the fulfilment of the individual expectation of salvation conditional on the abolishment of *singularitas*. They do not, however, demand the abolishment of individuality itself, which is embedded in God and defines itself through Him alone. On this point, let me cover the following aspects in detail.

Within rule-governed orders like monasteries, successful "communitisation" (*Vergemeinschaftung*) can be spoken of "if and so far" – quoting Max WEBER'S words – "the orientation of social action [...] is based on a subjective feeling of the parties, whether affectual or traditional, that they belong together"[24]. With regard to life in a monastery, the dictum *cor unum et anima una* ("one heart and of one soul") from the "Acts of the Apostles" (4, 32) postulated such a sense of belonging or stated its fulfilment again and again, like a leitmotif, as it were, and as a revival of the Early Church values of fraternity, of joint possession, and sharing.[25] Yet, first of all, this sense of belonging was also expressed as the idea of a homogeneity that existed in the community despite fundamental differences. Unity, however, demanded the contribution of individual qualities in order to strengthen the entirety: each and every convent, as for example Girardus de Arvernia pointed out in his admonitions to the Cluniacs, had to be like David's lyre, where harmony could only be achieved by the interplay of many different strings.[26] Self-limitation through humility (*humilitas*) and through love (*pietas*) for one's confreres meant the necessity to implement social virtues which made it possible to prevent hubris (*superbia*) and isolation (*singularitas*) as nuclei of discord. Monastic communities needed the *pax*, which is why, as was already programmatically laid down in early Premonstratensian and, later, also in Dominican texts, they required their members to realise a "unity that is to be maintained inwardly in the hearts" as well as a "uniformity that is maintained outwardly in the customs"[27] – thus, an inner and outer consonance.

---

24 Max WEBER, *Economy and Society*, vol. 1, 40. About basic structures of "communitisation" in medieval monasteries see Christina LUTTER, "Social Groups".

25 Cf. Klaus SCHREINER, "Ein Herz und eine Seele".

26 Cf. Gert MELVILLE, "Die 'Exhortatiunculae'", 212.

27 *Les statuts de Prémontré*, 1.

This structure, however, can only be achieved in a prolonged process of communal shaping. Benedict talks of the monastery as *officina*, as a workshop where "the tools of the spiritual art" can and shall be employed unceasingly night and day. Yet the prerequisite for the enjoyment of using these tools is the possession of the ability to use them, which, in turn, builds upon the corresponding conditioning and training of those involved. Thus, a monastery is always also a *schola* at the same time – a "school for the service to the Lord", as this reads in the Benedictine Rule.[28] Entering that school means cleansing oneself in so far as one sheds what represented one's own self-centred personality hitherto and gains a new one that is centred solely on God. When asked what ought to be taught to newcomers who were admitted as novices, Adam of Perseigne, for instance, answered that it was necessary to inform them carefully about the splendour of the new life, so that they could rightfully be called novices after having shed the old human being they have been till until now.[29] Petrus Cellensis put it in a more rigid way: "Upon joining a monastery, he shall give skin for skin and everything he has for his soul through shedding the old man and raising a new one, so that he might stride into the virtues of the new life and confess and leave behind the vices of his old one."[30] For what was required was, in the end, a *conversio totalis ad Deum* – a total devotion of the heart to God. This always also constituted an "alienation from oneself" (*alienatio a seipso*)[31], which, in theological terms, is to be named "kenosis" and thus meant the "self-emptying of one's own will and becoming entirely receptive to God's divine will".[32]

It is exactly this point which shows the difference between the two forms of the individual with which a monastic community was confronted. A monastic community was an institution of the transitory and of what led to individual salvation, and the monk who built an *interior domus* and also made it the dwelling of God by encountering his own conscience met all goals such a community aspired to for its members. As was pointed out in the beginning of this paper, it was possible to be absolutely obedient to this community and its representatives since one obeyed God in obeying the representatives and confreres. *Singularitas*, by contrast, was not transitory. Therefore, it did not want an alienation from oneself but wanted the self as final goal. Hence, it resulted in a discrepancy with the community, which – as we have seen – could only be mended with great effort or – as the very many cases of open rebellion in monasteries show – could not

---

28 Both quotations: *Benedicti regula*, Prologus, 45; ch. IV, 78.

29 Adam of Perseigne, *Lettres*, 112 (Lettre V, 45).

30 Petrus Cellensis, *De disciplina claustralis*, 192.

31 Anonymous, *Epistola cujusdam de doctrina vitae agenda*, col. 1187.

32 https://en.wikipedia.org/wiki/Kenosis

be mended at all. For in a manner of speaking, disobedience was inscribed into *singularitas*. Disobedience was called idolatry because *singularitas* meant placing oneself in God's stead.[33]

In monasteries, the relationship between community and individual was precarious at all times, had to be balanced continuously and could, despite all the optimism in the source texts, never be formed without compromises. However, the history of the Middle Ages shows that exactly because it required so much effort from both sides due to being both inescapable and necessary, the tension between monastic community and individual brought forth remarkable, innovative achievements in all sectors of religion, science, law, and society.[34] This system did not include *singularitas* as an individualism that was hostile to the community and remote from God. This problem probably was also less of a systemic but rather more of a psychological one that had to be ascribed to a general human weakness. This circumstance, however, did not render it less potent, which is why it always had to be taken into account.

Bibliography

Abelard, *Scito te ipsum*, edited by Rainer M. Ilgner (Corpus Christianorum, Continuatio Mediaevalis, 190). Turnhout, Brepols, 2001.

Adam de Perseigne, *Lettres*, edited by Jean Bouvet (Sources chrétiennes, 66). Paris, Du Cerf.

Anonymous, *De interiore domo*. Migne, Patrologia latina, vol. 184, col. 507–552.

Anonymous, *Epistola cujusdam de doctrina vitae agenda*. Migne, Patrologia latina, vol. 184, col. 1185–1190.

Anonymous, *Meditationes piissimae de cognitione humanae conditiones*. Migne, Patrologia latina, vol. 184, col. 485–508.

Anonymous, *Tractatus de statu virtutum*. Migne, Patrologia latina vol. 184, col. 791–812.

Antin, Paul, "Á la source de singularitas. Vie monastique", *Archivum Latinitatis Medii Aevi* 36 (1967/68), 111–112.

Bauer, Gerhard, *Claustrum animae. Untersuchungen zur Geschichte der Metapher vom Herzen als Kloster*, München, Fink, 1973.

Benedicti regula, edited by Rudolf Hanslik (Corpus scriptorum ecclesiasticorum Latinorum, 75). Wien, Hoelder-Pichler-Tempsky, 1960.

Bernard of Clairvaux, *De gradibus humilitatis et superbia tractatus*. In *Sancti Bernardi Opera*, vol. 3, edited by Jean Leclercq, and Henri-Maria Rochais. Roma, Ed. Cistercienses, 1963, 13–59.

Bernard of Clairvaux, *Epistola XI*. In *Sancti Bernardi Opera*, vol. 7, edited by Jean Leclercq, and Henri-Maria Rochais. Roma, Ed. Cistercienses, 1974, 52–60.

33 Anonymous, *Tractatus de statu virtutum*, col. 799.

34 Cf. Gert Melville, "The Innovational Power".

Bernard of Clairvaux, *In circumcisione, sermo 3,6.* In *Sancti Bernardi Opera*, vol. 4, edited by Jean Leclercq, and Henri-Maria Rochais. Roma, Ed. Cistercienses, 1966, 273–291.

Bernard of Clairvaux, *Parabolae 3.* In *Sancti Bernardi Opera*, vol. 6, edited by Jean Leclercq, and Henri-Maria Rochais. Roma, Ed. Cistercienses, 1970, 261–276.

Bernard of Clairvaux, *Sermo ad domenicam VI post Pentecosten, I.* In *Sancti Bernardi Opera*, vol. 5, edited by Jean Leclercq, and Henri-Maria Rochais. Roma, Ed. Cistercienses, 1968, 206–209.

Bertola, Ermenegildo, *Il problema della coscienza monastica del XII secolo*, Padua, CEDAM, 1970

Breitenstein, Mirko, "Die Verfügbarkeit der Transzendenz: Das Gewissen der Mönche als Heilsgarant". In *Innovationen durch Deuten und Gestalten. Klöster im Mittelalter zwischen Jenseits und Welt*, edited by Gert Melville, Bernd Schneidmüller, and Stefan Weinfurter, Regensburg, Schnell + Steiner, 2014, S. 37–56.

Breitenstein, Mirko, "Der Traktat vom 'Inneren Haus': Verantwortung als Ziel der Gewissensbildung". In *Innovation in Klöstern und Orden des Hohen Mittelalters. Aspekte und Pragmatik eines Begriffs*, edited by Mirko Breitenstein, Stefan Burkhardt, and Julia Dücker (Vita regularis. Abhandlungen, 48). Berlin, Lit, 2012, 263–292.

Bynum, Caroline W., "The Cistercian Conception of Community". In *Jesus as Mother. Studies in the Spirituality of the High Middle Ages*, edited by Caroline W. Bynum. Berkeley, University of California Press, 1984, 59–81.

Bynum, Caroline W., "The Spirituality of Regular Canons in the Twelfth Centruy". In *Jesus as Mother. Studies in the Spirituality of the High Middle Ages*, edited by Caroline W. Bynum. Berkeley, University of California Press, 1984, 22–58.

Chenu, Marie-Dominique, *L'éveil de la conscience dans la civilisation médiévale.* Montréal/Paris, Vrin, 1969.

Das Eigene und das Ganze. Zum Individuellen im mittelalterlichen Religiosentum, edited by Gert Melville, and Markus Schürer (Vita regularis. Abhandlungen, 16). Münster/Hamburg/London, Lit, 2002.

Diers, Michaela, *Bernhard von Clairvaux. Elitäre Frömmigkeit und begnadetes Wirken*, Münster, Aschendorff, 1991.

Erismann, Christophe, "*Singularitas.* Éléments pour l'histoire du concept: la contribution d'Odon de Cambrai". In *Florilegium Mediaevale. Études offertes à Jacqueline Hamesse à l'occasion de son éméritat*, edited by José Meirinhos, and Olga Weijers. Louvain-la-Neuve, Fédération Internationale des Instituts d'Études Médiévales, 2009, 177–183.

Füser, Thomas, *Mönche im Konflikt. Zum Spannungsfeld von Norm, Devianz und Sanktion bei den Cisterziensern und Cluniazensern* (Vita regularis. Abhandlungen, 9) . Münster/Hamburg/London, Lit, 2000.

Goffmann, Erving, *Asylums. Essays on the Social Situation of Mental Patients and other Inmates.* Chicago, Aldine, 1961.

Hübner, Wolfgang, "Der theologisch-philosophische Konservatismus des Jean Gerson", In *Antiqui und Moderni. Traditionsbewußtsein und Fortschrittsbewußtsein im späteren Mittelalter*, edited by Albert Zimmermann. Berlin, New York, De Gruyter, 1974, 171–200.

Hugo de Folieto, *De claustro animae.* Migne, Patrologia Latina, vol 176, col. 1017–1182.

Leinkauf, Thomas, "Die Bestimmung des Einzelseienden durch die Begriffe contractio, singularitas und aequalitas bei Nicolaus Cusanus", *Archiv für Begriffsgeschichte* 37 (1994), 180–211.

Les statuts de Prémontré au milieu du XIIe *siècle*, edited by Placide F. Lefèvre, and Wilfried Marcel Grauwen, Averbode, Praemonstratensia, 1978.
Lutter, Christina, "Social Groups, Personal Relations, and the Making of Communities in Medieval *vita monastica*" In *Making Sense as a Cultural Practice. Historical Perspectives*, edited by Jörg Rogge (Mainzer Historische Kulturwissenschaften, 18). Bielefeld, Transcript, 2013, 45–61.
Meinhardt, Helmut, "Singularität, Spontaneität, Subjektivität in Sein und Erkennen bei Nikolaus von Kues". In *Selbst - Singularität - Subjektivität. Vom Neuplatonismus zum deutschen Idealismus*, edited by Theo Kobusch, Burkhard Mojsisch, and Orrin F. Summerell, Amsterdam, Grüner, 2002, 231–240.
Melville, Gert, "Der Mönch als Rebell gegen gesatzte Ordnung und religiöse Tugend". In *De ordine vitae. Zu Normvorstellungen, Organisationsformen und Schriftgebrauch im mittelalterlichen Ordenswesen*, edited by Gert Melville (Vita regularis, Abhandlungen, 1). Münster, Lit, 1996, S. 153–186.
Melville, Gert, "Die 'Exhortatiunculae' des Girardus de Arvernia an die Cluniazenser. Bilanz im Alltag einer Reformierungsphase". In *Ecclesia et regnum: Beiträge zur Geschichte von Kirche, Recht und Staat im Mittelalter: Festschrift für Franz-Josef Schmale zu seinem 65. Geburtstag*, edited by Dieter Berg, and Hans-Werner Goetz, Bochum, Winkler, 1989, 203–234.
Melville, Gert, "Einleitende Aspekte zur Aporie von Eigenem und Ganzem im mittelalterlichen Religiosentum". In *Das Eigene und das Ganze. Zum Individuellen im mittelalterlichen Religiosentum*, edited by Gert Melville, and Markus Schürer (Vita regularis. Abhandlungen, 9). Münster/Hamburg/London, Lit, 2002, XI - XLI.
Melville, Gert, "The Innovational Power of Monastic Life in the Middle Ages". In *Monastic Culture. The Long Thirteenth Century. Essays in Honour of Brian Patrick McGuire*, edited by Lars Bisgaard, Sigga Engsbro, Kurt Villads Jensen, and Tore Nyberg. Odense, University Press of Southern Denmark, 2014, 13–32.
Melville, Gert, *The World of Medieval Monasticism. Its History and Forms of Life*. Collegeville, MN, Liturgical Press, 2016.
Petrus Cellensis, *De disciplina claustrali*, edited by. Gerard de Martel, Pierre de Celle, *L'école du cloître*, Paris, Du Cerf, 1977.
Schreiner, Klaus, "Ein Herz und eine Seele. Eine urchristliche Lebensform und ihrer Institutionalisierung im augustinisch geprägten Mönchtum des hohen und späten Mittelalters". In *Regula Sancti Augustini. Normative Grundlage differenter Verbände im Mittelalter*, edited by Gert Melville, and Anne Müller. Paring, Augustiner-Chorherren-Verlag, 2002, 1–48.
Straub, Christian, "Singularität des Individuums? Eine begriffsgeschichtliche Problemskizze". In *Individuum und Individualität im Mittelalter*, edited by Jan A. Aertsen, and Andreas Speer. Berlin, New York, De Gruyter, 1996, 37–56.
van 't Spijker, Ineke, *Fictions of the Inner Life. Religious Literature and Formation of the Self in the Eleventh and Twelfth Centuries*, Turnhoult, Brepols, 2004.
von Bredow, Gerda, "Participatio singularitatis. Einzigartigkeit als Grundmuster der Weltgestaltung", *Archiv für Philosophie* 71 (1989), 213–230.
Weber, Max, *Economy and Society*, vol. 1, translated by Guenther Roth, and Claus Wittich. Berkeley, Los Angeles, and London, University of California Press, 1978.

Gerd Schwerhoff

The Dark Side of Community – Early Modern German Witch Hunts

I

Rather than being innocent and purely value-neutral the German word for community, *Gemeinschaft*, is invariably normatively loaded. The term has been assigned positive connotations (in opposition to *Gesellschaft*, society) in political and scientific discourse since the nineteenth-century.[1] Incorporating the medieval horizons of meaning for the terms *allgemein* (global), *gemeinsam* (collective) and *zusammengehörig* (belonging together), *Gemeinschaft* describes a social group which is composed by individuals who not only live together in similar conditions but also hold common values. The term can also have an institutional dimension and has often been used to refer to specific social-political structural organizations. The closely related word *Gemeinde* denotes just such a corporate identity, being roughly equivalent to a parish or municipality. In the social sciences, Peter Blickle is credited with making the community or *Gemeinde* a point of recent historical focus with the analytical concept of communalism (*Kommunalismus*): leading to the rejection of one-sided, negatively biased representations of the past. Blickle argued against the prevailing narrative of a passive lower class and a particulary German subservient spirit. He shows that German peasantry actively participated in a continuous struggle for political autonomy. Between the years 1300 and 1800, according to Blickle, communalism blazed the trail for modern republicanism. This development involved the formation of associations of urban and rural authorities, who elected representative bodies, regulated their own affairs, passed bylaws and dispensed justice. All in all, the communities maintained conflict-laden partnerships with and struggled against their mostly noble lords. The latter generally had the upper hand, but occasionally the communities were able to resist. The socio-economic foundations of community building were the necessity of both economic cooperation and coordination of daily life so as to allow for peaceful coexistence in densely populated zones. Out of this need for solidarity arose the values and norms of communalism: the

1 Manfred RIEDEL, "*Gesellschaft, Gemeinschaft*", 801–862. For a differentiated discussion of the theoretical tradions and dimensions of meaning, see Hartmut ROSA and Lars GERTENBACH, *Theorien der Gemeinschaft*; Juliane SPITTA, *Gemeinschaft*.

common good (*bonum commune*), the *Hausnotdurft* (the right to secure one's own subsistence), peace and also justice and personal freedom.[2]

This is completely different from the image of the community, which two German ethnologists described forty years ago in their modern social history of a village. They used extant records from Württemberg's Kiebingen to exemplify the social psychology of village inhabitants. In fact large families and neighbors were able to build emergency alliances to better recover from shared catastrophes or – if called for – to stand together against enemies or territorial rulers. Outside of such situations, communal solidarity was a luxury, which few could command; instead individual interests were ruthlessly pursued. Marriages tended to be endogenous, meaning that one sought out partners who were local and from the same social class. Out of such marriage alliances arose large, multi-branched networks of relatives, who fought amongst each other. Clientelism reigned instead of formal politics. This required no formal communications or decision-making structures. All relevant information was public information anyways, everyone knew about each and every decision. Privacy in the bourgeois sense was unknown. Secrets did not exist, since everyone watched each other. In short, in their view the village life was an association of terror, a *Terrorzusammenhang*.[3] This empirical description could be supported by means of sociological analysis, which leads (albeit analytically distant) in the same direction. Following Max Weber, Rose Laub Coser sketched "The greedy nature of *Gemeinschaft*." Although writing in English, the Berlin-born Jewish emigrant consistently used the German term to emphasize that 'community' was not an exact translation for the highly charged notion of *Gemeinschaft*. As "structural property of greedy institutions" Coser identifies that

> ...they operate under conditions of restricted (or simple) role-sets. This is what appeals to many who yearn for a sense of belonging, and it is what helps to give the gemeinschaft character to such communities. The greediness consists in claims for priority in the allocation of material and emotional resources. The gemeinschaft is greedy to the extent that it absorbs individuals in unidimensional relationships, depriving them of the opportunity to confront multiple and contradictory expectations that would make them reflect about their roles. And in a gemeinschaft, role-sets are restricted.[4]

2 Basic concept in Peter Blickle, *Kommunalismus*, vol. 2. For broader context see Id., *Deutsche Untertanen*; Id., *Das Alte Europa*; see also Stollberg-Rilinger's review.

3 Utz Jeggle and Albert Illien, "Die Dorfgemeinschaft", 38–53; for the Kiebingen studies Robert Lorenz, *"Wir bleiben in Klitten"*, 22 sq.

4 Rose L. Coser, *In Defense of Modernity*, 74.

Coser's analysis focuses on the community as an obstacle on the way towards modern pluralism and individualism. But her research can also be used as the starting point for historical deliberation since she clearly articulated that her image of the *Gemeinschaft* is scarcely congruent with the romantic caricature of harmonious solidarity with which nineteenth-century scholars wistfully examined the past. In fact, the *Gemeinschaft* by no means excludes hostilities, fights or violence of any sort, since it is governed by rigid force and mutual suspicion.[5]

II

Rarely does a phenomenon bring out greed and hostility so vividly in a community as the persecution of witches and sorcerers. Nowhere on Earth have witch hunts assumed such excessive proportions as they did in early modern Europe. It is hardly possible to provide a quick overview of the complex phenomenon of witchcraft overall.[6] Both the belief in and persecution of witches are an intercultural and epoch-spanning phenomenon. Nonetheless, it was only in the cultural area of Latin Christendom that it resulted in legally authorized mass executions of alleged witches, based on the scholarship of Christian demonologists. It was thus assumed that sorcerers and witches relied upon a pact made with the devil when performing their own harmful magic (*maleficium*). They formed a secret sect of devil worshipers, who assembled at night to perform anti-religious rituals during a witches' sabbath – incidentally a special kind of imaginary 'community', which would be likewise worth some consideration.[7] According to serious estimates, between the fifteenth and eighteenth centuries approximately 50,000 – 60,000 people were convicted in Europe of being members of this imaginary sect of witches, sentenced to death and executed. About half of the victims were in the Holy Roman Empire of the German Nation, which is why Germany – reputed to be the "Heartland of the witch craze" – remains central to the discussion below.[8] The scale and intensity of persecution varied widely even within this core area. Regions with few or even no witch trials stood opposed to strongholds of witch hunting, where hundreds of victims were executed in large chain-reaction trials within short periods. Overall it seems that jurisdictions with large scale trials were located in central Europe, while noticeably fewer persecution activities were taking place on the northern and southern European peripheries.

---

5 Ibid., 74, 72.
6 For a good overview in English Wolfgang BEHRINGER, *Witches and Witch-Hunts*.
7 Elisabeth BIESEL, "'Die Pfeifer seint alle uff den baumen gesessen'", 290–302; Martine OSTORERO et al., eds., *L'Imaginaire du Sabbat*.
8 Hans C. Erik MIDELFORT, "Heartland of the witch craze", 113–119.

Furthermore, there were trans-regional peaks in witch hunting activity, occurring in Germany around 1590, 1630 and 1650/60, but continuing into the eighteenth century in Eastern Europe.

The early modern society was primarily rural. Thus, the witch hunts were likewise, for the most part, a rural phenomenon. But the cities were in no way immune to this persecution. Although there were scarcely any large waves of witch trials in the larger cities such as Nuremberg, Augsburg, Frankfurt or Hamburg, some of the smaller cities with between 1,000 and 3,000 inhabitants have proven to be catalysts for intensive witch hunting. This pattern includes residence cities such as Ellwangen or Bamberg, quasi-autonomous territorial cites such as Minden, Osnabrück or the proverbial 'witches nest' Lemgo, but also Imperial Free Cities such as Reutlingen or Esslingen with occasional intense waves of persecution with hundreds of victims. What is striking here is the small-scale structure. In the small Hansiatic city of Lemgo two thirds of the accused came from one of six quarters (*Bauernschaften*). In fact forty-three percent of all victims hailed from two streets: *Schusterstraße* and *Orpingstraße*. This was a rather poor neighborhood, out of which the accused tended to be among the relatively better off inhabitants.[9] This contrasts with Bamberg's *Lange Gasse*, where a portion of the upper and ruling classes of the city had settled and from which seventeen occupants and owners – some with their entire families – became victims of the witch hunts between 1626 and 1630. In this case, every third household was affected. These examples show that even within the city, smaller communities, neighborhoods and districts were important settings for persecution.

III

To explain the peculiar physiognomy of the witch hunts, it is necessary to consider the specific interactions of the three main actors: first there were the governing authorities, second the subjects and third those office holders and experts involved in the trials and acting as quasi intermediary authorities. For a long time historians have viewed the witch hunts as a campaign which was predominantly driven by the authorities "from above". Indeed, there can be no doubt that almost every Christian ruler considered it one of his primary duties to extirpate the atrocious witches' sect. More recently though, in connection with ethnological research into the problem of magic and witchcraft, the focus of

9 Ursula Bender-Wittmann, "Hexenprozesse in Lemgo", 235–266, chart 4; Britta Gehm, *Die Hexenverfolgung im Hochstift Bamberg*, 189 sq.

research has shifted to examine the roll of the general populace, and thus of the community. Gerhard Schormann coined a memorable phrase when he wrote that large waves of witch trials appeared in those places where the pent-up desire for persecution from below had received a 'green light' from above.[10]

Villages and small urban communities were thus the places, where suspicion against possible witches was generated and where desires for persecution were articulated. It is worth examining the context for this a little closer. Generally speaking, the belief in witches can be identified as a cultural pattern, which possesses explanatory power and creates concepts of meaning to reduce the contingencies of daily life. Hail destroyed the harvest, but spared the neighbor's field? The cow produces no more milk, while at the same time another's livestock is obviously thriving? Or the child died suddenly, with no apparent general epidemic or long period of illness? For early modern contemporaries it stood to reason that one should consider magic as a possible cause of such misfortune. Modern readers view this explanation as superstitious, irrational and incompatible with modern ideas of causality. But even the classical study of Evans-Pritchard of magic and witchcraft in Africa had already made it clear, that the witchcraft explanation has its own rationality. He observed the belief in a double causality among the Azande. His most famous example involved a granary which collapsed and crushed someone after being consumed by termites.[11] What to our eyes seems a tragic accident could be interpreted by the Azande as the work of witches. That is not to say that the Azande would not recognize the correlation between the termites and the granary's collapse, but rather that they would not accept this as a sufficient explanation: Why did the building collapse at precisely that time during which a particular person sat beneath it? In answering such questions, the explanation power of a magical belief system goes much further than modern logical reasoning. This type of scapegoat mechanism is naturally not restricted to small-scale communities, but it can become particularly effective in such locales.

What does the victim profile look like, who precisely was stigmatized as a witch? Here again, recent research has become highly dynamic, although no definitive understanding has been formed to date. The most likely hypothesis was for a long time that women and healers knowledgeable about magic would very quickly find themselves in the cross-hairs of witch hunters. Nonetheless, the opposite actually was the case. Although healers and other magical experts were criminalized by the authorities alongside witches, the former tended to be sanctioned with lighter sentences. Community members regularly consulted these

---

10 Gerhard SCHORMANN, *Hexenprozesse in Deutschland*, 56 sq.

11 Edward E. EVANS-PRITCHARD, *Witchcraft, Oracles and Magic*, 69 sq.

individuals and protected them from legal prosecutions. These 'experts' promised magical protection from witchcraft and acted as witch-finders.[12]

It is difficult to come up with the profile of a witch that is generalizable across time and space. There were, however, certain features prevalent among the victims. For instance, there were a disproportionate number of single women and widows over fifty years of age among those convicted of witchcraft in the Saar region. According to the analysis of Eva Labouvie, this group was singled out for particular behavioral problems, which could be classified as non-conforming or deviating from existing rules of cohabitation. More specifically, the cantankerous and shameless behavior of these women gave them a very high potential for conflict.[13] But it remains unclear to what extent such characteristics were disproportionately represented among accused witches with respect to the whole population – the majority of early modern people were poor and for each pattern discernible in one jurisdiction, there exists a counter-example somewhere else. Overall we know that four out of five of those burned as witches were female, but there were also many male victims in specific regions or in times of intense witch hunting.[14] In many places the trials specifically targeted the wealthy neighbors. This motive was first detected by Andreas Blauert in the 1500 trial of Oberhuserin, a woman from the Swiss town of Kriens, who attracted the envy and jealousy of her neighbors by virtue of being a rich newcomer. In the meantime, similar patterns have been documented in other research.[15] Furthermore, while some studies focus on the persecution of single old women, others have examined trials against children and male youths which apparently occurred with greater frequency in the later period.[16] Finally, there is a methodological problem in determining which non-conformist, deviant or even criminal behaviors sufficed for suspicions or accusations of witchcraft: The career of a witch can only be judged in retrospective. Thus, it is difficult to decide whether a significant pattern of behavior actually existed or whether the deviation from social norms instead resulted from a gradual process of attribution. Phrased differently: the reported non-conformism of the accused was not always the origin of suspicions of witchcraft, but it was often instead a product of those suspicions.

Could witchcraft be more easily uncovered if we understand it as the result of social conflicts and thus searched for potential root causes of discord? Carol

---

12 Walter RUMMEL, "'Weise' Frauen und 'weise' Männer'", 353–375.
13 Eva LABOUVIE, *Zauberei und Hexenwerk*, 176, 182.
14 Rolf SCHULTE, *Man as Witch*.
15 Andreas BLAUERT, "Hexenverfolgung in einer spätmittelalterlichen Gemeinde", 8–25; Ursula BENDER-WITTMANN, "*'Communis salutis hostis'*", 150–184.
16 Rainer BECK, *Mäuselmacher oder die Imagination des Bösen*.

Karlsen pursued this method of study in examining the well-researched New England witch trials. According to her observations, the attribution of abnormal behavior, and subsequently of witchcraft, could be seen to closely follow the logic of social conflicts. Thus, female heirs having no brothers or sons were at higher risk of being accused, since such situations went against the system of inheritance which was normally from man to man.[17] Numerous additional studies have affirmed this logic of conflicts, but the hope of finding a sort of master key for these conflicts was quickly quashed. Disputes about money, land, inheritance and honor are as likely to be found as economic competition and political antagonism.[18] At most, one common thread can be seen within a specific 'premodern' or 'rural' mentality which is rooted in the general shortage of resources of that era and in the intense feeling of jealousy connected with this shortage. It was anthropologist George M. Forster who developed the concept of 'limited goods' in his analysis of Mexico which was applied to discussions of Europe by Rainer Walz.[19] According to Forster, the peasant mindset is characterized by the belief that there exists a limited amount of (material and immaterial) goods, whether the goods in question are household items, livestock, friendship, love or good health. According to this logic, the acquisition of more goods by one member of the community always implies the loss of goods for others, giving rise to resentment or envy. Magic and witchcraft therefore offer attractive explanations in cases where there is no obvious reason for one neighbor to have more (cows, affection, good health, etc.) than another.

On the whole, witch trials can be seen as an extremely flexible multi-purpose instrument for resolving all manner of conflicts in very different social constellations and contexts.[20] In particular, the de facto fictionality of the sect of witches (in terms of the contemporaries: the secret, unobservable, only indirectly ascertainable character of magical practices) accounts for this incomparable elasticity of belief in witches, its universal applicability. On the one hand, no situation has been so difficult to objectify, verify and thus prove as witchcraft. On the other hand potential accusations existed which were even more difficult to invalidate, since it was equally difficult to furnish proof of innocence. Throughout the period of excessive witch hunting, this lack of provability often stood in inverse proportion to the riskiness of witchcraft accusations. The latter was therefore a sharp weapon, which appeared applicable without requiring a lot of training or risk. In

17 Carol F. KARLSEN, *The Devil in the Shape of a Woman.*

18 Ingo KOPPENBORG, *Hexen in Detmold*, 77 sq.; Johannes DILLINGER, *'Evil People'*, 79 sq.

19 George M. FORSTER, "Peasant Society", 293–315; Rainer WALZ, *Hexenglaube*, 52 sq.

20 Gerd SCHWERHOFF, "Hexerei, Geschlecht und Regionalgeschichte", 348 sq.

light of the actual existing conflicts, this overture was much too tempting not to be regularly used.

The notion of multi-purpose instrument, however, should not be misunderstood. One occasionally comes across the opinion, that accusers cynically exploited belief in witches for their own sordid materialistic purposes, without being personally convinced of the reality of witchcraft – an argument, which was especially popular among nineteenth-century historians. Upon closer reflection, this opposition between cynical functional rationalist on one side and pious but superstitious religious protagonist on the other side appears to be unproductive. The particular combination of both aspects directly influenced development of the belief in witches. Yet the specific power of witchcraft beliefs lay precisely in the notion of fulfilling completely material, thoroughly self-centered interests. The evil neighbor, with whom you had been fighting for decades, was a witch? That seemed completely logical. It explained her bad temper. In addition the public or private accusation that she was a witch opened up an option which promised a twofold win: victory over your opponent and the moral advantage of having freed the world from a devil worshiper. In the words of Johannes Dillinger:

> ... suspicion of witchcraft was directed at those who seemed to have proved themselves untrustworthy or who stood outside of community. The 'enemy image' of the witch was a category of interpretation that absorbed everything negative. It was thus possible to identify as a witch any person whom one experienced as adversarial [... as] 'Evil people (*böse Leute*)', a term that sources ... used as a synonym for "witches.[21]

IV

The logic of attribution and conflict, as described here, is compelling. But it implies a logic of causality which was clearly lacking in many witchcraft cases. For instance, a causal link between existing conflicts and witchcraft accusations can only be ascertained in around a quarter of all cases in the city of Minden.[22] One should bear in mind instead that the root causes of suspicion and the dynamics of witchcraft accusations could just as easily be a contingent phenomenon of communication. Using sources from the county of Lippe in the north of the Holy Roman Empire, Rainer Walz described the genesis and development dynamics of village witchcraft accusations in a vivid and analytically clear manner as 'magical

21 Johannes Dillinger, *'Evil People'*, 96.
22 Barbara Gross, *Hexerei in Minden*, 247 sq., 276.

communication'.[23] Distrust could originate from actual events or catastrophes, in situations where a person's words, gestures or mere presence made them seem suspicious. Also the rituals of popular witch-finders (*Hexenbanner*) could cast suspicions on a particular person. In the vast majority of witchcraft cases, however, speculation fell on persons who had already developed a reputation as a witch. But the accumulation of multiple incriminating incidents did not automatically lead to suspicions of witchcraft. Cursing is a good example: Excessive cursing – especially by women – was generally regarded as a sign that the person was lost to the devil. But often also the accusers and the witnesses had to admit in court that they themselves were known as cursers from time to time. Thus, general perceptions were selectively interpreted: The cursing of some people made them suspicious, others could use rough language without being suspected. A comparable difference can be observed with the abusive term 'witch' (*Hexe*). Frequently used in daily life, sometimes the word was used as a stereotypical insult, which did not imply any concrete suspicions and which could be remedied in the village's lower court through payment of a small fine. In other contexts, however, the term was used as a direct accusation with strong implications. Walz's study suggested that up to a certain point the creation of suspicion was rather coincidental. There was always gossip about the behavior and potential deviations of every member of the community. Nonetheless at a certain juncture, which is difficult to pinpoint, the rumors surrounding a particular person could reach a critical mass and they would henceforth be viewed as increasingly suspect. The veiled gossip about the suspect intensifies with the addition of innuendos in direct communication or plain-spoken suspicions and even 'face-to-face accusations'. From this point on there was a wide spectrum of aggressive behavioral patterns towards the accused witches. Shouts, threats, eviction demands and direct physical violence were intended to force the witch to reverse the magical harm. If these did not work, villagers then filed official accusations with the magisterial court. After this process reached a certain level the accused found themselves with dwindling options. A decisive reaction to a voiced suspicion or public accusation could help in the beginning, e. g. the "retorsion", the slinging back of an accusation onto the person who uttered it (person X "was considered to be a witch until he withdrew the allegation"), or a libel complaint to the court. Once too many rumors were circulating though, each possible option exacerbates suspicions – those who remained silent and failed to react to rumors implicitly acknowledged that the accusations were true, those who protested loudly contributed to the rumors, and exonerating factors were increasingly less accepted. This "paradoxical" form

23 Rainer WALZ, *Hexenglaube*, 188 sq.

of communication, identified by Walz, led many accused along an upward slope towards the funeral pyre lying in wait.

Rumor, gossip and scandal were, therefore, of central importance to a community's communication.[24] Gossip (*Gerücht*, *Gerede* or *Klatsch*) has certainly been given a certain importance in every historical period and all conceivable social formations. Nonetheless, in early modern small-scale communities it was given a special weight, in that the informal market of rumors continually tested the reputation of every single member and could be potentially damaging. Thus, having a bad reputation definitely became legally relevant as an indication for witchcraft. Above all, rumors produced continual behavioral uncertainties. To some extent they established their own reality behind the back of the suspect – a reality which quickly solidified and became nearly impossible to do away with.[25] In the extreme, rumors may have led certain people to rashly reference random statements or allusions to witches and thus generate suspicion in the first place like the following example from Lippe demonstrates. After a sick woman remarked in the presence of Lieseke Hase that evil people could be the cause of her illness, Lieseke erroneously assumed that the comment applied to herself. She soon afterwards returned to the sick woman's house and asked the husband directly if people considered her guilty. For his part the man was dismayed about this suggested magical accusation and wanted to call in the authorities. Lieseke tried to prevent this, in the hopes of putting an end to the formation of a rumor. Yet it was at this point that she made herself an actual suspect.[26]

Family members of suspected witches – especially their husbands – found themselves in a difficult situation. On the one hand both the village code of conduct and their own economical calculations demanded that they defend the family's honor. On the other hand the same code required that they definitively distance themselves from people with very bad reputations. Thus, most protagonists tended to follow social norms and the community's judgment became the family's imperative. Only a very few defended their relatives and also remained active during the advanced stages of the accusation. The majority of men recommended that their wives remain silent when suspicions were first articulated in the hopes that the rumors would not circulate very widely. As rumors intensified most then chose to distance themselves, leaving their relatives seeking to free themselves from the maelstrom of witchcraft suspicions. Such was the petition of several men in Amt Blomberg (Lippe), whose wives had been denounced as

24 Jörg R. BERGMANN, "Klatsch", 447–458; Regina SCHULTE, *Das Dorf im Verhör*, 166 sq.; Alexander COWAN, "Gossip and Street Culture", 113–133.

25 Barbara GROSS, *Hexerei in Minden*, 240 sq.

26 Rainer WALZ, *Hexenglaube*, 341.

witches by a previously banished sorceress. They demanded the women's immediate arrest so that their wives had a chance to defend themselves. The reason behind this petition was neither love nor attachment, but rather concern for their own family honor. The women were at the center of so many rumors, their husbands stated in the petition, that they could no longer be tolerated at home unless they could somehow successfully acquit themselves in court. If the women were found guilty, however, the men counseled that they should be burned at the stake as dictated by custom and law. Similar comments were made by another man: He stated that if his wife had actually mastered the art of witchcraft, then he wanted to help carry the wood with which she would be burned as required.[27]

The rumor mill of the community is consequently stronger than family ties, although the threatening character which an accusation of witchcraft potentially has for all family members must be taken into consideration. It was often claimed that condemned witches had likewise introduced their relatives (especially, but not exclusively daughters) to the devil's sect and the black arts. The consequences of this attribution are particularly evident, for example, in the persecution stronghold of Ellwangen, where a large majority of trials were focused against so-called witch families (*Hexenfamilien*) or at least against those who lived in the same household as an accused witch.[28] In light of this situation, there were scarcely any protests or resistance from relatives of the accused or even from bystanders who argued for 'common sense' against beliefs in witches and witchcraft suspicions. Thus it must be considered exceptional that a certain Tönnies Strohmeyer in Lippe 1653 defended his accused wife by claiming that he sensed nothing in her besides piousness, love, honor and virtue. He then demanded that authorities severely punish her devilish slanderers.[29] Doubts about rightness of witch-trial procedures admittedly surfaced in some places in the course of massive-scale chain-reaction witch-trials. These doubts could even lead to a 'crisis of confidence' (Erik Midelfort), by means of which the trials would prematurely come to an end.[30]

V

Until now the discussion has considered the community as a whole along with individual protagonists who were members of it. Nonetheless, rural and urban

27 Ibid., 365 sq., esp. 367, 369; cf. Ingo Koppenburg, *Hexen in Detmold*, 141.

28 Wolfgang Mährle, "'O wehe der armen seelen'", 404 sq.

29 Rainer Walz, *Hexenglaube*, 368.

30 Edward Bever, "Witchcraft Prosecutions", 263–293.

communities in the Early Modern era often were not structureless entities, but rather well organized legal and political bodies. They also appeared in the witch-hunts as just such organized groups, generally with the goal of pressuring authorities and their representatives into taking decisive action on the matter of witches. The common communication channel between citizens and authorities – the supplication (*Bittschrift*) – was mostly used for this purpose. Thus, residents of Amt Brühl appealed to the privy council of Electoral Cologne in March 1631 with the complaint that the extirpation (*Ausrottung*) of witches was not proceeding as required in Amt Brühl, with the result that the use of harmful magic increased day by day and much damage occurred.[31] A second example comes from the village of Dieburg in Electoral Mainz, where the decisive initiative of an organized community stood at the start of a large witch-hunt, which claimed almost 150 victims in the following years. In March 1626 Dieburg's bailiff reported to the Elector that the citizenry requested him unrelentingly and imploringly with regards to the eradication of the abominable vices of witchcraft. They conveyed to him a further supplication containing the names of suspected witches.[32] Many other villages behaved similarly.

As the example from Dieburg made clear, communities did not merely perform as supplicants or as groups making demands, but instead sometimes also took hold of the reigns of persecution themselves. How far they succeeded with that in multiple regions has been made apparent by the groundbreaking studies of Eva Labouvie und especially Walter Rummel for the Saar and Mosel regions.[33] In the villages and small cities of this region (but not only there) there developed formal witchcraft committees (*Gemeindeausschüsse*) which systematically conducted witch-hunts. Members of this new variety of 'village inquisition'[34] were elected and ritually appointed at a community meeting.[35] All heads of households gathered, for example, under the village's linden tree, in city hall or under the high altar in the Church of Our Lady as was done in the small Eifel city of Bitburg. After swearing an oath, they elected the committee members. Nonetheless, by and by all citizens could in turn enjoy the benefits of this dubious office, even those whose family members were rumored to practice magic. The function of the com-

31 Shigeko KOBAYASHI, "Kommissar und Bittschrift", 83; Rita VOLTMER and Shigeko KOBAYASHI, "Supplikationen und Hexereiverfahren", 247–269.

32 Source at http://www.hexenprozesse-kurmainz.de/quellen/stadtarchiv-mainz/anzeigen-und-obrigkeitliche-korrespondenz/lage-6-draengen-der-dieburger-nach-hexenprozessen.html; i.e. Herbert POHL, *Zauberglaube*, 127 sq.

33 Walter RUMMEL, "Communal Persecutions", 201–203; summarized research Rita VOLTMER, "Monopole", 5–67; EAD., "Konspiration ", 213–244.

34 Eva LABOUVIE, *Zauberei und Hexenwerk*, 82.

35 Rita VOLTMER, "Konspiration", 225 sq.

mittee was to initialize and to forward a witch-hunt, in concrete terms to collect clues, evidence and witness statements and also to arrest suspects. Secondly, they had to stand surety for the costs of expensive trials, which is why Rita Voltmer wrote about 'private accusation syndicates' (*Klagekonsortien*). If they did not want to be liable for the costs themselves, they had to arrange for the confiscation of the goods of the condemned. They operated in no way fully autonomously, but were required to coordinate and agree with representatives of the court and functionaries of the authorities. Indeed the territories in which the community witchcraft committees were active are characterized by a strong political fragmentation and tended to have 'weak' state structures. Such was the case in Rummel's study of the *Vogtei* of Winningen, which belonged to the County (*Hintere Grafschaft*) of Sponheim, which was situated isolated in the middle of the territory of the Electorate of Trier. Other courts were governed *kondominial*, that means by two or more rulers at the same time, who may have had different confessions. Against this background the community committees were able to usurp the on-site manorial functions and become supremely important. The magisterial courts sporadically simply 'ratified' the incriminating evidence provided by community committees. In contrast, local representatives of the authorities may have allied themselves with the committees. As traveling experts in matters of witchcraft, the so-called witch commissioners (*Hexenkommissare*) were equally likely to aid the committees, since they could draw tangible profits and prestige from the trials. Both of these persecution alliances of village committees and local authorities imparted a cataclysmic dynamic to the witch-hunts.[36] Separately they could nevertheless adopt very different forms. Thus, a witchcraft committee could, for example, adopt a positively rebellious form in opposition to the authorities as happened after 1593 in the small Mosel city of Cochem.[37] Made up of artisans, the witchcraft committee succeeded in disempowering the mercantile council elite and tyrannize the city for a while with witch trials primarily conducted against members of the old ruling class. Armed bailiffs dragged suspects from their homes to be interrogated and cruelly tortured. The reeve (*Vogt*) colluded with the committee. Electoral Trier's bailiff avoided assuming responsibility by means of his pointed absence. A commission of inquiry was set up in Cochem only after a victim was able to contact the territorial lord. The commission put an end to the witchcraft committee's activities. In appeal proceedings, the manorial court in Koblenz deemed the trials to be invalid and imposed the costs on the community's accusers.

---

36 Walter RUMMEL, *Bauern, Herren und Hexen*, 80, 114. The roll of local office holders or traveling specialists is very important, but cannot be expanded upon here, see Johannes DILLINGER, *Hexen und Magie*, 97 sq.
37 Walter RUMMEL, "Soziale Dynamik", 26–55.

The mechanism observed here can be generalized. Trials often stagnated or ended when protagonists from 'outside' (i.e. central manorial courts or rulers) broke open the small-scale persecution milieu and start asking critical questions. Having inherently established a greater physical distance between the location of events and decision makers, larger territories with stronger, centralized and formalized justice systems (i.e. Electoral Bavaria, Electoral Saxony or the Duchy of Württemberg) had relatively few witch-trials and, most importantly, no excessive mass persecutions.[38] This also reflects the relationship between community and the witch-hunts, albeit *ex negativo*.

The villagers' anti-witch supplications and the witchcraft committees' activities are not isolated factors, but must instead be considered in the context of the political-social structures of rural and urban communities. However, there were vast regional variations and the ideal-typical structure can only be characterized with broad brush strokes. These communities were in no way identical in terms of the number of inhabitants. Usually only those owning land or property were fully entitled to the benefits of citizenship. They filled the community assemblies, elected the local leader (the *Schultheiß*, *Ammann*, *Zender*, or similar), enacted local commands and prohibitions and were also responsible for local justice (N*iedergerichtsbarkeit*). Described another way, they had partial autonomy to govern from within over day-to-day matters of communal life on the one hand and were responsible for communicating with feudal authorities on the other. Thus, the community's political structure was also particularly constrained by economic realities. The inhabitants of the villages were obliged to rely on cooperation in order to survive, most notably the shared cultivation of common land. For masters and princes the existence of authorized representatives, able to speak for all local inhabitants, was quite practical. Nevertheless, the community also functioned as a spokesperson for the same people. They brought forward the interests and viewpoints of the inhabitants, generally by means of the much used instrument of petition (*Supplication*) which was used to air grievances and demand concrete improvements. If these voices were not heard, then peasants could actively resist the authorities' agents. In the extreme case, the community could become the central nucleus of violent disturbances and rebellions.[39]

Thus the witch hunts were not an exotic foreign element, but to some extent rather a logical expression of communal life in the same way that peasant or

---

38 Johannes DILLINGER, *Hexen und Magie*, 106 sq.; i.e. for Saxony Gerd SCHWERHOFF, "Zentren und treibende Kräfte", 61–100.

39 Heide WUNDER, *Die bäuerliche Gemeinde*; for social unrest Peter BLICKLE, *Unruhen in der ständischen Gesellschaft*; for supplications Cecilia NUBOLA and Andreas WÜRGLER, eds., *Bittschriften und Gravamina*.

middle class opposition to lordship was. Nonetheless, to what extent witchcraft committees can be understood in a specific sense as phenomenon of Blickle's communalism is a highly contentious issue. In particular Dillinger has advocated for this interpretation.[40] Contrary to him Rummel and Voltmer argued that the witchcraft committees may have availed themselves of a communal structure and pattern of behavior (i.e. in the form of a ritual oath), but their function and content did not have anything to do with communalism. It was not about defending collective values (i.e. the common good), nor about representing common interests against authorities, but rather that the committees were normally an instrument of a particular fraction of inhabitants and used to resolve personal enmities and fulfill selfish interests.[41] These objections are substantial, but refer in part to the weaknesses or rather the idealization of the concept of communalism which portrays the image of peasants united in resistance, but with their inner frictions and enmities hidden.[42]

VI

Even beyond the conceptual aggregation of communalism and beyond the formal witchcraft committees we have collected enough evidence to show that suspicions and accusations of witchcraft were anchored to the early modern community. In small-scale face-to-face early modern communities conflicts possessed a completely different quality than they do in modern, functionally differentiated societies where the differentiation of roles is greater and it is easier to keep out of someone's way. In comparison, according to Rainer Walz again in connection with the sociologist Niklas Luhmann, characteristic for early modern communities was a form of 'agonal' communication, whereby every interactant fearfully tried to defend his or her honor and mistrustfully gauged each utterance and action of others for possible assaults to that honor. In the event that such an attack was detected, it was valid to immediately react so as not to lose face.[43] Conflicts could scarcely be avoided or ignored in early modern communities. The medium of witchcraft suspicions and the possibility of being formally accused enhanced this setting in an often dramatic way. The imaginary sect of witches virtually formed the anti-type of the community, the orientation towards the *bonum commune* and neighborly solidarity. Thus, each attribution of negative behavior or injury

40 Johannes Dillinger, *'Evil People'*, 147 sq., 195 sq.

41 Walter Rummel, "Communal Persecutions", 11 sq.; Rita Voltmer, "Konspiration", 226, 234.

42 Heide Wunder, "Hexenprozesse und Gemeinde", 62.

43 Rainer Walz, "Agonale Kommunikation", 232 sq.

had the potential to be interpreted as an expression of attitudes prejudicial to the community and to be criminalized as indicator of witchcraft. Witchcraft suspicions and witch trials could be understood as extreme variants of village (and small city) communities of terror. In addition, Coser's sociological analysis of the community as a greedy institution could find its most impressive confirmation in the persecution of witches. In these cases the greed of the community often went so far that many of its members were devoured.

Bibliography

Beck, Rainer, *Mäuselmacher oder die Imagination des Bösen. Ein Hexenprozess 1715–1723.* München, C. H. Beck, 2011.

Behringer, Wolfgang, *Witches and Witch-Hunts. A Global History* (Themes in History). Cambridge, Polity Press, 2004.

Bender-Wittmann, Ursula,"'Communis salutis hostis'. Die Kauffrau Anna Veltmans". In *Biographieforschung und Stadtgeschichte. Lemgo in der Spätphase der Hexenverfolgung*, edited by Gisela Wilbertz and Jürgen Scheffler (Studien zur Regionalgeschichte, 13). Bielefeld, Verlag für Regionalgeschichte, 2000, 150–184.

Bender-Wittmann, Ursula, "Hexenprozesse in Lemgo (1628–1637). Eine sozialgeschichtliche Analyse". In *Der Weserraum zwischen 1500 und 1650: Gesellschaft, Wirtschaft und Kultur in der Frühen Neuzeit* (Materialien zur Kunst- und Kulturgeschichte in Nord- und Westdeutschland, 4). Marburg, Jonas-Verlag, 1993, 235–266.

Bergmann, Jörg, "Klatsch". In *Historisches Wörterbuch der Rhetorik*, vol. 10: *Nachträge A–Z*, edited by Gert Ueding, Darmstadt 2012, 447–458.

Bever, Edward, "Witchcraft Prosecutions and the Decline of Magic", *The Journal of Interdisciplinary History* 40 (2009), 263–293.

Biesel, Elisabeth, "'Die Pfeifer seint alle uff den baumen gesessen'. Der Hexensabbat in der Vorstellungswelt einer ländlichen Bevölkerung". In *Methoden und Konzepte der historischen Hexenforschung*, edited by Gunter Franz and Franz Irsigler (Trierer Hexenprozesse, 4). Trier, Spee, 1998, 290–302.

Blauert, Andreas, "Hexenverfolgung in einer spätmittelalterlichen Gemeinde. Das Beispiel Kriens/Luzern um 1500", *Geschichte und Gesellschaft* 16 (1990), 8–25.

Blickle, Peter, *Das Alte Europa. Vom Hochmittelalter bis zur Moderne*. München, C. H. Beck, 2008.

Blickle, Peter, *Deutsche Untertanen. Ein Widerspruch*. München, C. H. Beck, 1981.

Blickle, Peter, *Kommunalismus. Skizzen einer gesellschaftlichen Organisationsform*, vol. 2: *Europa*. München, Oldenbourg, 2000.

Blickle, Peter, *Unruhen in der ständischen Gesellschaft 1300–1800* (Enzyklopädie deutscher Geschichte, 1). Munich, Oldenbourg, [2]2010.

Coser, Rose Laub, *In Defense of Modernity. Role Complexity and Individual Autonomy*. Standford, Standford University Press, 1991.

Cowan, Alexander, "Gossip and Street Culture in Early Modern Venice", *Journal of Early Modern History* 12 (2008), 113–133.

Dillinger, Johannes, *'Evil People'. A Comparative Study of Witch Hunts in Swabian Austria and the Electorate of Trier* (Studies in Early Modern German History). Charlottesville, University of Virginia Press, 2009.

Dillinger, Johannes, *Hexen und Magie. Eine historische Einführung* (Historische Einführungen, 3). Frankfurt a. M., Campus, 2007.

Evans-Pritchard, Edward E., *Witchcraft, Oracles and Magic Among the Azande.* Oxford, Clarendon Press, 1937.

Forster, George M., "Peasant Society and the Image of Limited Good", *American Anthropologist* 67 (1965), 293–315.

Gehm, Britta, *Die Hexenverfolgung im Hochstift Bamberg und das Eingreifen des Reichshofrates zu ihrer Beendigung* (Rechtsgeschichte, Zivilisationsprozeß, Psychohistorie, 3). Hildesheim, Olms, 2000.

Groß, Barbara, *Hexerei in Minden. Zur sozialen Logik von Hexereiverdächtigungen und Hexenprozessen (1584–1684)* (Westfalen in der Vormoderne, 2). Münster, Aschendorff, 2009.

Jeggle, Utz and Albert Illien, "Die Dorfgemeinschaft als Not- und Terrorzusammenhang. Ein Beitrag zur Sozialgeschichte des Dorfes und zur Sozialpsychologie seiner Bewohner". In *Dorfpolitik. Fachwissenschaftliche Analysen und didaktische Hilfen*, edited by Hans-Georg Wehling (Analysen, 22). Opladen, Leske & Budrich, 1978, 38–53.

Karlsen, Carol F., *The Devil in the Shape of a Woman: Witchcraft in Colonial New England.* New York, Norton, 1987.

Kobayashi, Shigeko, "Kommissar und Bittschrift in der Kurkölnischen Hexenverfolgung", *Veröffentlichungen des Japanisch-Deutschen Zentrums Berlin* 63 (2011), 77–85 [http://www.jdzb.de/fileadmin/Redaktion/PDF/veroeffentlichungen/tagungsbaende/D63/11%20p1409%20kobayashi.pdf].

Koppenborg, Ingo, *Hexen in Detmold. Verfolgung in der lippischen Residenzstadt 1599–1669* (Sonderveröffentlichungen des Naturwissenschaftlichen und Historischen Vereins für das Land Lippe, 57). Bielefeld, Verlag für Regionalgeschichte, 2004.

Labouvie, Eva, *Zauberei und Hexenwerk. Ländlicher Hexenglaube in der frühen Neuzeit* (Fischer-Taschenbücher Geschichte, 10493). Frankfurt a. M., Fischer-Taschenbuch-Verlag, 1991.

Lorenz, Robert, *"Wir bleiben in Klitten". Zur Gegenwart in einem ostdeutschen Dorf* (Europäische Ethnologie, 8). Berlin, LIT, 2008.

Mährle, Wolfgang, "'O wehe der armen seelen'. Hexenverfolgungen in der Fürstpropstei Ellwangen (1588–1694)". In *Zum Feuer verdammt. Die Hexenverfolgungen in der Grafschaft Hohenberg, der Reichsstadt Reutlingen und der Fürstpropstei Ellwangen*, edited by Johannes Dillinger, Thomas Fritz and Wolfgang Mährle (Hexenforschung, 2). Stuttgart, Steiner, 1998, 325–500.

Midelfort, Hans C. Erik, "Heartland of the witch craze". In *The Witchcraft Reader*, edited by Darren Oldridge. London, Routledge, 2002, 113–119.

Nubola, Cecilia and Andreas Würgler, eds., *Bittschriften und Gravamina. Politik, Verwaltung und Justiz in Europa (14.–18. Jh.)* (Schriften des Italienisch-Deutschen Historischen Instituts in Trient, 19). Berlin, Duncker & Humblot, 2005.

Ostorero, Martine, et. al., eds., *L'Imaginaire du Sabbat. Édition critique des textes les plus anciens (1430 c. – 1440 c.)* (Cahiers lausannois d'histoire médiévale, 26). Lausanne, Université de Lausanne, 1999.

Pohl, Herbert, *Zauberglaube und Hexenangst im Kurfürstentum Mainz. Ein Beitrag zur Hexenfrage im 16. und beginnenden 17. Jahrhundert* (Hexenforschung, 3). Stuttgart, Steiner, 1998.

Riedel, Manfred, "Gesellschaft, Gemeinschaft". In *Geschichtliche Grundbegriffe. Historisches Lexikon zur politisch-sozialen Sprache in Deutschland*, vol. 2, edited by Otto Brunner, Werner Conze and Reinhart Koselleck. Stuttgart, Klett-Cotta, 1975, 801–862.

Rosa, Hartmut and Lars Gertenbach, *Theorien der Gemeinschaft zur Einführung* (Zur Einführung, 367). Hamburg, Junius, 2010.

Rummel, Walter, "'Weise' Frauen und 'weise' Männer im Kampf gegen die Hexerei. Die Widerlegung einer modernen Fabel". In *Europäische Sozialgeschichte. Festschrift für Wolfgang Schieder*, edited by Christof Dipper et al. (Historische Forschungen, 68). Berlin, Duncker & Humblot, 2000, 353–375 [online at https://www.historicum.net/purl/7pzyl/].

Rummel, Walter, "Communal Persecutions". In *Encyclopedia of Witchcraft. The Western Tradition*, vol. 1, edited by Richard M. Golden. Santa Barbara, ABC-Clio, 2006, 201–203.

Rummel, Walter, "Soziale Dynamik und herrschaftliche Problematik der kurtrierischen Hexenverfolgungen. Das Beispiel der Stadt Cochem (1593–1595)", *Geschichte und Gesellschaft* 16 (1990), 26–55.

Rummel, Walter, *Bauern, Herren und Hexen. Studien zur Sozialgeschichte sponheimischer und kurtrierischer Hexenprozesse 1574–1664* (Kritische Studien zur Geschichtswissenschaft, 94). Göttingen, Vandenhoek & Ruprecht, 1991.

Schormann, Gerhard, *Hexenprozesse in Deutschland* (Kleine Vandenhoeck-Reihe, 1470). Göttingen, Vandenhoeck & Ruprecht, 1981.

Schulte, Regina, *Das Dorf im Verhör. Brandstifter, Kindsmörderinnen und Wilderer vor den Schranken des bürgerlichen Gerichts, Oberbayern 1848–1910.* Reinbek bei Hamburg, Rowohlt, 1989.

Schulte, Rolf, *Man as Witch. Male Witches in Central Europe* (Palgrave Historical Studies in Witchcraft and Magic). Basingstoke, Macmillan Palgrave, 2009.

Schwerhoff, Gerd, "Hexerei, Geschlecht und Regionalgeschichte – Überlegungen zur Erklärung des scheinbar Selbstverständlichen". In *Hexenverfolgung und Regionalgeschichte. Die Grafschaft Lippe im Vergleich*, edited by Gisela Wilbertz, Gerd Schwerhoff and Jürgen Scheffler (Beiträge zur Geschichte der Stadt Lemgo, 4). Bielefeld, Verlag für Regionalgeschichte, 1994, 325–353.

Schwerhoff, Gerd, "Zentren und treibende Kräfte der frühneuzeitlichen Hexenverfolgung – Sachsen im regionalen Vergleich", *Neues Archiv für sächsische Geschichte* 79 (2008), 61–100.

Spitta, Juliane, *Gemeinschaft jenseits von Identität? Über die paradoxe Renaissance einer politischen Idee.* Bielefeld, Transcript, 2013.

Stollberg-Rilinger, Barbara, Review on Peter Blickle: *Das Alte Europa. Vom Hochmittelalter bis zur Moderne.* München, 2008 [H-Soz-Kult, 16.09.2008, <http://www.hsozkult.de/publicationreview/id/rezbuecher-11174>].

Voltmer, Rita, "Konspiration gegen Herrschaft und Staat? Überlegungen zur Rolle gemeindlicher Klagekonsortien in den Hexenverfolgungen des Rhein-Maas-Mosel-Raumes". In *Staatsbildung und Hexenprozess*, edited by Johannes Dillinger and Jürgen-Michael Schmidt (Hexenforschung, 12). Bielefeld, Verlag für Regionalgeschichte, 2008, 213–244.

Voltmer, Rita, "Monopole, Ausschüsse, Formalparteien: Vorbereitung, Finanzierung und Manipulation von Hexenprozessen durch private Klagekonsortien". In *Hexenprozesse und*

Gerichtspraxis, edited by Herbert Eiden and Rita Voltmer (Trierer Hexenprozesse, 6). Trier, Spee, 2002, 5–67.

Voltmer, Rita and Shigeko Kobayashi, “Supplikationen und Hexereiverfahren im Westen des Alten Reiches – Stand und Perspektiven der Forschung”, *Kurtrierisches Jahrbuch* 51 (2011), 247–269 [online at http://www.jdzb.de/fileadmin/Redaktion/PDF/veroeffentlichungen/tagungsbaende/D63/11%20p1409%20kobayashi.pdf].

Walz, Rainer, “Agonale Kommunikation im Dorf der Frühen Neuzeit”, *Westfälische Forschungen* 42 (1992), 215–251.

Walz, Rainer, *Hexenglaube und magische Kommunikation im Dorf der Frühen Neuzeit. Die Verfolgungen in der Grafschaft Lippe* (Forschungen zur Regionalgeschichte, 9). Paderborn, Schöningh, 1993.

Wunder, Heide, “Hexenprozesse und Gemeinde”. In *Hexenverfolgung und Regionalgeschichte. Die Grafschaft Lippe im Vergleich*, edited by Gisela Wilbertz, Gerd Schwerhoff and Jürgen Scheffler (Beiträge zur Geschichte der Stadt Lemgo, 4). Bielefeld, Verlag für Regionalgeschichte, 1994, 61–70.

Wunder, Heide, *Die bäuerliche Gemeinde in Deutschland* (Kleine Vandenhoeck-Reihe, 1483). Göttingen, Vandenhoek & Ruprecht, 1986.

Laura S. Carugati[1]

Considerations on the Role of Translation in the Building of *Symphilosophy*-Community by the Early German Romantics

Philosophieren heißt die Allwissenheit ***gemeinschaftlich*** *suchen.*

To philosophize means to jointly search for universal wisdom.
(Friedrich Schlegel, *Athenaeum*, *Fragmente*, Nr. 344)[2]

This paper attempts to contribute to the reflections on community as a challenge of life in view of the programmatic proposal of the early German romantics, considered a new generation who developed their theory and practice of poetry, that is to say art, philosophy, religion, and literature, based on the historical consciousness of the past in the present. The "aesthetic revolution" proposed by Friedrich and A. W. Schlegel and their circle consists of a program which seeks the transcendental transformation in the history of German literature.[3] I will not go into too much detail here on the different conceptions of art and literature that these theorists formulated from the answers they reached when they pondered each of the "ancient," the "classic," the "modern," and the "romantic." I would like to point out that the romantic program of the Circle of Jena can only be understood as an attempt to revolutionize art and literature through a historical perspective which makes a new philosophy of art possible. The young romantics built their hopes for an aesthetic revolution in the coming future "with a series of reasonings based both in aesthetics and in the philosophy of history"[4] transferring the revolutionary principles to poetology. The revolution shall not be limited, according to Friedrich Schlegel, to a national revolution, but will encompass all humankind because it is an intellectual one, a cultural revolution for humankind that yearns for the consummation of infinite plenitude (*unendliche Fülle*), which can only be understood and guessed via a historical basis that does not consider history from a historicist perspective, a model to be copied, simulated or restored. We could say that the Schlegelian circle's programmatic proposal is derived from

---

1 Translated by www.Aiki-Translations.com.

2 Friedrich Schlegel, *Charakteristiken und Kritiken*, 226. Henceforth referred to as KFSA, vol 2. Unless otherwise noted, the translations into English of the citations are by Aiki-Translations.

3 Cf. Hans Robert Jauss, *Literaturgeschichte als Provokation*, 65–99.

4 Ibid., 68.

an attempt to appropriate and resignify the past within the common horizon of the sense of meaning, a common horizon that was necessary for the creation of this, and perhaps all, living communities, and possible only when the ultimate task of *Bildung* is for man to own his transcendental self as a condition of possibility for the complete understanding of himself, without which he will never be able to truly understand others.[5]

I will now present, on one hand, *symphilosophy*, which was considered by Friedrich Schlegel and Novalis as the only possible way to philosophize on religion, poetry, and art, and a necessary condition to universal progressive poetry, and on the other hand, the role that translation played within this program, in which philosophizing is a joint search for truth, in the becoming of community through *symphilosophy.* I will attempt to elaborate on the ways in which the community creating nature of translation—the hermeneutic practice of re-creation of meaning based in the unavoidable dialectic of "comprehension/non-comprehension"—emerges once it is understood as "an infinite practice of approximation,"[6] which is to say, as an experience of meaning, a continuous and infinite resignification in which meaning is allocated not by the semantic character of what is translated, but through consideration of the underlying motivation of translation. This motivation, for the early romantics, is none other than the creation of a universal, cosmopolitan community, founded "in the desire for a collective activity, a true 'community' life"[7] which was born in the journal *Athenaeum* and whose "project will not be a literary project, and it will not cause a crisis *in* literature, but a crisis and a general criticism (social, moral, religious, political: all these aspects are found in *Fragments*) with literature or literary theory as the privileged place of expression."[8]

***

In the *Vorerinnerung* of *Athenaeum,* brothers August and Friedrich Schlegel, editors and authors of the magazine published between 1798 and 1800, outlined the purpose and the spirit that had brought them to this joint undertaking. *Athenaeum* would be the organ and medium of expression for the work of a very

---

5 In the words of Novalis: "Die höchste Aufgabe der Bildung ist, sich seines transcendentalen Selbst zu bemächtigen, das Ich seines Ichs zugleich zu seyn. Um so weniger befremdlich ist der Mangel an vollständigem Sinn und Verstand für Andere. Ohne vollendetes Selbstvertändniß wird man nie andere wahrhaft verstehen lernen. Novalis, "Blüthenstaub", 78.

6 Hans-Georg Gadamer, "Lesen ist wie Übersetzen", 283.

7 Philippe Lacoue-Labarthe and Jean-Luc Nancy, *L'absolu littéraire*, 14. Consulted version: *El absoluto literario*, 27.

8 Ibid., 23.

diverse group of men and women, united by a shared conviction and passion: to revolutionize the philosophy of art and literature. With respect to the themes and issues that would occupy them throughout the six issues of the magazine, they announced that, in everything directly related to *Bildung*, they aspired to the greatest degree of generality possible and to the freest form of expression for what they called "a brotherhood of knowledge and abilities."[9] A naïve interpretation could make this look like a brotherly project in terms of its aim and interest, but in reality it was the proclamation of a broader common project, a spiritual revolution in the area of knowledge and a radical transformation in the theory of art and literature that would call into question much more than the concept of art as *mimesis*. In the words of Philippe Lacoue-Labarthe and Jean-Luc Nancy, it was:

> [...] a romantic project, which is to that say a brief, intense, and brilliant *moment of writing* (barely two years, a few hundred pages) that by itself inaugurated a whole era, but which exhausts itself in its inability to grasp its own essence and aim—and that will ultimately find no other definition than a place (Jena) and a journal (*Athenaeum*). [...] Its founders, as we all know, are the Schlegel brothers: August Wilhelm and Friedrich. They are philologists. [...] They are perceived as politically 'advanced' (which during this period means 'revolutionary,' 'republican,' or 'Jacobin'), [...] they are involved in the 'literary' and social circles of Berlin (the 'Jewish' salons of Rahel Levin or Dorothea Mendelsson-Veit) which makes them, according to the French model of the period, perfect 'intellectuals.' [...] It is within this milieu that the *Athenaeum* begins to take shape. What initially begins to take shape is the group—a close-knit and relatively closed circle, which was founded, at least at the beginning, on intellectual fraternity and friendship, and on the desire for a collective activity, for a certain 'community' life as well. It is by no means the 'committee' of a journal; nor was it a simple circle of friends or a group of intellectuals. It was more like a kind of 'cell,' marginal (if not totally clandestine), equivalent to the nucleus of an organization called to develop itself as a 'network' and as a model of a new way of life. [...] Friedrich, the most adept at this form of community in which he would be the true driving force, would cherish, perhaps, the utopia of an 'alliance' or of a 'league' of artists of which *Athenaeum* constituted the embryo.[10]

The programmatic proposal of the Circle of Jena, developed in *Athenaeum* and known as universal progressive poetry, consists in breaking down the limits between literature and life, in romanticizing the world. In reference to this program Rüdiger Safranski affirms that

> Friedrich Schlegel and Novalis established for this Project the concept of romanticization. Each and every one of the activities of life must be charged with poetic significance, must sensitively express a strange beauty and demonstrate a conforming force that at the same

9 August Wilhelm SCHLEGEL and Friedrich SCHLEGEL, „*Vorerinnerung*, III", *Athenaeum* 1, 1.
10 Philippe LACOUE-LABARTHE and Jean-Luc NANCY, *El absoluto literario*, 25.

> time has its own 'style' as does any artistic product in the strict sense. [Schlegel and Novalis] didn't see art as a product, but as an event that could occur at any time and in any place where man carries out his activities with creative force and vital impulse.[11]

Therefore, in universal progressive poetry as a programmatic project to break down the limits between life and poetry, the romanticization of man and the world, what can be seen is the "clear consciousness of eternal agility," agility which facilitates the continuous alteration of expansion and contraction, the perpetual play between thesis and antithesis, which results in the synthesis of antagonistic elements, that is, the romanticization not only of art, philosophy, and religion, but of the world and man as such. This romanticization of the world and of man is only possible in poetry in the Schlegelian sense. In the opening of *Conversation on Poetry,* poetry is presented as a common lover that links those who love her with unbreakable bonds, configuring a universal community in which the determining factor is not what each may seek in their individual lives, but the greater magical force that unites their community. This magical force is nothing other than poetry itself:

> Poetry connects and unites with unbreakable bonds all spirits that love her. Even if they pursue in their individual lives the most disparate things, even if one totally disregards that which for the other is the most sacred, even if they underestimate or ignore one another and remain eternal strangers, nevertheless, in this region they will all unite in peace, by virtue of a higher power. Each muse searches for the other and finds it, and so all the currents of poetry meet in the great universal sea.[12]

The programmatic proposal is nothing more than the theory (and practice) of progressive universal poetry, it has as a point of departure the response of the intellectuals of the Schlegelian circle to the *Quarrel of the ancients and the moderns*, reopening the debate which had occurred eighty years prior in France, to evaluate once again, through a historical resignification, what is understood as the *antiqui–moderni* antinomy. The romantic revolution proposed in *Athenaeum* consisted of a new aesthetic-literary program which proposed the romanticization or total poetization of the world. The best-known definition of romantic poetry as universal progressive poetry is offered by Friedrich Schlegel in *Athenaeum* fragment 116[13] according to which poetry (*Poesie*) encompasses philosophy, theology,

11 Rüdiger SAFRANSKI, *Romantik. Eine deutsche Affäre,* 58–59.

12 Friedrich SCHLEGEL, KFSA, vol. 2, 284.

13 Friedrich SCHLEGEL, KFSA, vol. 2, 182, Fragment 116: "Romantic poetry is a progressive universal poetry. Its destiny is not merely to reunite all of the different genres and to put poetry in touch with philosophy and rhetoric. Romantic poetry wants to and should combine and fuse

and even natural sciences. It is, in the end, universality that will overcome and break down the hierarchical structure of genres as well as techniques and styles. The universality dealt with here is not an abstract, formal one, but a universality that tends to construct itself on what is common to diversity, which extends to diversity and encompasses it, thus, to peculiarities. The objective is to take what is common as a starting point, even though it will always be a provisional one. The universality of universal progressive poetry is based on, and at the same time strives for what is common, the creation of community without suppressing or annihilating multiplicity.

Following Fichte's concept of reflection, the fundamental problem that universal poetry seeks to solve is to unite the system and the world, the finite and the infinite. This synthesis is represented as the absolute. Therefore the objective of the romantic program is to assure the union or, more precisely, the synthesis of antagonistic elements, achieving infinite fullness (*unendliche Fülle*), the absolute which is nevertheless intrinsically unattainable. In the words of Novalis:

> Through the voluntary renunciation of the absolute, infinite free activity arises in us—the only possible absolute which can be given to us, and which we find only through our incapacity

---

poetry and prose, genius and criticism, art poetry and nature poetry. It should make poetry lively and sociable, and make life and society poetic. It should poeticize wit and fill all of art's forms with sound material of every kind to form the human soul, to animate it with flights of humor. Romantic poetry embraces everything that is purely poetic, from the greatest art systems, which contain within them still more systems, all the way down to the sigh, the kiss that a poeticizing child breathes out in an artless song. Romantic poetry can lose itself in what is represented to the extent that one might believe that it exists solely to characterize poetic individuals of all types. But there is not yet a form which is fit to fully express an author's spirit. Thus many artists who only wanted to write a novel ended up presenting a kind of self-portrait. It alone is able to become a mirror of the entire surrounding world, an image of their age in the same manner as an epic. And yet it is Romantic poetry which can best glide between the portrayer and what is portrayed, free from all real and ideal interests. On the wings of poetic reflection, it can raise that reflection to a higher power and multiply it in an endless row of mirrors. Romantic poetry is capable of the highest and most comprehensive refinement [*Bildung*] – not merely from the inside out, but also from the outside in. In everything that should be a whole among its products, it organizes all parts similarly, through which a vision of an infinitely expanding classicism is opened. Romantic poetry is to the arts what wit is to philosophy and what society, company, friendship, and love are in life. Other kinds of poetry are finished and can now be fully analyzed. The Romantic form of poetry is still in the process of becoming. Indeed, that is its true essence, that it is always in the process of becoming and can never be completed. It cannot be exhausted by any theory, and only a divinatory criticism would dare to want to characterize its ideal. Romantic poetry alone is infinite, just as it alone is free and recognizes as its first law that the poetic will submits itself to no other law. The Romantic kind of poetry is the only one which is more than a kind – it is poetry itself. For, in a certain sense, all poetry is or should be Romantic." Translation: Jonathan Skolnik (http://germanhistorydocs.ghi-dc.org/sub_document.cfm?document_id=368).

> to arrive at and recognize an absolute. This absolute which is given to us may only be recognized negatively, in that we find through no action we may arrive at that which we seek.[14]

Thus, only through voluntary renunciation of the absolute, a form of the absolute which consists of the infinitely free activity of representing the unrepresentable arises in us, i.e. the unachievable absolute. The result of this free attempt at representing the unrepresentable, which should be considered as an ongoing production, would be what Friedrich Schlegel and Novalis called the romantic novel, the absolute book, which they conceived as a "system of books" in which each work should be understood as a monadic manifestation of the whole, as an individual unit that contains and remits to the book of books, simultaneously constituting itself as peculiar and universal.[15] In the words of Gerhard Poppenberg:

> The concept of universal progressive poetry and Schlegel's concept of the novel became fused together, making the true novel a project for the future, a book yet to be written, that would be quintessential to all literature as the union of philosophy and poetry. It would be an absolute book, a book that would encompass all books. Novalis shared with him this utopian ideal of the absolute book and both would identify it with the project of a Bible. The idea of a new Bible as the ideal of the book.[16]

It is worth recalling here the well-known quote by Novalis on the novel in *Preliminary Works* of 1798: "Life should not be seen as a novel given to us, but as a story written by us."[17] The project of the absolute novel is founded on this conception of the novel. This notion emerges from *symphilosophy* which goes beyond the individuality of authors to a "symphony of brotherhood" that produces an "authorless" text or a text of "collective authorship." From this oscillation between anonymity and *symphilosophy*, a distinctive trait of romantic writing,[18] we may attempt to comprehend the universality which universal progressive poetry[19] strives for as well as August Wilhelm Schlegel's later development of what could be considered the first "comparative grammar"[20] and the hermeneutics of Wilhelm von Humboldt as an attempt to comprehend man's constant attempts to understand himself and orient himself in the world, in the realm of thought, and

14 NOVALIS, "Merckwürdige Stellen", 269 sq.
15 Cf. Friedrich SCHLEGEL, KFSA, vol. 2, 265, Fragment 95. Also, Monika SCHMITZ-EMANS, *Einführung in die Literatur der Romantik*, 52–53 and Ernst BEHLER, *Friedrich Schlegel in Selbstzeugnissen*, 69–71 and Ernst BEHLER "Athenaeum. Die Geschichte einer Zeitschrift", 18 and sq.
16 Gerhard POPPENBERG, "El libro de los libros", 221–240.
17 NOVALIS, [Vorarbeiten], 352, Fragment 187.
18 LACOUE-LABARTHE and NANCY, *El absoluto literario*, 56.
19 Cf. Friedrich SCHLEGEL, KFSA, vol. 2, 168, Fragment 22.
20 George STEINER, *Nach Babel*, 91.

in the realm of action, which is only possible in and through language. The emergence of the program of a poetry with universal reach gave way to Humboldt's hermeneutics,[21] unparalleled in history, which maintain that in language as well as in the consciousness of the nation "the concentration is placed on the spirit of the individual and at the same time on the vision of the community."[22] This community is based in the notion that "language, conceived in its true essence, is something constant and provisional in every instance. [...] It is not a work (*ergon*), but an activity (*energeia*)," which is to say, a constant and infinite resignification. Perhaps it's in language understood as *energeia* that the dialectic of the whole and the parts plays out in life regarding poetry. Where according to Novalis: "The individual lives in the whole and the whole lives in the individual [and] through poetry the highest sympathy and coactivity emerges, the most intimate *community* of the finite and the infinite."[23]

In order to comprehend the scope and aims of this non-systematized programmatic system, this proposal for the romanticization of the world which is nothing more than the unattainable objective of *symphilosophy*[24] that the *Athenaeum* community did not succeed in achieving,[25] it is important to consider the role of translation as the medium where intercultural dialogue takes place, the bridge between cultures. The theory and practice of the art of literature developed in community by the authors and editors of *Athenaeum*, as well as

---

21 Cf. Wilhelm VON HUMBOLDT, "Über die Verschiedenheit", 46.

22 George STEINER, *Gedanken dichten*, 91.

23 NOVALIS, [Vorarbeiten], 322, Fragment 31.

24 Cf. Friedrich SCHLEGEL, KFSA, vol. 2, 161, Fragment 122: "The analytical writer observes the reader, he observes him as he is. Then he makes his calculation, utilizes his machines to obtain the appropriate effect in the reader. The synthetic writer constructs and creates the reader for themselves, as they should be. He doesn't imagine him tranquil and dead, but vital and resistant. He lets the one whom he has created gradually yield before his eyes or incites the reader himself to invent it. He does not desire to produce a determined effect in the reader, but enters with him into the sacred relationship of the most intimate *symphilosophy and sympoetry.*"

25 With respect to the early dissolution of the circle of Jena, Nancy and Labarthe (The critics) affirm that "it is understood that the anecdote is here even more 'essential' (if possible) than in any other moment of history: the dispersion of the group of Jena *truly* implies the suspension of the romantic project. It is the effective fragmentation of its 'fragment of the future,' it's the confirmation by the romantics of the 'absence of romanticism' of their period. And as it is known today, it is therefore, in a twisted in ineluctable way, the confirmation (paradoxical, ironic, critical?), of 'Romanticism' itself." The writing *On incomprehensibility* with which *Athenaeum* concludes is the ironic reconfirmation of the successful failure of the journal which had already announced the impossibility of the realization or conclusion of the early romantic program. Cf. Friedrich SCHLEGEL, KFSA, vol. 2, 367: „Über die Unverständlichkeit".

the work of August W. Schlegel as professor and editor[26] in the decades that followed, are characterized by an unusual opening-up that goes beyond the limits of the German letters. An example can be found in the transcriptions of folk songs (*Volkslieder*) compiled and edited by Achim von Arnim and Clemens Brentano as well as the folk tales (*Volksmärchen*)[27] compiled by the brothers Grimm which included literary works from other cultures in response to the cosmopolitan interest of the romantics who wanted to construct a transnational community based in intercultural dialogue. The creation of such a community is only possible through translation, conceived as a hermeneutic practice of re-creation in which "the challenge fundamentally consists in registering the difference between the original and the recreation and, therefore, in placing the attention on the forms of critical cultural transference proposed by A. W. Schlegel, forms that lay the foundation of the European model of civilization."[28] The end goal was the creation of a community that embraces diversity, that is to say, sameness as well as alterity. This dialectic of the whole and the parts makes it possible to realize the horizon of a universal community created through communication in "literary practice based in the transculturalism of August W. Schlegel."[29] Therefore

> together with philological works, lessons, and extensive correspondence, the magnitude of the translation project can be considered as the base on which the pretension of a universal poetry was founded and made possible through communication across historical and national differences. An ambitious literary translation with a high degree of technical self-reflection is nothing less than a prominent example of the scope of cultural transference.[30]

Finally, I'd like to present the concept of translation as a hermeneutic practice of recreation of meaning based on the dialectic of comprehension/non-comprehension as a dimension that functions as a basis for the undertaking of ambitious literary translation through the program of universalization of the Circle of Jena, resulting from *symphilosophy*. For this, some considerations about translation as

---

26 August Schlegel was the editor of the Indische Bibliothek (1823–30), of the Bhagavad Gita (1823) and the Ramayana (1827–45).

27 It is interesting to note that more than sixty stories of the Grimm brothers' compilation were translations of the stories of the Italian author Giambattista Basile and the French author Charles Perrault, which they accessed through Dorothea Viehmann, whose Huguenot ancestors had moved to Kassel. Cf. Christina SCHLAG, Christoph OTTERBECK, and Harm-Peer ZIMMERMANN (eds.), *Echt hessisch? Land – Leben – Märchen*, 34–35.

28 York-Gothart MIX and Jochen STROBEL, eds., *Der Europäer August Wilhelm Schlegel*, 4.

29 *Op. cit.*, 5.

30 *Op. cit.*, 5.

a hermeneutic task of re-creation are necessary, recurring to notions of Gadamerian hermeneutics.

If we conceive of translation as a hermeneutic process of re-creation of meaning, translation does not differ qualitatively from the hermeneutic task in general,[31] in both cases the objective is not the overcoming of difference, but the search for "the best solution, which will never be more than a compromise."[32] As expressed by Günter Figal in his homage to Hans-Georg Gadamer,

> the hermeneutic practice doesn't deal with searching for undefined or colorless compromises, but in developing an art of mediation in which the other becomes understandable to myself and vice versa without producing an assimilation of both. It is more of an 'intermediation' (*Vermitteln*) in which nothing is considered in isolation, but everything remains in its peculiarity [...] which permits the discovery of co-belonging, the truly common experience.[33]

The pretense of totality and, therefore, of unification that generally hides behind an apparent understanding of the other, does not take into consideration the alterity and insupressible difference between the self and the other. Therefore, "Comprehension—according to the Gadamerian conception—should be shaped in such a way that difference is maintained. This means that only then the other may have the possibility to remain, to some extent, under his or her own subjectivity. If I wish to understand the other under the pretense of totality, then there is no longer any difference between myself and the other."[34] If the difference between the self and the other is not maintained, then there is no way to establish the boundary or delimitation between the two, and therefore, there is no way to extend a "bridge" between the two terms, that is to say, between two cultures. The elimination of difference aims for the totalization that seeks unification of the self with the other, the foreign, the alien, the unification of the self with the self of the other. This does not allow us to conceive of translation as a rewriting of another's literary material, an interpretation that constitutes a new work, the work of the other mediated through the self in the self. The translation that does not conserve difference and distinction between the source and target languages and, therefore, the referential framework in which each work exists, is conceived merely as a unification and uniformity resulting from a mechanical practice of replacement, a mere word for word transposition that does not contemplate the specificity of the emerging moments of each text. Only by maintain-

31 Cf. Hans-Georg GADAMER, *Wahrheit und Methode*, 391.

32 Ibid., 392.

33 Günter FIGAL, "Hans-Georg Gadamer", XI.

34 Hans-Martin SCHÖNHERR-MANN, *Hermeneutik als Ethik*, 182.

ing difference, understanding the other without assimilation or absorption, is it possible to conceive of translation as a comprehensive interpretation that results in a work of integration that recognizes the linguistic and cultural boundaries as it simultaneously maintains them linked yet separate. Translation understood as hermeneutic intercultural experience should conserve the foreign as well as the familiar as it not only extends a bridge between two cultures, but becomes the bridge itself between two languages, between two worldviews that initiate a hermeneutic conversation, characterized mainly by the openness necessary to understand the other while conserving its alterity in the process of appropriation of the foreign that is produced in translation. This dialectic of the familiar and the foreign responds to the hermeneutic principle which establishes that "the other is not only the counterpart of the self, but that it belongs to the intimate constitution of the meaning of the self".[35] The task of translation consists precisely in appropriating something that is foreign to the self, which is enclosed in a culture distinct from one's own, extending a bridge between the two. However, the translator must exercise caution to keep from falling prey to the naïve belief in the possibility of overcoming those borders between cultures due to the apparent full comprehension of one culture on the part of another. Quite the opposite, there should be a constant awareness of the limits of comprehension between cultures and people, highlighting difference as the only possible route to communication and comprehension between the people of different historical periods, cultures, or religions. Therefore although it initially appears to be a deficiency, on closer inspection, the impossibility of total and complete comprehension acquires a positive character in the context of exchange between cultures which is only possible thanks to translation.

In any event, considering translation as a hermeneutic practice of re-creation of meaning from the perspective of the Gadamerian concept of comprehension, the provisional nature of the hermeneutic experience can be extended to translation. One of the principles of Gadamer's hermeneutics is that every attempt to put something into words is and must be provisional, since otherwise it would never be possible to say the same thing in more than one way. Extending this principle to translation, every attempt to translate something is and must be provisional, so all translation is provisional in that it is always possible to translate the same thing in a different way. Therefore, the provisional nature of a translated text demonstrates the possibility and even the need to translate the same text infinite times, to update the text infinite times incorporating the previous re-creations, almost in the way outlined in the program of the early romantics. The provisional

35 Paul Ricoeur, *Soi-Même comme un Autre*, 380: "[...] l'Autre n'est pas seulement la contrepartie du Même, mais appartient à la constitution intime de son sens."

nature of translation represents the inexhaustible (*Unerschöpflichkeit*) potential of meaning that is refreshed in the translator's constant and unending hermeneutic practice of re-creation.

Within this context I am interested in presenting incomprehension or non-understanding in order to ponder this notion as the insuppressible moment of understanding, of interpretation, of translation as the hermeneutic practice of resignification. For this I will make reference to some aspects of non-understanding or incomprehensibility in Friedrich Schlegel that allows us to consider it as the *factum* on which comprehension as experience of meaning is based.[36]

For Schlegel, incomprehensibility is not merely an occasional individual-subjective misunderstanding or lack of comprehension that occurs in a given moment. He does not see incomprehensibility as an instance characterized by a deficiency that must be suppressed through a corrective hermeneutics that establishes the concepts and rules needed to comprehend or to learn to comprehend another person perfectly and completely – whether it be an interlocutor, a "you" as a "not-me,"[37] as an "other," or a discourse or work of writing[38] – but as what can be considered a verification, manifestation or awareness of the configuration of human life conceived as the provisional and infinite process of establishment of meaning, as the infinite process of interpretation[39] or resignification that according to Friedrich Schlegel creates meaning from nonsense. Comprehension is created successively and progressively in the dialectic of comprehension/non-comprehension. Non-comprehension includes in itself confrontation with alterity or awareness of alterity, confrontation with the foreign element without which the very experience of comprehension would not be possible. Friedrich Schlegel states that the dialectic of "comprehension/non-comprehension" does not have the objective of clarification of the pronounced or written, but instead aims to obscure or relativize what is supposed to be understood: "In order to understand someone who only partially understands himself, you first have to

---

36 In the aphorisms of 1805 and 1809–10 Friedrich Schleiermacher formulates that non-understanding is the *factum* upon which hermeneutics is based ("Die Hermeneutik beruht auf dem Factum des Nicht-Verstehens der Rede"), cf. Friedrich D. E. SCHLEIERMACHER, *Hermeneutik*, 31–49.

37 Cf. Manfred FRANK, *Philosophische Grundlagen*, 55 sq.: "Wer 'Ich' sagt, faßt einen bestimmten Gedanken, also einen, dessen Verständlichkeit auf der Abgrenzung von (wenigstens) einem Oppositionsterminus ('Nicht-Ich') besteht. Schon darum ist es ganz abwegig, das Ich für ein Prinzip der Philosophie zu halten oder ihm einen Vorrang vor dem Gedanken des Nicht-Ich einzuräumen."

38 Cf. Johann Martin CHLADENIUS, *Einleitung zur richtigen Auslegung*.

39 Cf. Friedrich A. KITTLER, *Aufschreibesysteme 1800–1900*, 87 sq.

understand him completely and better than he himself does, but then only partially and precisely as much as he does himself."[40]

Friedrich Schlegel elaborates a new concept of the understanding which abandons old gnoseological structures, founded in comprehension as paradoxical to non-comprehension. According to Manfred Frank: "Out of the paradox comes an 'infinite agility,' a reciprocal affirmation and negation, out of which a unidirectional desire is released toward a 'progressive, ironic dialectic' as the 'true method.'"[41] The dialectic of "comprehension/non-comprehension" is one of the ways the tension between peers and opposites manifests itself, remitting to the final opposition on which the philosophy of Friedrich Schlegel is based on, that is to say the dialectic of finite/infinite which constitutes the two modes of being of conscience (finiteness and infiniteness) that meet in an opposition not only temporal, but also qualitative. For Friedrich Schlegel the concept of finiteness is dialectically linked to that of infiniteness: neither can be considered in isolation because it is ultimately a matter of the tension between the two opposites, not of each of them considered as elements that subsist by themselves.[42] Incomprehensibility is the insuppressible instance that emerges from the lack of irony and causes us to consider, to think (it could almost be said to "experience") each of the elements of these pairs of opposites separately, as subsisting on their own, and not as a pair of opposites in permanent tension. Novalis states in the first fragment of *Blüthenstaub* "We seek the absolute everywhere and only ever find things."[43] Incomprehension can only be "obliterated" through the romanticization of the world, that is to say, through the infinite, progressive movement of synthesis of opposites that does not cancel them out, but precisely avoids their mutual annihilation because they are necessary for the relationship to subsist. It could be said about Schlegel that he was interested in suppressing the (Kantian) dualism between awareness and understanding, nature and spirit, practice and theory, postulating an ultimate unity that is the unattainable totality that results from the synthesis of opposites. Out of the original chaos pairs of opposites already exist, making it a less chaotic chaos. Incomprehensibility emerges from the fact that man, being finite, cannot definitively and completely achieve unity, he will never obtain more than a finite representation of the infinite, not immediate, but progressive, continuously approaching the unity that constitutes totality. The oft-mentioned romantic *Sehnsucht* is precisely this permanent tendency

---

40 Friedrich Schlegel, KFSA, vol. 2, 241, Fragment 401.

41 Manfred Frank, *Philosophische Grundlagen*, 80.

42 Cf. Friedrich Schlegel, *Philosophische Vorlesungen*, 334: "Das eigentlich Widersprechende in unserm Ich ist, daß wir uns zugleich endlich und unendlich fühlen [...]".

43 Novalis, "Blüthenstaub", 70.

of man towards the absolute, despite the fact that he can only express himself and think in a finite way. Therefore life cannot help but be a *Rätsel*, the game of exchange and oscillation played by irony. Through the dialectic of "comprehension/non-comprehension," incomprehensibility ceases to be an element or moment of deficiency. According to Petra Rennecke, in opposition to the conventional readings of Friedrich Schlegel, frustrated communication does not need to be seen in terms of what is missing, but instead as failure, as the "most solid and pure incomprehensibility" that guarantees the success of aesthetic communication.[44] To which we could perhaps add that the most solid and pure incomprehensibility is what guarantees the success of translation as an art of interpretation of human finiteness.

Based on the considerations presented up to now on *symphilosophy* and the role of translation in the creation of the community, perhaps Friedrich Schlegel's fragment 125 of *Athenaeum* can be understood with greater clarity:

> Perhaps there would be a birth of a whole new era of the sciences and arts if *symphilosophy* and *sympoetry* became so universal and heartfelt that it would no longer be anything extraordinary for several complementary minds to create communal works of art. One is often struck by the idea that two minds really belong together, like divided halves that can realize their full potential only when joined. If there were an art of amalgamating individuals, or if a wishful criticism could do more than merely wish—and for that there are reasons enough—then I would like to see Jean Paul and Peter Leberecht combined. The latter has precisely what the former lacks. Jean Paul's grotesque talent and Peter Leberecht's fantastic turn of mind would, once united, yield a first-rate romantic poet.[45]

Therefore, the creation of the community of *symphilosophy* as envisioned by Schlegel invites us to reflect on community as a continuous call to reinterpret, resignify, and recreate the foreign through and via the self without erasing differences, but, on the contrary, allowing the foreign to rule on it. This is only possible by virtue of the opening up those results from the hermeneutic Gadamerian principle according to which the fact that the other may be correct is taken seriously.

---

44 Petra RENNECKE, "Das große Lalula", 228.

45 August Wilhelm SCHLEGEL and Friedrich SCHLEGEL, *Athenaeum* 1, 209 and Friedrich SCHLEGEL, KFSA, vol. 2, 185. English version at http://www.carolineschelling.com/letters/volume-1-index/letter-195b/.

Bibliography

Behler, Ernst, *Friedrich Schlegel in Selbstzeugnissen und Bilddokumenten* (Rowohlts Monographien, 123). Hamburg, Rowohlt, 1966.
Behler, Ernst, "Athenaeum. Die Geschichte einer Zeitschrift". In *Athenaeum. Eine Zeitschrift von August Wilhelm Schlegel und Friedrich Schlegel*, vol 3. Darmstadt, Wissenschaftliche Buchgesellschaft, 1992, pp. 1–64.
Chladenius, Johann Martin, *Einleitung zur richtigen Auslegung vernünftiger Reden und Schriften* (Instrumenta philosophica, Series hermeneutica, 5). Düsseldorf, Stern-Verlag Janssen, 1969 (photomechanic reprint after the edition of 1742).
Figal, Günter, "Hans-Georg Gadamer", *Internationales Jahrbuch für Hermeneutik* 1 (2002), IX-XII.
Figal, Günter, *Gegenständlichkeit. Das Hermeneutische und die Philosophie*. Tübingen, Mohr Siebeck, 2006.
Frank, Manfred, "Philosophische Grundlagen der Frühromantik", *Athenäum. Jahrbuch der Friedrich Schlegel-Gesellschaft* 4 (1994), 37–130.
Gadamer, Hans-Georg, "Lesen ist wie Übersetzen". In Id., *Gesammelte Werke*, vol. 8: *Ästhetik und Poetik I. Kunst als Aussage*. Tübingen, Mohr, 1993, 279–285.
Gadamer, Hans-Georg, *Gesammelte Werke*, vol. 1: *Hermeneutik I. Wahrheit und Methode. Grundzüge einer philosophischen Hermeneutik*. Tübingen, Mohr, [6]1990.
Humboldt, Wilhelm von, "Über die Verschiedenheit des menschlichen Sprachbaues und ihren Einfluss auf die geistige Entwicklung des Menschengeschlechts". In *Wilhelm von Humboldts Werke*, vol. 7, part 1, edited by Albert Leitzmann (Wilhelm von Humboldts Gesammelte Schriften I: Werke). Berlin, De Gruyter, 1968, 1–344 (photomechanic reprint after the edition of 1907).
Jauss, Hans Robert, *Literaturgeschichte als Provokation* (Edition Suhrkamp, 418). Frankfurt a. M., Suhrkamp, 1970.
Kittler, Friedrich A., *Aufschreibesysteme 1800–1900*. München, Wilhelm Fink Verlag, 1985.
Lacoue-Labarthe, Philippe and Jean-Luc Nancy, *El absoluto literario: teoría de la literatura del romanticismo alemán*, translated by Cecilia González, and Laura S. Carugati. Buenos Aires, Eterna Cadencia, 2012 (*L'absolu littéraire. Théorie de la littérature du romantisme allemand*. Paris, Éditions du Seuil, 1978)
Mix, York-Gothart and Jochen Strobel, eds., *Der Europäer August Wilhelm Schlegel. Romantischer Kulturtransfer – romantische Wissenswelten* (Quellen und Forschungen zur Literatur- und Kulturgeschichte, 62). Berlin/New York, De Gruyter, 2010.
Novalis, "Merckwürdige Stellen und Bemerkungen bey der Lectüre der Wissenschaftslehre". In *Novalis. Schriften. Die Werke Friedrich von Hardenbergs*, vol. 2: *Das philosophische Werk I*, edited by Richard Samuel. Stuttgart, W. Kohlhammer, [2]1965, 268–274.
Novalis, [Vorarbeiten zu verschiedenen Fragmentsammlungen 1798]. In *Novalis. Werke, Tagebücher und Briefe Friedrich von Hardenbergs*, vol. 2: *Das philosophisch-theoretische Werk*, edited by Hans-Joachim Mähl. Darmstadt, Wissenschaftliche Buchgesellschaft, 1999, 311–424.
Novalis, "Blüthenstaub". In *Athenaeum. Eine Zeitschrift von August Wilhelm Schlegel und Friedrich Schlegel*, vol. 1. Darmstadt, Wissenschaftliche Buchgesellschaft, 1992, pp. 70–106.

Poppenberg, Gerhard, "El libro de los libros. Prolegómenos a una lectura del *Quijote*". In *Discursos explícitos e implícitos en el* Quijote, edited by Christoph Strosetzki. Pamplona, EUNSA, 2006, 221–240.
Rennecke, Petra, "Das große Lalula. Friedrich Schlegels Konzept einer progressiven Universalpoesie", *Athenäum. Jahrbuch der Friedrich Schlegel-Gesellschaft* 20 (2012), 211–228.
Ricoeur, Paul, *Soi-Même comme un Autre*. Paris, Editions du Seuil, 1990.
Safranski, Rüdiger, *Romantik. Eine deutsche Affäre*. München, Carl Hanser Verlag, 2007.
Schlag, Christina, Christoph Otterbeck, and Harm-Peer Zimmermann, eds., *Echt hessisch? Land – Leben – Märchen*. Marburg, Jonas Verlag, 2013.
Schlegel, August Wilhelm and Friedrich Schlegel, eds., *Athenaeum. Eine Zeitschrift von August Wilhelm Schlegel und Friedrich Schlegel*, 3 vols. Darmstadt, Wissenschaftliche Buchgesellschaft, 1992.
Schlegel, Friedrich, *Philosophische Lehrjahre 1796–1806 nebst philosophischen Manuskripten aus den Jahren 1796–1828*, vol. 1, edited by Ernst Behler (Kritische Friedrich-Schlegel-Ausgabe, 18). München/Paderborn/Wien, Ferdinand Schöningh, 1963.
Schlegel, Friedrich, *Philosophische Vorlesungen [1800–1807]*, vol. 1, edited by Jean-Jacques Anstett (Kritische Friedrich-Schlegel-Ausgabe, 12). München/Paderborn/Wien, Ferdinand Schöningh, 1964.
Schlegel, Friedrich, *Charakteristiken und Kritiken*, vol. 1, edited by Hans Eichner (Kritische Friedrich-Schlegel-Ausgabe, 2). München/Paderborn/Wien, Ferdinand Schöningh, 1967.
Schleiermacher, Friedrich D. E., *Hermeneutik*, edited by Heinz Kimmerle (Abhandlungen der Heidelberger Akademie der Wissenschaften, Philosophisch-Historische Klasse, 1959/2). Heidelberg, Carl Winter Universitätsverlag, [2]1974.
Schmitz-Emans, Monika, *Einführung in die Literatur der Romantik*. Darmstadt, Wissenschaftliche Buchgesellschaft, 2004.
Schönherr-Mann, Hans-Martin, ed., *Hermeneutik als Ethik*. München, Wilhelm Fink Verlag, 2004.
Steiner, George, *Nach Babel. Aspekte der Sprache und des Übersetzens* (Suhrkamp-Taschenbuch Wissenschaft, 2125). Berlin, Suhrkamp Verlag, 2014.
Steiner, George, *Gedanken dichten*. Berlin, Suhrkamp Verlag, 2011.

Cultural Identities

Christof Zotter*

The Bonds of the Liberated: On Community among Hindu Ascetics

In his seminal and influential work on Indian asceticism, the French social anthropologist Louis Dumont observes that,

> By renunciation a man can become dead to the social world, escapes the network of strict interdependence [...] and becomes to himself his own end as in social theory of West, except that he is cut off from the social life proper. That is why I have called this person, this renouncer, an individual-outside-the-world.[1]

As Max Weber, who distinguished "other-wordly" asceticism (under which he groups the practices of Indian ascetics and their escapism)[2] from "this-wordly" asceticism (as underlying the vocative ethic of Protestantism),[3] Dumont opposes the Indian renouncer, the "individual-outside-the-world", to the social world. For him the most excellent form of the renouncer's conceptual counterpart, the "man-inside-the-world", is the Brahman who occupies the top position in the hierarchy of the castes, which is – according to Dumont (and Weber) – the most fundamental institution of Hindu society. Dumont is aware of the fact that the renouncer who subsists on alms and remunerates generous householders by preaching, does not, in fact, leave society but adopts a new role vis-à-vis the social world and that it is the dialogue between the renouncer and the Brahman that brought into being what is called Hinduism today.

Although Dumont's interpretation has been repeatedly subjected to criticism, the 'ideal' motif of the lone renouncer free from social bonds can be found in many recent works on Hinduism.[4] A contribution on the community among Hindu ascetics might appear paradoxical,[5] but the situation is more complex then idealized views suppose.

As the anthropologist Richard Burghart has observed, Dumont's structuralist model is remarkable, given the fact that at the time when it was proposed no

---

* I wish to thank Axel Michaels and Astrid Zotter for their comments on earlier drafts and to Terry Kleeman for revising my English.

1 Louis DUMONT, *Homo Hierarchicus*, 184 sq.

2 Max WEBER, "Die Wirtschaftsethik der Weltreligionen. II. Hinduismus und Buddhismus".

3 ID., "Die protestantische Ethik und der Geist des Kapitalismus".

4 For references, see: Matthew CLARK, *The Daśanāmī-Saṃnyāsīs*, 8 n. 36.

5 Cf. Véronique BOUILLIER, "The Ambiguous Position of Renunciants", 199.

(or only very little) published ethnographic data on renunciation was available.[6] Over the last decades, several articles and monographs have helped fill this gap and confirm that, as Burghart phrases it, "no simple dichotomy can describe the relation between Brahman householder and renouncer."[7] In response to Dumont's model, ethnographers collected case-types to illustrate that there was a spectrum of possible intermediate positions between the extreme of the man of caste and that of the solitary renunciant,[8] or argued that these two persons situate themselves in different conceptual universes.[9] The present paper will not enter these discussions. Instead, it focuses on a fact that has often been neglected in simplistic portrayals of Hindu ascetics, but figures prominently in ethnographers' accounts. The accepted form of renouncing the world is to subordinate oneself to an ascetic guru, and therewith, to enter a spiritual lineage or order. In 'real life', to take initiation as an ascetic usually means to take up new social and ritual obligations.[10]

But why, then, is the idea of the lone renouncer leaving caste and ritual so persistent in the literature on Hinduism? One reason is surely, that "indigenous Brahmanical theory itself depicts renunciation as the antithesis of caste system".[11] Before taking up the material on 'real life' ascetics, it is worthwhile to have a brief look at the influential Brahmanical model of asceticism and address an issue that, on the one hand, can support what Dumont writes about the important role ascetics played in the formation of Hinduism but, on the other hand, blurs the categorical opposition between the ascetical 'individual-outside-the-world' and the householder 'man-in-the-world'.

The Brahmanical model of renunciation

Indian asceticism has its roots in the distant past.[12] In the late Vedic period (from ca. 850 BCE), different kinds of ascetics appeared, seemingly in great numbers;

---

6 Richard BURGHART, "Renunciation in the Religious Traditions", 636.
7 Ibidem.
8 See e.g. Véronique BOUILLIER, "The Ambiguous Position of Renunciants".
9 Richard BURGHART, "Renunciation in the Religious Traditions" and ID., "Wandering Ascetics".
10 See e.g. ibidem, 165; Matthew CLARK, *The Daśanāmī-Saṃnyāsīs*, 81; Robert L. GROSS, *Sādhus of India*, 156.
11 Kirin NARAYAN, *Storytellers, Saints, and Scoundrels*, 75.
12 For a summary of the controversy on the origin of Indian asceticism and its early forms, see: Johannes BRONKHORST, *The Two Sources of Indian Asceticism*, 1–9; Matthew CLARK, *Daśanāmī-Saṃnyāsīs*, 5 sq. n. 19; Axel MICHAELS, *Die Kunst des einfachen Lebens*, 132–134.

among them were the Buddha Śākyamuni and Mahāvīra, the founder of Jainism. The image of the lone wanderer who has abandoned the social and ritual obligations of family and caste and undertakes austerities in order to liberate himself from the bonds of transient existence, which casts "an interpretative shadow"[13] on the discussion of Hindu ascetics up to the present day, is typically related to a Brahmanical model of the social world, known as *varṇāśramadharma*, that developed in the early centuries BCE.[14] According to its classical exposition, the life as a renouncer (in later sources often called *saṃnyāsin*)[15] is reserved for the last of four, successively passed life-stages (*āśrama*) and it is not open to all four classes (*varṇa*) of men. In some sources, it is prescribed exclusively for the Brahmans. Furthermore, the life stage of renunciation (*saṃnyāsa*) can be entered only after having discharged three kinds of debts (*ṛṇa*) of central importance to the Brahmanical way of life, the debt:

1) to the ancient seers (*ṛṣi*), by studying the Veda as a celibate student,
2) to the gods, by performing rituals and
3) to the forefathers, by the procreation of a son.[16]

Evidently, the sources promoting this model represent the viewpoint of the Brahmanical householders (and not the ascetics) on what asceticism is. The renouncer is styled as antithesis to the man who follows the rules and regulations of caste and ritual, but the system also offers a conclusion or synthesis to mediate these two positions.

Another detail is of interest here. The three life-stages preceding *saṃnyāsa* in the *varṇāśramadharma* model[17] are also suffused with ascetical ideas of self-restraint. The student lives a celibate life in service of a preceptor. The householder has to follow ritual obligations that often involve temporal ascetical regimes (such as vows of fasting, sexual abstinence, etc.) meant to guarantee the purity required for the performance of rituals. Even the procreation of a son is regulated by so many rules (concerning timing, preparations, etc.) that the importance of self-control for the householders' life is obvious here, too. So, there is a "domestication of asceticism" or a "domestic asceticism".[18]

---

13 Matthew CLARK, *The Daśanāmī-Saṃnyāsīs*, 8.

14 For the development of the *āśrama* system, see: Johannes BRONKHORST, *The Two Sources of Indian Asceticism*; Patrick OLIVELLE, *The Āśrama System*.

15 For other denominations, see Matthew CLARK, *Daśanāmī-Saṃnyāsīs*, 6 sq.

16 For the development of the doctrine of debts, see: Patrick OLIVELLE, *The Āśrama System*, 47–53.

17 I.e. the life as a student of the Veda (*brahmacārin*), as a married householder (*gṛhastha*) and as an anchorite living in the forest (*vanaprastha*).

18 See e.g. Patrick OLIVELLE, *Rules and Regulations*, 12–26.

Taking this observation into account and referring to Geoffrey G. Harpham's thesis that asceticism "is the 'cultural' element in culture",[19] Patrick Olivelle[20] and, following him, Oliver Freiberger[21] distinguish for heuristic purposes three levels, or grades, of asceticism. Both authors are aware of the pitfalls of such a broad conception of asceticism, but it enables them to emphasize that "the ascetic is at the very root of the cultural, and it is this deep association with culture that gives the extraordinary forms of asceticism their extraordinary power over human society and over human imagination".[22] The levels, which they call 'root' and 'cultural' asceticism – i.e. a kind of unconsciously underlying "operating system" of culture and its specific cultural manifestation or "application", respectively – would be of interest in the context of the present volume, too. This paper will however focus on the groups of religious virtuosi who practise an extraordinary, more radical form of self-control and self-constrain; the 'professional' or (in the terminology of Olivelle and Freiberger) 'elite' ascetics.

As ethnographers report, in Hinduism, the 'elite' ascetics do not necessarily live very close to the model of the Brahmanical *saṃnyāsin* traced above. Still, they live in opposition. As ascetics in other religions traditions, they demonstrate that they are extraordinary and demarcate the difference to the common people in their behaviour, dress, hair style, eating habits and so on.[23] (Less ascetical) householders, if they acknowledge the extraordinariness of an ascetic,[24] ascribe to him special qualities, sometimes including supernatural powers, and try to use these for their own purposes, not only in spiritual matters but also in profane problems and homely affairs (see below).[25]

For a Brahman bound by his ritual obligations and caste restrictions the 'ideal' ascetic may indeed live beyond the world of ritual and caste. However, if a Hindu ascetic departs from the ideal of the Brahmanical *samnyāsin*, this does not necessarily mean that he cannot be considered a 'true' ascetic. There are other ideal types of ascetics found in Indian literature, e.g. the tantric "hero" (*vīra*) or "perfected one" (*siddha*) who is believed to have obtained supernatural powers by passing through complicated rituals, or the "devotional ascetic", an ecstatic

19 Geoffrey G. HARPHAM, *The Ascetic Imperative*, xi.

20 Patrick OLIVELLE, "The Ascetic and the Domestic", 26 sq., 29 and *passim*.

21 Oliver FREIBERGER, "Introduction", 6, and ID., *Der Askesediskurs*, 252.

22 Patrick OLIVELLE, "The Ascetic and the Domestic", 40; cf. Oliver FREIBERGER, *Der Askesediskurs*, 252.

23 Cf. Axel MICHAELS, *Die Kunst des einfachen Lebens*.

24 Even inside one specific cultural context, the question what practice is extraordinary or 'ascetical' is an issue of discourse: Oliver FREIBERGER, *Der Askesediskurs*, 252 sq.

25 For this pragmatic aspect of the "domestication of asceticism", see: Christof ZOTTER, "The Cremation Ground", 61.

mystic subordinating everything to the direct experience of the union with his tutelary deity.[26]

'Real life' ascetics

In 'real life', an ascetic might be labelled as *sannyasī* (Skt. *saṃnyāsin*, renouncer) or as *yogī* (Skt. *yogin*, the one engaged in *yoga*, which is in its original meaning an act of yoking). More commonly he is known as *sādhu* (the good or righteous one). Entering the wide field of Indian "holy men", a few general observations on the differences between the lifestyle of the *sādhu*s and normal life in householder communities are appropriate.

Hindu ascetics are aiming at an individual release from the transient world (*saṃsāra*) and the cycle of death and rebirth during their lifetime.[27] What is the most effective, the most appropriate or the only path to liberation might be conceived differently by different sectarian groups and is an important issue in inter-ascetical debates.[28] Generally speaking, in the life style of *sādhu*s an exclusive pursuit of religious experience is emphasized.[29] Hindu householders, too, are often deeply engaged in religious activities, but for the ascetics "pivoting around spiritual concerns"[30] is a full-time job. Ideally, ascetics are celibate. Some *sādhu*s however maintain what they call 'celibacy in marriage'. They do not equate the transient world that has to be overcome with the social world and lay stress on the attainment of a desireless state.[31] Hindu ascetics are, at least in theory, non-productive in an economic sense.[32] They are expected to have no possessions, except for a few ritual paraphernalia, and – again in contrast to the locally rooted householders – many (not all, see below) *sādhu*s lead an itinerant existence. As in case of 'ascetical' couples (who may rear children[33]), a clear distinction between ascet-

---

26 Cf. Id., "Ascetics in Administrative Affairs".

27 For the difference to the Christian concept of assurance of salvation (*certitudo salutis*), see: Axel Michaels, *Die Kunst des einfachen Lebens*, 160sq.

28 Cf. Richard Burghart, "Renunciation in the Religious Traditions", 642. Burghart criticises Dumont's model for ignoring these inter-ascetical discourses: ibidem, 641 and *passim*. See also Kirin Narayan, *Storyteller, Saints, and Scoundrels*, 78.

29 Cf. Robert L. Gross, *The Sādhus of India*, 111.

30 Kirin Narayan, *Storyteller, Saints, and Scoundrels*, 77.

31 Cf. Richard Burghart, "Renunciation in the Religious Traditions", 643.

32 For a discussion of this issue and the differences with ascetical traditions in other religions, see: Axel Michaels, *Die Kunst des einfachen Lebens*, 50–72.

33 See, for example, the case of a "naga baba" who lived with his wife and six children as "full-fledged Shiva family" at a cremation ground in Benares: Surajit Sinha and Baidyanath Saraswati, *Ascetics of Kashi*, 145 sq.

ics and householders can appear to be blurred. Nonetheless, *sādhus* usually live in clear opposition to the ordinary way of life.

For a person suffering the constraints of communal life this opposing form of life can appear as an alternative option. Max Weber has characterized Hindu asceticism as a valve offering a way out of the static social world.[34] Combining Louis Dumont's approach with Victor Turner's concept of liminality and anti-structure,[35] Robert L. Gross draws a similar conclusion. He understands the lifestyle of the *sādhus* as a state of "institutionalized liminality"[36] providing an outlet for those "who are either incapable of dealing with life in any other way or who do not wish to live within the narrow confines of the domestic setting of family, kin, caste and village".[37] The ascetic's path is sanctioned by tradition and therefore, as has been repeatedly stressed, "a viable alternate life style".[38] For many it is the only realistic alternative.[39]

Ideally, the motivation for renunciation should be a strong spiritual inclination, but people have become *sādhu* or have assumed the guise of a holy man for a number of other reasons, too. Some tried to "escape from a criminal charge or some painful personal experience",[40] others to gain minimal financial security and make a living by collecting alms,[41] and yet others simply "to smoke hashish in a fraternity-like setting".[42] In Sanskrit and vernacular literature the notion of 'false' and 'fake' ascetics is a common cliché;[43] the modern *sādhus*, too, are normally treated with respect, but can be ambivalent figures encountered not without suspicion. As there are different kinds of ideal ascetics (see above), there are also different criteria by which one determines whether an ascetic is a 'true' or a 'fake' one.[44]

As mentioned above, *sādhus* are typically itinerants.[45] A small number of them roam around in the wilderness of the forests of the subcontinent subsisting

---

34 Max WEBER, "Die Wirtschaftsethik der Weltreligionen. II. Hinduismus und Buddhismus", 122 and *passim;* cf. Axel MICHAELS, *Die Kunst des einfachen Lebens*, 60.

35 Victor TURNER, *The Ritual Process.*

36 Cf. Robert L. GROSS, *The Sādhus of India*, 300–302, 461 sq. and *passim.*

37 Ibidem, 418.

38 Ibidem, 415; Sondra L. HAUSNER, *Wandering with Sadhus*, 44.

39 Robert L. GROSS, *Sādhus of India*, 416 sq.

40 Ibidem, 99.

41 Ibidem, 133 and *passim*; cf. Axel MICHAELS, *Die Kunst des einfachen Lebens*, 82.

42 Sondra L. HAUSNER, *Wandering with Sadhus*, 19.

43 Cf. e.g. Maurice BLOOMFIELD, "On False Ascetics and Nuns"; Monika HORSTMANN, "Approaching Sant Satire".

44 Cf. Christof ZOTTER, "The Cremation Ground", 53–56, 67 sq. and *passim.*

45 Many *sādhus* wander at least during their "learning stage" and later on settle down at an hut, an ashram or a monastery: cf. Sondra L. HAUSNER, *Wandering with Sadhus*, 100–103, 125 and *passim.*

on what they find there.[46] For some householders these 'invisible' renouncers are (besides the legendary hermits meditating hidden in remote caves) the only 'true' ascetics. Most of the wandering *sādhus*, however, have a different pattern of itinerancy[47] and they definitely have a public existence. On their route from one religious pilgrimage site to another they wander (alone or in small groups,[48] rarely in larger regiments) through villages and cities begging for alms from the householders. There is also a cyclic temporal pattern underlying their movement. Throughout the year, a number of religious festivals related to certain religious sites provide occasions to get together with other ascetics (of one's own tradition or of another, allied or rival). Of special importance for many *sādhus* is the Great Kumbha Melā, a festival held every twelve years at a confluence of rivers in Allahabad for which millions of people gather at the bank of the river. Several ascetic traditions initiate new recruits or appoint their officers (see below) on this occasion. Another reason for ascetics to visit such events is to meet pious and open-handed lay pilgrims,[49] who often come to meet holy men.

Ascetics and Householders

Given the focus of the present volume on the challenges community provides for the individual, only a few remarks on the complex interrelation between *sādhus* and householders should be made here. As mentioned, most ascetics subsist on alms.[50] Householders support them for different reasons. They may consider it as a pious act, or they do so because this is the way to deal with *sādhus*. They might be also afraid of the scorn of the suppliant or simply want to get rid of him.[51] Ascetics bless their supporters (or at least do not curse them).[52] Furthermore, they can use such interactions for religious teaching or other transmissions of cultural values, sometimes with the obvious aim of getting further support or of recruiting new members to their lineage. But the contact between ascetics and household-

46 Robert L. Gross, *Sādhus of India*, 134 sq.

47 See e.g.: Richard Burghart, "Wandering Ascetics"; Robert L. Gross, *Sādhus of India*, 125–131.

48 Often a guru and his disciple(s).

49 For the 'temporal asceticism' of pilgrims, see: Axel Michaels, *Die Kunst des einfachen Lebens*, 82 sq.

50 For a discussion on the non-reciprocal and therefore extraordinary character of alm giving, see: ibidem, 68–72.

51 Cf. ibidem, 67.

52 For a legendary example of a *sādhu*'s curse, see: Christof Zotter, "The Cremation Ground", 63 sq.

ers has many more facets. People approach *sādhus* in search of help, or an ascetic showing up in the village can be entertainment. In his monograph on the *sādhus* of India, Robert L. Gross provides examples of how ascetics provide medical and psychological advice, engage in healing and exorcism, mediate in conflicts, assist in arranging marriages, bless business ventures or use their network of lay followers to find someone employment, advise and intervene in legal and criminal proceedings, counsel village, state or even national political leaders, entertain by spreading local news and gossips, and so on.[53] Kirin Narayan is critical of the theoretical approach of Gross (see above)[54] but confirms, that a "*sādhu*, in short, is someone who may be turned to in time of need, and who serves as spiritual advisor, doctor, lawyer, political commentator, councillor, entertainer, and psychotherapist all rolled into one".[55] A 'true' *sādhu* is a spiritual and moral authority. He provides psychological assurance by "showing sympathy for suffering even as he emphasizes that this is a precondition for life in this world".[56] He may also help in material terms and, not least, he is often ascribed extraordinary powers which common people hope to use for their own benefit.

Asceticism has a social function for the householder. It offers solutions to his problems which are caused by the endemic stress of community. It constitutes a safety valve for social pressure, not just because it offers a permanent outlet into an "alternative lifestyle" (see above), but also because it creates an alternative community between the *sādhu* and the householders who gather around him. Victor Turner would call it a *communitas*.[57] This community can be temporal and very informal. A wandering *sādhu* makes halt in a pilgrimage place or village, establishes his sacred fire (*dhūnī*) and, before he is on the road again, the people can approach him with their problems, including issues they cannot address elsewhere. This community can also take a permanent form (see below) and, while becoming more institutionalized, it often becomes more formalized. But even the informal gatherings just sketched are not without rules and regulations. The *dhūnī* of the ascetic is a sacred fire and the space around it, where people can gather, listen to religious teachings, ask for help or simply gossip, is a ritual one.[58]

---

53 Robert L. GROSS, *Sādhus of India*, esp. 176–198; cf. also Kirin NARAYAN, *Storyteller, Saints, and Scoundrels*, 78–81.
54 Ibidem, 75.
55 Ibidem, 79.
56 Ibidem, 81.
57 Cf. Axel MICHAELS, *Die Kunst des einfachen Lebens*, 77. For the usefulness and limits of Turner's theoretical approach, see also PETER VAN DER VEER, "Strucure and Anti-Structure in Hindu Pilgrimage to Ayodhya".
58 Cf. Robert L. GROSS, *Sādhus of India*, 89. For the *dhūnī* as the center of the ascetics' ritual activities, see: ibidem, esp. 357–362; cf. also Sondra L. HAUSNER, *Wandering with Sadhus*, 120–124.

Ascetics and Ascetics

As indicated at the beginning of the paper, there is another form of community that needs to be considered; namely the community of ascetics. Before coming to the intra-sectarian organisation and the challenges it poses for the individual, let us consider institutionalisation and inter-sectarian relationships.

As seen above, ascetics entertain relations with householders and might be supported by them with some coins and a handful of food, but also with large money donations or land grants. A typical story goes: A charismatic ascetic impresses the locals, a rich landlord, or even the king of the state by working miracles or wondrous deeds. The impressed give so much support that the ascetic can establish a hut, an ashram, a temple, or a monastery.[59] In such institutions, other *sādhus* of the tradition are given food and shelter, lay followers can gather to listen to the teachings or to consult a (resident or visiting) ascetic with their personal problems, etc. The temporary community between householder and ascetics sketched above is getting a more permanent form, and it overlaps with the community of ascetics (see below).

Véronique Bouillier has exemplarily shown how such institutions can grow by recruiting disciples and founding sub-establishments, but also how they can decline due to quarrels over inheritance or the transmission of privileges.[60] Such conflicts frequently end up in legal disputes[61] and can also occur because an ascetic marries and wants to pass the properties he administered on to his son.[62]

At the peak of their influence, such institutions can accumulate enormous wealth that can be reinvested, e.g. in obtaining further land-holdings to be rented out to tenants. There are many instances of rich *mahant*s, or "abbots", working as bankers, money-lenders, traders, etc.[63] Naturally, such an institutionalisation also goes along with an increase in bureaucracy (regulating both internal affairs and interactions with the surrounding society), and with the creation of different offices.[64] Besides the *mahant* there might be a treasurer, an armed guard, a priest, a cook, etc.

If *sādhus* do not meditate in a cave, and have not settled down in a hut or a bigger institution of their order, they are wandering; often from one pilgrimage

---

59 Nowadays, ascetics also open social institutions, such as schools, hospitals, etc.

60 Véronique Bouillier, "Growth and Decay" and Id., "The King and His Yogī".

61 Cf. Matthew Clark, *Daśanāmī-Saṃnyāsīs*, 44.

62 Cf. Véronique Bouillier, "The Ambiguous Position of Renunciants", 216.

63 See Matthew Clark, *Daśanāmī-Saṃnyāsīs*, 256–262; Bernhard S. Cohn, "The Role of the Gosains"; Dirk H.A. Kolff, "Sannyasi Trader-Soldiers".

64 Cf. Christof Zotter, "Ascetics in Administrative Affairs".

site to another. It is therefore no wonder that these places are preferable spots for the establishment of institutions. Although the routes of ascetic groups can differ along sectarian lines—for instance worshippers of Śiva (*śaiva*) tend to visit different places than worshippers of Viṣṇu (*vaiṣṇava*)—there are several overlaps in the itinerant pattern and, as history shows, numerous disputes over the control of pilgrimage routes and centers have arisen. During the seventeenth and eighteenth century, inter-sectarian rivalries achieved institutional form with the formation of militant regiments (so-called *akhāṛā*s, or 'wrestling rings') and led to outright battles.[65] Conflicts arose over precedence in religious activities, most prominently the bathing priority at a pilgrimage site, but they were also caused by the competition of the different orders in their financial and commercial activities (such as collecting taxes from pilgrims).[66] The military organisation of the ascetics also opened a new field of activities; namely to hire oneself and one's disciples out as mercenary warriors.[67] The British colonial administration banned these military activities, but, until today, occasional violent eruptions of hostilities remind us of the militant past.[68]

Several of these *akhāṛā*s still exist and they constitute important units in the social organisation of a number of ascetic orders.[69] The example of the *daśanāmī*s, the order of the "Ten names", will illustrate how intricate such an organisational structure can be. Mathew Clark distinguishes three types of ascetics in this group, which are organised according to different, partly overlapping models:[70]

1) The *daṇḍin*s, or *daṇḍadhārī*s, the 'stick holders', who are in general of high-caste origin, take initiation from a *daṇḍin* monastery and tend to regard only themselves as 'true *saṃnyāsī*s'. They are sometimes also referred to as *śāstradhārī*s ('scripture holders') and they constitute the monastic wing of the order.

---

65 See Richard BURGHART, "The Foundation of the Ramanandī Sect", 126 sq.; Matthew CLARK, *Daśanāmī-Saṃnyāsīs*, 61–65; Robert L. GROSS, *Sādhus of India*, 62–75; Sondra L. HAUSNER, *Wandering with Sadhus*, 78–80; John N. FARQUHAR, "The Fighting Ascetics of India"; David N. LORENZEN, "Warrior Ascetics in Indian History".

66 Cf. Richard BURGHART, "The Foundation of the Ramanandī Sect", 137 n. 3; Matthew CLARK, *Daśanāmī-Saṃnyāsīs*, 61 sq.

67 For the famous example of *gosain* Anūp Giri's army fighting for the Marathas, the Mughals, but also for the British, see: Matthew CLARK, *Daśanāmī-Saṃnyāsīs*, 249–251; William R. PINCH, *Warrior Ascetics and Indian Empires*.

68 Cf. Matthew CLARK, *Daśanāmī-Saṃnyāsīs*, 65; Sondra L. HAUSNER, *Wandering with Sadhus*, 87.

69 See Matthew CLARK, *Daśanāmī-Saṃnyāsīs*, 53–57; Robert L. GROSS, *Sādhus of India*, 61, 74 and *passim*.

70 Cf. Matthew CLARK, *Daśanāmī-Saṃnyāsīs*, 40–52.

2) The *paramahaṃsa*,[71] too, usually reside in a monastery but they (or their guru, or their guru's guru) received *saṃnyāsa* initiation from a high-ranking official (the *mahāmaṇḍaleśvara*) of one of the seven *daśanāmī ākhāṛas* and they are therefore included in the 'military wing' of the order, although they are not fully involved in its activities, because this would require another initiation, namely that of
3) the *nāgas* ("naked ones"), who are also known as *astradhārīs*, 'weapon-holders'. They are full-fledged members of an *akhāṛā* and eligible to vote for the representatives who oversee the affairs of this institution.

Besides these three basic divisions, there are several overarching structures and numerous sub-divisions.[72] For other traditions, too, different types of ascetics and organisational patterns can be distinguished. Furthermore, ascetic traditions can attract and institutionalize householder followers.[73]

The *akhāṛās* have been prominently mentioned here because they can also illustrate another point of relevance for the present study. Being part of a 'military unit' obviously requires discipline and subordination.[74] It has been argued that the *akhāṛās* of the *daśanāmī nāgās* are in some respect democratic and non-hierarchical institutions because the representatives are elected by vote,[75] but as Clark has argued, they are "essentially hierarchical in terms of authority and decision-making",[76] in spiritual as well as practical affairs (see below).[77]

Whatever the organisational pattern of an ascetical order might be, the backbones are segmentary lineages "which perpetuate themselves by spiritual initiation rather than sexual reproduction".[78] Initiation creates a "personal, unique and irrevocable"[79] bond between a disciple and a guru who represents an eternal knowledge that originates – transmitted through time by the chain of gurus – from the tutelary deity of the lineage. As Richard Burghart points out, strictly speaking, the candidate ascetic is not initiated into a sect but "the initiation con-

---

71 The name refers to a mythological bird that is said to be able to filter milk out of water with his beak.
72 Ibidem, 53–80.
73 See e.g. the intra-sectarian differences among the *rāmānandīs*: Richard BURGHART, "Regional Circles and the Central Overseer", 645–648, Robert L. GROSS, *Sādhus of India*, 153–156.
74 Cf. Matthew CLARK, *Daśanāmī-Saṃnyāsīs*, 231.
75 For references see ibidem, 50 sq.
76 Ibidem, 51, 80.
77 Ibidem, 51.
78 Richard BURGHART, "The Founding of the Rāmānandī Sect", 126.
79 ID., "Renunciation in the Religious Traditions", 650.

ducts the candidate outside transient existence"[80], or, as he phrases it elsewhere, "into an eternal situation".[81] As a residue the initiate enters a "network of fictive kin-relationships"[82] which is modelled on the extended joint family.[83] Other disciples of the same preceptor become "guru brothers," the "guru brothers" of the guru, the respected "uncles" and so on. As the example of the *daṇḍins* of the *daśanāmīs* has demonstrated (see above), even caste distinctions can be replicated in the social body of the sect.

The procedures of the initiation by which a novice enters a lineage depend on the sect. While for a future *saiva* ascetic often the separation from the old life is marked by celebrating his death ritual,[84] *vaiṣṇava* orders rather stress the forging of a (new) relationship with the redeeming tutelary deity of the lineage during initiation.[85] As general rule, the initiate is given a new name (often indicating his new affiliation) and he is revealed information that is important for his spiritual practice under the guidance and supervision of his guru. He might be also given a kind of identification code. Among the *daśanāmī nāgās*, for example, such a code includes:

> ...the ascetic's name, his guru's name, his akhara and dawa [a subdivision of an *akhāṛā*, CZ], the name of one of the four cardinal Dasanami maths [monasteries, CZ] to which his akhara is affiliated, his *istadevata* (here, the tutelary deity of his akhara), his guru-mantra, a specific planet associated with his akhara, and several other esoteric attributes.[86]

Given the great number of 'fake' ascetics, this is an effective precaution.[87]

The new 'communal' identity obtained by initiation also involves new forms of social practice and discipline. In particular, the novice must serve the guru and the other 'advanced' members of the lineage[88] and the 'period of apprenticeship' may range up to twelve years.[89] Besides the subordination to the guru who structures and supervises the religious training, the *sādhu* also has to obey the rules of the institution he belongs to. The latter provides shelter and food, and

80 Ibidem.
81 Id., "Regional Circles and the Central Overseer", 165.
82 Robert L. Gross, *Sādhus of India*, 96.
83 See e.g. ibidem, 142; Sondra L. Hausner, *Wandering with Sadhus*, 76 sq.
84 See Véronique Bouillier, "The Ambiguous Position of Renunciants", 108–110.
85 Richard Burghart, "Renunciation in the ReligiousTraditions", 644 and 650; Id., "Regional Circles and the Central Overseer", 165.
86 Robert L. Gross, *Sādhus of India*, 150.
87 Cf. ibidem, 55.
88 Cf. ibidem, 98, 158 and *passim*.
89 Ibidem, 117.

encourages well-behaving members to make use of it,[90] but its officers are also charged with the responsibility to maintain the rank and the reputation of their establishment. Nonconformity to the rules is punished. After reporting several cases of public and private beating of *sādhus* by a higher official of the order, Sondra Hausner writes:

> The internal rules of social *sādhu* life are strict, and the structures of renunciation do not brook misbehavior even from the wildest and idiosyncratic individuals. Hierarchies are firmly in place to prevent anyone getting too big for his britches, or using the relative freedom of renouncer life to the wrong ends.[91]

During his ascetic career a *sādhu* may take further initiations from different gurus of his tradition[92] which results in additional personal bonds and subordinations[93] but also increases his status until he himself eventually becomes a preceptor initiating disciples.[94]

Conclusion

This paper has dealt less with the question of which challenges a community poses for the individual but rather has portrayed Hindu asceticism as an institution that helps to overcome them. For householders who themselves may follow a temporary ascetic regime in conducting their domestic rituals or undertaking a pilgrimage, professional asceticism can, among other functions, constitute a social valve. I have shown it functions in two ways: It helps to reduce the 'communal' pressure on the individual by offering temporary or permanent access to another community in which an ascetic is using his powers (and other skills and means) to solve the problems of devotee suppliants. The *sādhu*, although detached, shows sympathy for their suffering and thereby provides psychological assurance.

If this does not suffice, it is a permitted option for a householder – although in many cases a rebellious one – to permanently adopt the *sādhu*'s way of life. As

90 Sondra L. HAUSNER, *Wandering with Sadhus*, 83.

91 Ibidem.

92 As a rule, *sādhus* do not change gurus or sect affiliations: Robert L. GROSS, *Sādhus of India*, 117.

93 E.g. the *nāgas* of the *dasanāmīs* have to obey five gurus: Matthew CLARK, *Daśanāmī-Saṃnyāsis*, 51.

94 Richard BURGHART, "Renunciation in the Religious Traditions", 650.

shown in the last part of the paper, entering the ascetic's path in a proper way implies not only cutting off bonds with one's community but also entering into new bonds with another one. The new community is extraordinary and consists of spiritual lineages originating from a deity, however, its organisation is modelled on the ordinary extended joint family.

Such communities, too, impose sanctions on a person. How can the ascetic escape? He is initiated into an 'eternal' knowledge and a solitary spiritual practice that is meant to dissolve sufferings by relinquishing the attachment to the 'transient' world and to the ego identity which suffers it. This process is described as the merging of the individual soul (*ātman*) into the cosmic soul (*brahman*), or, more theistically, as discovering the union between, or identity of, worshipper and worshipped. For a 'true' Hindu ascetic aiming at or having reached this state, it is less important to be separated from the social community than to belong to a different kind of community, the spiritual community of the divine and his preceptors.

Bibliography

Bloomfield, Maurice, "On False Ascetics and Nuns in Hindu Fiction", *Journal of the American Oriental Society* 44 (1924), 202–242.

Bouillier, Véronique, "Growth and Decay of a Kanphata Yogi Monastery in South-West Nepal", *The Indian Economic and Social History Review* 28 (1991), 151–170.

Bouillier, Véronique, "The King and His Yogī: Prthivi Nārāyaṇ Śāh, Bhagavantanāth and the Unification of Nepal in the Eighteenth Century". In *Gender, caste and power in South Asia. Social status and mobility in a transitional society*, edited by John P. Nelson. New Delhi, Manohar, 1991, 3–21.

Bouillier, Véronique, "The Ambiguous Position of Renunciants in Nepal: Interrelations of Asceticism and the Social Order" in: *Journal of the Nepal Research Centre 8* (1985), 199–229.

Bronkhorst, Johannes, *The Two Sources of Indian Asceticism*. Delhi, Motilal Barnarsidass, ²1998.

Burghart, Richard, "Regional Circles and the Central Overseer of the Vaishnavite Sects in the Kingdom of Nepal". In *Changing South Asia*, vol. 1: *Religion and Society*, edited by Kenneth Ballhatchet and David Taylor. London, Centre of South Asian Studies, 1984, 165–179.

Burghart, Richard, "Renunciation in the Religious Traditions of South Asia", *Man N. S.* 18 (1983), 635–653.

Burghart, Richard, "Wandering Ascetics of the Rāmānandī Sect", *History of Religion* 22 (1983), 361–390.

Burghart, Richard, "The Foundation of the Ramanandī Sect", *Ethnohistory* 25 (1978), 121–139.

Clark, Matthew, *The Daśanāmī-Saṃnyāsīs. The Integration of Ascetic Lineages into an Order* (Brill's Indological Library, 25). Leiden, Brill, 2006.

Cohn, Bernhard S., "The Role of the Gosains in the Economy of the Eighteenth and Nineteenth Century Upper India", *The Indian Economic and Social History Review* 1 (1964), 175–182.
Dumont, Louis, *Homo Hierarchicus. The Caste System and Its Implications. Complete Revised English Edition.* Chicago/London, University of Chicago Press, 1980.
Farquhar, John N., "The Fighting Ascetics of India", *Bulletin of John Rylands Library* 9 (1925), 431–452.
Freiberger, Oliver, *Der Askesediskurs in der Religionsgeschichte. Eine vergleichende Untersuchung brahmanischer und frühchristlicher Texte* (Studies in Oriental Religions, 57). Wiesbaden, Harrassowitz, 2009.
Freiberger, Oliver, "Introduction. The Criticism of Asceticism in Comparative Perspective". In *Asceticism and its Critics. Historical Accounts and Comparative Perspectives*, edited by Oliver Freiberger. New York, Oxford Unversity Press, 2006, 3–21.
Gross, Robert L., *The Sādhus of India. A Study of Hindu Asceticism*. Jaipur/Delhi, Rawat Publications, 1992.
Harpham, Geoffrey G., *The Ascetic Imperative in Culture and Criticism*. Chicago, University of Chicago Press, 1987.
Hausner, Sondra L., *Wandering with Sādhus. Ascetics in the Hindu Himalayas*. Bloomington, Indiana University Press, 2007.
Horstmann, Monika, "Approaching Sant Satire". In *Indian Satire in the Period of First Modernity*, edited by Monika Horstmann and Heidi R.M. Pauwels (Khoj, 9). Wiesbaden, Harrassowitz, 2012, 95–115.
Kolff, Dirk H.A., "Sannyasi Trader-Soldiers", *Indian Economic and Social History Review* 8 (1971), 213–218.
Lorenzen, David N., "Warrior Ascetics in Indian History", *Journal of the American Oriental Society* 98 (1978), 61–75.
Michaels, Axel, *Die Kunst des einfachen Lebens. Eine Kulturgeschichte der Askese* (Beck'sche Reihe, 1600). München, C.H. Beck, 2004.
Narayan, Kirin, *Storytellers, Saints, and Scoundrels. Folk Narrative in Hindu Religions Teaching*. Delhi, Motilal Banarsidass, 1992.
Olivelle, Patrick, "The Ascetic and the Domestic in Brahmanical Religiosity". In *Asceticism and its Critics. Historical Accounts and Comparative Perspectives*, edited by Oliver Freiberger. New York, Oxford Unversity Press, 2006, 25–42.
Olivelle, Patrick, *Rules and Regulations of Brahmanical Asceticism*. Albany, State University of New York Press, 1995.
Olivelle, Patrick, *The Āśrama System. The History and Hermeneutics of a Religious Institution*. Oxford, Oxford University Press, 1993.
Pinch, William R., *Warrior Ascetics and Indian Empires* (Cambridge Studies in Indian History and Society, 12). Cambridge, Cambridge University Press, 2006.
Sinha, Surajit and Baidyanath Saraswati, *Ascetics of Kashi. An Anthropological Exploration*. Varanasi, N.K. Bose Memorial Foundation, 1978.
Turner, Victor, *The Ritual Process. Structure and Anti-Structure*. Chicago, Aldine Publishing Company, 1969.
Van der Veer, Peter, "Structure and Anti-Structure in Hindu Pilgrimage to Ayodhya". In *Changing South Asia*, vol. 1: *Religion and Society*, edited by Kenneth Ballhatchet and David Taylor. London, Centre of South Asian Studies, 1984, 59–67.

Weber, Max, "Die protestantische Ethik und der Geist des Kapitalismus". In *Gesammelte Aufsätze zur Religionssoziologie*, edited by Max Weber, vol. 1. Tübingen, Mohr, ²1920, 17–206.

Weber, Max, "Die Wirtschaftsethik der Weltreligionen. II. Hinduismus und Buddhismus". In *Gesammelte Aufsätze zur Religionssoziologie*, edited by Max Weber, vol. 2. Tübingen, Mohr, ²1923.

Zotter, Christof, "The Cremation Ground and the Denial of Ritual. The Case of the Aghorīs and Their Forerunner". In *The Ambivalence of Denial. Danger and Appeal of Rituals*, edited by Ute Hüsken and Udo Simon. Wiesbaden, Harrassowitz, 2016, 43–79.

Zotter, Christof, "Ascetics in Administrative Affairs. Documents on the *maṇḍalāī* of the *jogīs* and the *mahantamaṇḍalāī* of the *sannyāsīs*". In *Studying Documents in South Asia and Beyond*, edited by Simon Cubelic, Axel Michaels and Astrid Zotter, Heidelberg, Universitätsverlag Winter, forthcoming.

Terry F. Kleeman

Individual and Community in Early Daoism

Diffuse and Organized Religion in Chinese History

When Westerners first encountered China, they found its religious life perplexing.[1] Religion did not fit into neat categories and the symbols of various religious traditions were jumbled together at sacred sites and in the practice of the people. The Jesuits, seeking their proper counterparts in Chinese society, decided that the closest parallel was not the representative of any faith, but rather officials of the theocratic Chinese state. By contrast, in nearby Japan, they identified with Buddhist priests. In this paper, I will sketch out the Chinese religious world as it developed through time and space, describing first the diffuse common religion and its extension into the world of politics, the state cult, then examine the processes by which organized religions have taken form in China.

Chinese history begins with the oracle bones of the Shang dynasty, dating from roughly 1250–1045 BCE.[2] They record divinations made by the king and other high officials of the Shang state, subsequently etched onto the carapaces of turtles or the shoulder bones of bovines. These inscriptions are constrained in many ways: the topics addressed are limited, the speakers come from a select group, and the language is laconic and still not fully understood. Still they give us an idea of religious practice at the time. At the center was the practice of sacrifice, involving the offering of foodstuffs, drink, and occasionally other valuable objects, to dead ancestors and a handful of nature spirits.

A prime concern of the oracle bone inscriptions is the happiness and well-being of the ancestors. When something unfortunate struck the Shang king, diviners inquired systematically as to which of the ancestors of the Shang royal line were displeased, and exactly what sort and scale of offering would bring them back into harmony. The dominant model is feeding. There is a sort of high god, called Di, who cannot be approached directly through sacrifice, but must rather be entertained by one of the king's deceased ancestors. This god seems to have had special authority over issues of import like where to locate the capital and the success of the harvest, but otherwise is not recorded as interfering directly in the relationship between the Shang ruler and his forebears.

---

1 On the West's encounter with China, see Lionel M. JENSEN, *Manufacturing Confucianism*; Timothy H. BARRETT, "Chinese religion in English guise".

2 For an introduction to the world of the Shang dynasty and oracle bones, see David N. KEIGHTLEY, *The Ancestral Landscape*.

We know little of the religion of the common folk during this period, but it was probably much the same: worship through sacrifice to familial ancestors and a small group of local gods, mostly nature spirits associated with mountains, springs, wells, rivers, etc.[3] Ancestors would have exerted an influence over the lives and well-being of their descendants. Village life centered on planting and harvest festivals in the second and eighth lunar months (usually our March and September) celebrated by a communal banquet. Villagers pooled resources to provide this feast, which was duly sacrificed to the local deities and then consumed communally as a way to share the blessings of the village's supernatural protectors, which had been infused into the food.

The Chinese common religion developed from this foundation. It is a diffuse religion with no clear hierarchy or organization, made up of a million individual cults to gods of various sorts, including dead human beings recognized for singular achievements as well as indigenous tutelary deities associated with ancient sites of religious power. These were originally solely local in character, but beginning in the Song dynasty (960–1279), we begin to see the rise of first regional then national deities who respond to the entreaties of supplicants over a larger geographical area.

A trend toward specialization shaped this cult expansion. Originally, most local gods were all-purpose deities to whom any sort of entreaty could be addressed. Gods that gained renown and a cult over a larger area typically did so by developing a reputation for efficacy in dealing with one or more specific types of problems. Thus one god might be particularly adept at dealing with illness while another assured fertility and a third might be useful to guaranteeing victory, or at least survival, in war. Gods also became associated with specific professions, as patron deities who watched over the participants in that trade. Thus by late imperial times, there were specific gods associated with carpenters, printers, merchants, tailors, physicians, potters, and a hundred other professions, even to prostitutes, beggars, and thieves.

Worshippers make use of these gods eclectically.[4] A family would often have a tradition of worship of one specific deity, who would hold a cherished place on the home altar and would be the first recourse for most entreaties. But this would in no sense deter the family from calling upon a different deity, perhaps one never before worshipped, if that god were said to be particularly efficacious in dealing with the problem at hand. Such an action would constitute no break of fealty or feelings of disloyalty toward the primary deity of the family. All gods

3 Mu-chou Poo, *In Search of Personal Welfare*.
4 See the classic anthropological study of David K. Jordan, *Gods, Ghosts, and Ancestors*.

work together toward the common good and expect the same sort of attitude and conduct from the faithful.

Similarly, there would be no set religious professional to whom the family would turn. Instead, they would again choose from a variety of practitioners representing different approaches to deity and charges different amounts for their services.[5] Today in Chinese communities we regularly encounter a range of male and female ritual practitioners, who approach deity orally or through simple written forms. The lowest status belongs to illiterate spirit ladies who communicate directly with the dead, followed by spirit mediums who practice spirit possession and often ritual violence, as well as ritual masters who claim to control rather than serve the deities.

Over the centuries and across the breadth of Chinese, such religious professionals have been known by a welter of shifting names, but most can be subsumed under the two traditional categories of spirit medium (*wu*) and priest (*zhu*).[6] The spirit mediums were originally woman who danced and brought the spirits down into their body, so that they could speak directly with mortals. Today, they are found throughout China, both male and female. By embodying deity, they are able to sanctify and imbue with divine power a variety of objects and documents, which worshippers can bring home to relieve their problems. The priests often work with the spirit mediums, overseeing large-scale or communal ritual to the deity. The priest knows the proper forms for making offerings and entreaties to the god, whereas the spirit medium confirms that the gods have indeed come to and partaken of the sacrifice.

The Chinese state is to a degree just the family writ large. The royal ancestors are feted with greater ceremony and often for more generations back (up to nine), and the offerings are larger, but not different in kind. The Emperor as Son of Heaven is also the son of a line of dead kings who require constant ritual service.[7]

The state also oversees a ritual program involving the deities of the earth and sky, and ultimately the universe. Representatives of the central government in each locality offer blood sacrifice on specific dates (generally, the planting and harvest dates plus a special day designated as the god's "birthday") to the officially recognized deities of their region. This in fact replicates an injunction in

5 Vincent GOOSSAERT, "Mapping Charisma".

6 On spirit mediums in Early China, see Lothar VON FALKENHAUSEN, "Reflections on the Political Role of Spirit Mediums".

7 For an account of the Chinese state religion, see, Jan J. M. DE GROOT, *The religious system of China*.

the ritual texts concerning the proper recipients and offerings for each class of the traditional nobility:[8]

> The Son of Heaven sacrifices to Heaven and Earth; the feudal lords sacrifice to the gods of soil and grain; the Grand Ministers sacrifice to the five tutelary cults. The Son of Heaven sacrifices to the famous mountains and great rivers throughout the empire. . . . the feudal lords sacrifice to the famous mountains and great rivers within their domains.

The emperor sits atop this sacrificial pyramid, making offerings to the cosmic forces of Heaven and Earth and the Five Sacred Peaks (*wuyue* 五嶽), with the aid of a cohort of officials.

Thus this state religious system takes on some of the characteristics of an organized religion, but there are key differences. First, it is a religion with no laity, with no direct relationship to the great part of the populace. Moreover, the goal is the health and welfare of the state rather than the individuals who make up the quasi-clergy for this religion. Finally, the system is routinized to a high degree, so that it is not possible to have a charismatic leader or significant internal movements.

The diffuse religion of the Chinese people described above is really rather similar to what we call Hinduism. It differs most significantly in that there was never a colonial overlord of China ready to call it by one name. Instead we founder around with awkward terms like "popular religion" or "folk religion" even though it is not in any sense limited to the folk. Part of the problem is that China did eventually have organized religions, with distinctive deities and ritual programs, which were appropriated by the Chinese common religion. This has led many to dismiss it as a derivative hodgepodge, but deities deriving from Buddhism or Daoism, for example, and effectively shorn of their sectarian nature and all treated in roughly the same manner as the gods on the popular pantheon.

The Rise of Organized Religion in China

The nature of sources surviving from Early China makes us wonder whether or not something has been elided, but we so far see no evidence of anything like an organized religion before the turn of the first millennium of our era. Two centuries into the Han dynasty, there was a prolonged failure of leadership and an attempted usurpation accompanied by profuse use of portents, signs, and other purported revelations.

8 *Liji zhengyi*, 12/16a-b.

Two phenomena arose in response. The first was a sort of millennial movement centering on a figure called the Grandmother of the West. Together with the Grandfather of the East, she was frequently depicted in tomb art of the period, where she represented the night, and death, to the Grandfather's day and life. In 3 B.C. a prophecy circulated that the Grandmother would appear in this world at a certain date. Sources record mass pilgrimages to meet the Grandmother at that time, with whole families taking to the roads.[9]

The second was a political rebellion of a group called the Red Eyebrows (*chimei* 赤眉), because of the distinctive way they colored their eyebrows with cinnabar. They worshipped a long-dead prince of the Han royal house, the Prince of Jingyang, who had been instrumental in suppressing an attempted usurpation of the Han throne by the agnates of the sitting emperor. We know little of their beliefs, which were clearly tied to a restoration and revitalization of the Han dynasty, but we do know a bit about their structure, because it was recorded by contemporary historians. They traveled in large groups encompassing entire families, and were led by a female shamanic figure. It is likely that she functioned in part as a spirit medium, relaying communications with the dead prince.

As the Han dynasty continued its slow disintegration, the second century became a breeding ground for a new type of movement that combined revelation, charismatic leadership, and social organization to produce China's first recorded organized religions. One indicator of the new attitude toward divine revelation can be seen in a isolated scripture produced by a failed group. The *Scripture of the Transformations of Laozi* (*Laozi bianhua jing*) survives in a single manuscript exemplar from the Dunhuang hoard of manuscripts discovered at the beginning of the 20th century and studied by the eminent French Sinologist Anna Seidel.[10] This scripture was produced by a cult that worshiped the ancient figure of Laozi as an incarnating deity who has returned repeatedly to aid the Chinese people and state by serving as a counselor to rulers. His last such avatar was in the mid-second century, and his followers clearly looked forward to his return. This apocalyptic millenarianism reflected by this text is manifest on the historical stage with two other groups from this period.

The Yellow Turbans (*huangjin* 黃巾) are still poorly understood, despite their very significant role in the events at the end of the second century that eventually led to the final fall of the Han.[11] They were led by an enigmatic figure named Zhang Jiao 張角, who proclaimed himself the Great Wise and Virtuous Master

9 Homer H. DUBS, "An Ancient Chinese Mystery Cult".

10 Anna K. SEIDEL, *La divinisation de Lao Tseu.*

11 For information on the Red Eyebrows and Yellow Turbans, see Barbara HENDRISCHKE, "Early Daoist Movements", 134–164.

(*da xian liang shi* 大賢良師) and send eight disciples out to proclaim his message in all directions, "using the Way of Goodness to teach and convert the realm."[12] Within ten years, his followers numbered over one hundred thousand. He was inspired by possession of portions of a voluminous revealed text, the *Scripture of Great Peace* (*Taiping jing* 太平經).[13] Now the *Scripture of Great Peace* itself introduced a number of ideas that were important in movements of this time, including elaborating upon the idea of a coming age of great peace and social equity that fed millenarian aspirations and an account of a supernatural record-keeping system that recorded mortal deeds, good and evil, and adjusted their fates on this basis, but it does not seem to have been tied originally to a social group. Zhang Jiao added ritual practices like the confession of sins, followed by the administration of secret written formulae in the form of talismans, which the sufferer consumed. The healing was said to be effective only if the sufferer had true faith in the group's message.

The Yellow Turbans were organized into thirty-six regionally-based units (*fang*), each nominally of ten thousand faithful, actually varying from 6,000–12,000. All planned to rise in rebellion as a group in 184, an important year symbolically because it marked the first year of the sixty-year cycle by which Chinese traditionally counted time, thus an appropriate time for new beginnings, but their plans were revealed. The scale of the movement was great, with battles extending through much of the North China plain in the east, including all or portions of modern Hebei, Shandong, Anhui, Jiangsu, Zhejiang, Jiangxi, and Hubei, and extending even into the far western areas.[14] It was suppressed, in the absence of effective military response by the central government, only by the mobilization of much of the elite of East China, leaving behind local warlords who independent power bases signaled the end of Han rule.

Roughly contemporary with the Yellow Turbans, another religion took form in the Chengdu plain of modern Sichuan province. There in 142 CE, it was said, the Supreme Lord Lao descended to Cranecall Mountain to bestow upon a man named Zhang Ling the title of Celestial Master, forging with him the Way of the Correct and Unitary Covenant with the Powers (*zhengyi mengwei zhi dao*).[15] The covenant, encapsulated in the Pure Bond: "The gods do not eat or drink; the master does not accept money," proclaimed a new relationship between the divine and the people of this world. Heretofore, they had worshipped the profane spirits of the Six Heavens, heroic or violent dead who demanded offerings from

12 See *Hou Hanshu* 71/2299–2300.

13 Barbara HENDRISCHKE, *The Scripture on Great Peace.*

14 Werner EICHHORN, "Bemerkungen zum Aufstand des Chang Chio".

15 On the founding of the Celestial Masters, see Terry F. KLEEMAN, *Celestial Masters*, ch. 1.

the living in return for their divine protection, in other words, the gods of the common religion we introduced above.

Zhang Ling revealed a higher Three Heavens, filled with pure deities who subsisted on the pneumas of the Dao and had no need or use for the bloody sacrifices of mortal beings. Instead, these transcendent deities determined the fate of humankind solely on the basis of their conduct, rewarding the good and punishing the evil. They could not be entreated through offerings or swayed by words of praise; only through exemplary conduct defined as obedience to a set of precepts or commandments (*jie* 戒) could they be approached. All failures must be confessed through a formal petition to the heavens, followed by penance, or illness and misfortune would befall one. Moreover, if having once been cured, one falls ill again, this is a sign that the repentance was not sincere.

The Celestial Master set up a network of twenty-four parishes around the Chengdu plain, centers of missionary evangelization where priests called libationers (*jijiu* 祭酒) ministered to the faithful on the days of the Three Assemblies , on the seventh of the first and seventh months, and on the fifth day of the tenth (1/7, 7/7, 10/5).[16] Each family contributed an annual tithe of five pecks of rice (roughly 9 liters), which was used to provide communal feasts called kitchen-feasts. In this dangerous age of rampant demons and unforeseen perils, participation in Daoist communities brought protection from supernatural onslaught. Each member underwent a ritual to become a Daoist citizen (*daomin* 道民), invested with a protective talisman and a document called a register (*lu* 籙), which imparted the authority to call directly upon the Daoist Heavens for aid, but also imposed precepts limiting their conduct. Their names were recorded on the local libationer's Fate Roster, a detailed record of all the faithful under his or her care that would be memorialized up to Heaven on the Assemblies, there to be duly recorded in the cosmic Fate Roster so that the Daoist network of deities and their servants would know to protect and look out for them.

A Daoist citizen family could seek higher status by sending a son or daughter to train as a register student (*lusheng* 籙生) or novice.[17] The novice was first invested with the One General Register, conveying a single divine general and his retinue of clerks and soldiers, roughly 120 in number, to take up residence in their body and protect again all dangers. Subsequent ordinations conveyed 10, then 75, then 150 Generals, dramatically increasing the novice's ritual power. With each new ordination came also the right to use a progressively greater number of scriptures as well as the imposition of a progressively greater and more severe set of precepts. The generals, clerks, and soldiers that protected the wearer also

16 Franciscus VERELLEN, "The Twenty-four Dioceses".

17 On these ordination ranks, see Kristofer M. SCHIPPER, "Taoist Ordination Ranks".

oversaw the precepts and punished the individual when he or she sinned. Eventually, the novice would ideally rise to the rank of libationer, ready to take on a flock of citizens and novices of his or her own. Status within the church was determined solely by merit, hence even women and individuals of mean birth could rise to leadership. But the ultimate arbiter was still the Celestial Master, a lineal descendant of the first, Zhang Ling.

The Celestial Masters were able to achieve high social cohesion within their communities, in part by teaching a millennial doctrine that promised hardship and travail before salvation. The accumulated sins of the demonic Six Heavens had to be cleansed from the earth, and this could be done only through fire, war, disaster, and disease. The Daoists were warned by their leaders, "If you want the dawn, the sun must first set. If you want Great Peace, there must first be chaos."[18]

Salvation was offered to the Seed Citizens (*zhongmin* 種民), a select group chosen from among the Daoists who had performed the Merging the Pneumas sex ritual and further been found of irreproachable conduct. They would survive all the calamities and disasters of the end times and live to see the golden age of Great Peace.[19] It seems early church members thought to see the advent in their lifetimes, but as the founding events receded, charismatic leaders continued to arise. For centuries, the name of the savior, Li Hong, was invoked again and again by local rebels with a utopian dream.

For most Celestial Master Daoists of the early medieval period, the Daoist church provided a secure environment of friends and co-religionists who shared protection of fierce supernatural defenders. They lived in communities where their conduct was overseen by human priests and well as divine supervisors, sharing bounty and hardship through rites of communal feasting and public works. It is difficult to trace such Daoist communities in the historical record much beyond the end of the Six Dynasties (ca. 600 CE), but the modern Yao ethnic group of South China and mainland Southeast Asia maintains a communal Daoism very similar in some respects to that of the early Celestial Masters. Every member of a village undergoes Daoist ordination and holds a Daoist rank, females only marry males of similar rank, and social status is determined by rank of ordination.[20]

Beginning in the second half of the fourth century CE, we begin to see more individualistic forms of Daoism that permitted the individual seeker to quest for the divine within the Daoist sacred world. Such quests had long been part of the

---

18 This quote is from a spirit revelation to the community on the First Assembly of the year, 255 CE. See Stephen R. Bokenkamp, *Early Daoist scriptures*, 172.

19 Anna K. Seidel, "Taoist Messianism".

20 On Yao Daoism, see Hirota Ritsuko, "Chūgoku Konanshō no Yaozoku no girei ni midasu Dōkyō no eikyō".

Chinese world. Lore about divine transcendent beings called *xian* 仙, who lived in the remote mountains and subsisted on mist, became popular toward the end of the Warring States period (406–221 BCE), with epic poems like "Encountering Sorrow" (*Lisao* 離騷) and "Distant Roaming" (*Yuanyou* 遠遊) providing poeticized accounts of astral travels to distant realms in search of the divine. During the Han, diverse programs of self-cultivation sought to extend life and transcend this realm through physical exercises, macrobiotic, herbal, and mineral regimens, sexual practices, meditation and visualization on the body and the universe, and external alchemy.[21] After the fall of the Han, North China was evangelized by Celestial Master Daoists, who rejected divination and medicine as attempts to evade the proper moral fruits of one's actions, and the Arcane Learning (*xuanxue* 玄學) philosophical movement also eschewed esoteric explanations for the world, but these Han traditions continued to flourish in southern China.

The Sima 司馬 family briefly united China under the Jin dynasty in 265, but they quickly turned to fratricide, and by the beginning of the fourth century, North China came under severe pressure from horse-riding nomads like the Xiongnu (Huns), Xianbei (Sarbee), and Qiang (Tibetans). The western capital was sacked in 311 and the north fell finally, the emperor captured, in 317. A mass migration to the south ensued, with whole villages moving as groups and settling in "guest" (*qiao* 僑) towns and counties with the same name. For the next three hundred years, North China was lost to Chinese control and the center of Chinese civilization shifted decisively to the South.

Evangelizing Daoists converted individuals with deep roots in these southern occult arts, and a new synthesis of Daoism emerged, that was to reshape the faith. In a series of spirit revelations during the years 364–370, one of the great old southern gentry families, the Xus, received word of new, more exalted realms associated with the Supreme Purity heavens. In addition to precious lore concerning their own personal fates in Heaven and that of their ancestors, they were granted new knowledge of solitary meditation and visualization methods that extracted energies from the sun, moon, and other astral bodies. Moreover, these ethereal deities, often in the form of young women, promised transcendental trysts with the divine in supernatural unions called "pairing the phosphors" (*oujing* 偶景) that were said to far transcend the all-too-fleshly unions of the Celestial Masters Joining the Pneumas rite.[22]

In sum, the Supreme Purity revelations provided an elite, individually-oriented Daoism within the Celestial Master framework of orthodoxy. This new form

---

21 David HAWKES, "The Quest of the Goddess"; ID., *The Songs of the South*.

22 On Supreme Purity Daoism, see Michel STRICKMANN, *Le taoïsme du Mao Chan*; Isabelle ROBINET, *La révélation du Shangqing*.

of Daoism flourished among the high literati of the subsequent Tang dynasty (618–907) and informed the inner alchemy traditions that took form in the Song (960–1279). Xie An 謝安, a devout Celestial Master Daoist who served as Prime Minister became famous as the "minister in the mountains" because he spent so much time in retreat. From this time on, famous Daoist priests are awarded hermitages in the mountains, first called "offices" (*guan* 館), later "belvederes" (*guan* 觀), where the solitude and quiet are thought to promote reflection and spiritual growth. Eventually, these belvederes transformed into quasi-monastic institutions on the model of Buddhist monasteries. This development reached full fruition with the institution of Complete Perfection (*quanzhen* 全真) Daoism in the 12th century. Complete Perfection Daoists remain a major part of the Daoist world today. They do perform rituals for the local communities where they are based, but the primary orientation of the Complete Perfection priests is on personal self-cultivation through a rigorous program of meditation as well as physical exercises related to mastering and controlling the pneumas (*qi* 氣). Moreover, the ideas first formulated in the scriptures of the Supreme Purity movement, as well as the visualization practices, were important in the development of Inner Alchemy practices that became dominant among elite practitioners in the Ming and Qing dynasties.

Another new revelation from this period was to permanently change the nature of priesthood in Daoism. The Numinous Jewel (*Lingbao* 靈寶) scriptures started to appear in the 390s, combining traditional lore about celestial writing and sacrifices to the traditional gods with new ideas that had arrived with Buddhism, like the hells and reincarnation.[23] Whereas Celestial Master petition rites had sought the safety and well-being of specific ancestors who were threatened in the other world, Numinous Jewel texts followed Buddhism in arguing that all the dead suffered the punishments of their inevitable karmic errors, and thus needed salvation from cosmic figures rather that the simple tinkering with divine records offered by the Celestial Master priests.

The Numinous Jewel texts provided large-scale rites for the salvation of the dead or of the state as a whole in the Yellow Register and Golden Register Rites. Even when a Numinous Jewel ritual was in fact sponsored by a single family, and intended primarily for the salvation of a specific recently-deceased member of the sponsor's family, the rite in fact saved all the dead, including all relatives of the sponsor's neighbors, and ended with a feast to which all were welcome. Here is an example from the Southern Song dynasty (1127–1289) of such a large-scale

23 Stephen R. BOKENKAMP, "Sources of the Ling-pao Scriptures".

ritual. Despite being a mass ritual, each individual has a personal encounter with their deceased relative.[24]

> On the 16th day of the second month of 1198, the Abbey of Celestial Auspiciousness in Raozhou 饒州天慶觀 held a Yellow Register Great Offering. They solicited people to put forward their dead, preparing on behalf of each individual one thousand two hundred cash. A thousand people came to participate in the assembly. As the rite was about to conclude, Fu San saw his mother, who had recently died, wearing the clothes she had worn while alive. She was wet top and bottom, and approached from a distance, entering the public feast area. Seeing her, Fu suffered unbearably, and could not abide. Tears streaming, he turned to home. His mother followed him all the way home, chatting and asking him questions just like always. When she said, "Thanks for all your efforts in in seeing to me after my death.", her voice was just as when she was alive. When Fu went to investigate more closely, she disappeared.
>
> 庆元四年二月十六日，饶州天庆观设黄箓大醮，募人荐亡，每一位为钱千二百，预会者千人。将毕事，市侩傅三，见近所亡母，著生前衣服，上下皆湿，自远而来，入供筵中。傅瞻视悲痛不堪处，垂泣遽还。母冉冉随至家，语话问讯，一如常时。且云："荷汝追修之力。"声音全与生前不异。傅欲审叩其所以然，奄忽而没。

In large-scale rituals like this, priests could minister to the needs of large groups at the same time through a single ritual.

The rise of national deity cults, beginning again in the Song, marked the final major stage in the development of the modern Chinese religious world.[25] China had always worshipped a plethora of local deities, each different and distinct to his or her local setting. On the local level, these deities might have specific associations but they were, in general, all-purpose deities responded to all sorts of entreaties. As they grew beyond the local level, becoming noticed by worshippers in surrounding communities and eventually throughout the region, it was usually because of reported efficacy in one specific respect. Thus one god might develop a reputation for healing, while another was better at amassing wealth; a third might be a martial figure who protects in battle or enforces oaths. Some gods became so specialized that they developed into occupational deities worshipped by guilds, like Lu Ban for carpenters or Shennong for herbalists. An even smaller group was adopted into Daoism as cosmic figures who aid in the survival of the world and the salvation of humanity from the kalpic disasters engendered by human evil.

This rise of popular god cults also marked the final evolution of the Daoist priest from communal pastor to a member of a guild of religious profession-

---

24 HONG Mai, *Yijian zhi*, 3/1319; Edward L. DAVIS, *Society and the Supernatural*, 179 sq.
25 Terry KLEEMAN, "The Evolution of Daoist Cosmology".

als open to employment by any member of the population.[26] Since Song times, Daoists have been just one member of a religious cohort from which supplicants draw as needed. They have, to be sure, a privileged position as the most educated and ritually most elaborate religious professional to deal with the popular religious world. When a major ritual confirming the god in his place is held, this Offering or Rite of Cosmic Renewal is the sole province of the Daoist priest. No common, vernacular priest could master the multi-day liturgical program in literary Chinese. These elaborate rituals have been performed continuously since at least the 12th century.[27]

Thus Daoist priests have assumed a unique role with regard to the common religion, regulating ritual observances to local gods and subordinating them to the cosmic system of Daoism. People are free to make use of the full range of religious specialists, who appeal to the authority and sacred power of the common, Daoist, and/or Buddhist pantheons. At any given time, certain specialists will have a reputation for efficacy that trumps their affiliation. But in general, for the most intractable or significant problems, the final resort is usually the Daoist priest.

This then is the normal state of the Chinese religious world: a public world of religious devotion to local and national gods, worshipped in public temples on an ad hoc basis. Certain gods (typically national gods of high moral purpose) might enjoy regular devotion, but most deities are only addressed when a problem arises they are well equipped to resolve. Daoist priests and Buddhist monks form independent guilds of religious professionals, interacting with the public primarily as clients who call upon the priests or monks to deal with major life transitions and other weighty matters.

One might think that this current situation is a permanent answer to the question of organized versus diffuse religion, but in fact the Chinese religious world is constantly developing new organized religions. The Complete Perfection Daoist movement of the Southern Song is in some senses a new religion, built upon a synthesis of Daoism with other faiths.[28] The characteristics of new religions in Early Modern and Modern China has been an overt syncretism fusing elements of different origin, a rejection of the traditional sacrificial common religion, and exclusive identification with the new faith. In the twentieth century, we saw the rise of the Unity Sect (*Yiguan dao* 一貫道), sometimes called the Way of Heaven (*Tiandao* 天道), which is now a pervasive element of Chinese society outside of

26 For the training and life of a modern Daoist priest, see Kristofer M. SCHIPPER, *Le corps taoïste*.
27 On the Daoist offering ritual, see John LAGERWEY, *Taoist Ritual in Chinese Society and History*.
28 Vincent GOOSSAERT, "The Invention of an Order".

the PRC.[29] More recently, we have seen the phenomenal rise of the Falungong 法輪功 organization, which is now found throughout the world. In this sense, the Sinophone world is an active breeding ground for new organized religions, at the same time that is also is characterized by a diffuse religious tradition with which the older, more established organized religions of Buddhism and Daoism have evolved an elaborate and satisfying *modus vivendi*. This is what makes the Chinese religious realm so uniquely dynamic.

Bibliography

Barrett, Timothy H., “Chinese religion in English guise: The history of an illusion”, *Modern Asian Studies* 39 (2005), 509–533.

Bokenkamp, Stephen R., “Sources of the Ling-pao Scriptures”. In *Tantric and Taoist Studies in Honour of R. A. Stein*, vol. 2, edited by Michel Strickmann (Mélanges chinois et bouddhiques, 21). Bruxelles, Institute Belge des Hautes Etudes Chinoises, 1983, 434–486.

Bokenkamp, Stephen R., *Early Daoist scriptures* (Taoist classics, 1). Berkeley, University of California Press, 1997.

Bosco, Joseph, “Yiguan Dao: Heterodoxy and Popular Religion in Taiwan”. In *The Other Taiwan: 1945 to the Present*, edited by Murray A. Rubinstein. Armonk, NY/London, M.E. Sharpe, 1994, 423–444.

de Groot, Jan J. M., *The religious system of China. Its ancient forms, evolution, history and present aspect, manners, customs and social institutions connected therewith*. 6 vols. Leiden, E.J. Brill, 1892–1910.

Davis, Edward L., *Society and the Supernatural in Song China*. Honolulu, University of Hawaii Press, 2001.

Dubs, Homer H., “An ancient Chinese mystery cult”, *Harvard Theological Review* 35 (1942), 221–240.

Eichhorn, Werner, “Bemerkungen zum Aufstand des Chang Chio und zum Staate des Chang Lu”, *Mitteilungen des Instituts für Orientforschung* 3 (1955), 291–327.

von Falkenhausen, Lothar, “Reflections on the Political Role of Spirit Mediums in Early China: The *wu* Officials in the *Zhou li*”, *Early China* 20 (1995), 279–300.

Goossaert, Vincent, “The Invention of an Order: Collective Identity in Thirteenth-Century Quanzhen”, *Journal of Chinese Religions* 29 (2001), 111–138.

Goossaert, Vincent, “Mapping Charisma among Chinese Religious Specialists”, *Nova Religio* 12 (2008), 12–28.

Hawkes, David, “The Quest of the Goddess”, *Asia Major NS* 13 (1967), 71–94.

Hawkes, David, *The Songs of the South. An ancient Chinese anthology of poems*. Harmondsworth, Penguin, ²1985.

Hendrischke, Barbara, “Early Daoist movements”. In *Daoism Handbook*, edited by Livia Kohn, Leiden, E.J. Brill, 2000, 134–164.

29 On the Unity Sect see Joseph BOSCO, “Yiguan Dao”.

Hendrischke, Barbara, *The Scripture on Great Peace: The Taiping jing and the beginnings of Daoism* (Daoist classics series 3). Berkeley, University of California Press, 2007.
Hirota Ritsuko 広田律子, "Chūgoku Konanshō no Yaozoku no girei ni midasu Dōkyō no eikyō", *Tōhō shūkyō* 110 (2007), 57–81.
Hou Hanshu. Beijing, Zhonghua, 1963.
Hong Mai, *Yijian zhi*. Beijing, Zhonghua, 1981.
Jensen, Lionel M., *Manufacturing Confucianism: Chinese Traditions and Universal Civilization*. Durham, N.C./London, Duke University Press, 1997.
Jordan, David K., *Gods, Ghosts, and Ancestors: The folk religion of a Taiwanese village*. Berkeley, University of California Press, 1972.
Keightley, David N., *The Ancestral Landscape: Time, Space, and Community in Late Shang China (ca. 1200–1045 B.C.)* (China Research Monograph, 53). Berkeley, Institute of East Asian Studies, 2000.
Kleeman, Terry F., "The Evolution of Daoist Cosmology and the Construction of the Common Sacred Realm", *Taiwan Journal of East Asian Studies* 2 (2005), 89–110.
Kleeman, Terry F., *Celestial Masters: History and Ritual in Early Daoist Communities* (Harvard-Yenching Institute Monograph Series, 102). Cambridge, Harvard East Asia Institute, 2016.
Lagerwey, John, *Taoist Ritual in Chinese Society and History*. New York, Macmillan, 1987.
Liji zhengyi. Taipei, Yeewen, 1974, 12/16a-b.
Poo, Mu-chou, *In Search of Personal Welfare: A View of Ancient Chinese Religion* (SUNY Series in Chinese Philosophy and Culture). Albany, State University of New York Press, 1998.
Robinet, Isabelle, *La révélation du Shangqing dans l'histoire du taoïsme* (Publications de l'École Française d'Extrême-Orient, 137). 2 vols. Paris, École Française d'Extrême-Orient, 1984.
Schipper, Kristofer M., *Le corps taoïste: corps physique, corps social* (L'espace intérieur, 25). Paris, Fayard, 1982.
Schipper, Kristofer M., "Taoist Ordination Ranks in the Tunhuang Manuscripts". In *Religion und Philosophie in Ostasien. Festschrift für Hans Steininger*, edited by Gert NAUDORF, Karl-Heinz POHL and Hans-Herrman SCHMIDT, Königshausen, Neumann, 1985, 127–148.
Seidel, Anna K., *La divinisation de Lao Tseu dans le taoïsme des Han* (Publications de l'École Française d'Extrême-Orient, 71). Paris, École Française d'Extrême-Orient, 1969.
Seidel, Anna K., "Taoist Messianism", *Numen* 31 (1984), 161–174.
Strickmann, Michel, *Le taoïsme du Mao Chan: Chronique d'une revelation* (Mémoires de l'Institut des Hautes Études Chinoises, 17). Paris, Collège de France, 1981.
Verellen, Franciscus, "The Twenty-four Dioceses and Zhang Daoling: The Spatio-liturgical Organization of Early Heavenly Master Taoism". In *Pilgrims, Patrons, and Place: Localizing Sanctity in Asian Religions*, edited by Phyllis Granoff and Koichi Shinohara (Asian Religions and Society). Vancouver, University of British Columbia Press, 2003, 15–67.

Rodrigo Laham Cohen

Languages, Names and Images

Community and Identity Markers in Ancient Jewish Epigraphy of Western Europe

Introduction

It is not easy to understand the relationship between the individual and communal life during Late Antiquity and the Early Middle Ages. Given the paucity of sources, it is very difficult to identify the thoughts, ideas and concerns of people in ancient times. This situation is even true for well-known men who wrote their own biographies such as Julius Caesar, Flavius Josephus or Augustine of Hippo.

But here I will tackle an even more complex scenario: Western European Jews of the first millennium. After Flavius Josephus, no text written by Jews survived before the 9th century[1]. In fact, we have to wait until Shabbethai Donnolo to see another Jewish work in Europe[2]. There are, also, three texts that are considered Jewish by some scholars, but there is still no clear consensus about them[3]. In this context, the Jewish epigraphical record of Western Europe offers an opportunity to explore the relationship between the individual and the communal. It is not an easy path, but I will try to present guidelines with which to obtain some information on an intricate topic.

First of all, the area of enquire will be confined to Jewish religious communities. It is important to stress that there were different Jewish communities in the period: the rabbinic communities of Palestine and Babylon were clearly differentiated from the European communities of the first part of the millennium; even each of these Jewish European communities was different from the other.

Secondly, three are the main questions that I will try to answer here: 1) To what extent was an individual able to live outside a community or, in other words,

---

1 Shaye Cohen and Joshua Schwartz, *Studies in Josephus*; Louis Feldman, "Josephus (CE 37 – c. 100)", 901–921; Zuleika Rodgers, *Making History*.

2 Andrew Sharf, "Shabbetai Donnolo", 160–177; Paul Skinner, "Conflicting Accounts", 1–15.

3 There are several controversies around the authorship of the following texts: *Collatio legum Mosaicarum et Romanarum*; *Epistola Anne ad Senecam*, and *Liber Antiquitatum Biblicarum* (also known as pseudo-Philo). Some scholars affirm that the texts were written by Jews but others suggest that they are Christian works. Frederick Murphy, *Pseudo-Philo*; Alfredo Rabello, "La datazione della *Collatio legum Mosaicarum et Romanarum*", 411–422; Ilaria Ramelli, "L'*Epistola Anne ad Senecam*", 25–50.

was it possible to escape from communal life? 2) To what extent did the binding to a community allow otherness or diversity, which could lead to the fracturing of identity or to its alteration, respectively? 3) Could an alternative community influence an individual against the compulsion of his own community?

Jewish Epigraphy in Western Europe

There are two main sets of Jewish inscriptions in Western Europe: Rome and Venosa. More than six hundred Jewish inscriptions – dated between the 2nd and the 5th century – were found in Rome[4]. In Venosa there are almost ninety five inscriptions dated between the 2th and the 9th century[5]. The entire epigraphical record of Western Europe reaches approximately nine hundred registers. The big majority of these inscriptions consist of epitaphs found in Jewish catacombs.

In what follows I will focus on the Italian Peninsula and mainly in the next issues: languages, names, art, and religious offices. I will try to detect communal patterns and, when possible, individual deviations. We will see that the communal patterns can be understood as group identity markers altered by individuals that adopted alternative identity markers, generally belonging to another community. It is important to highlight that, in epigraphy, it is sometimes hard to discern communal compulsion from individual decisions. The absence of contemporary Jewish texts does not help us to reconstruct the dynamic. However, some Christian sources provide useful insight on this situation.

Last but not least, I will deal with individuals that broke some community patterns but remained nevertheless inside the community. If a Jew abandoned his community and was buried in a non-Jewish catacomb or cemetery without any Jewish identity marker in his grave, it would be impossible to recognize his tomb. Because of that – I insist – the evidence only contains individuals that did not abandon their communities.

Let's begin now with languages.

---

4 David Noy, *Jewish Inscriptions*, vol. 2 (= JIWE II).

5 Inscriptions dated between 2nd and 6th century: Id. *Jewish Inscriptions*, vol. 1 (= JIWE I). Dated between 6th and 9th century: Umberto Cassuto, "Le iscrizioni ebraiche", 99–120; Jean Baptiste Frey and Baruch Lifshitz, *Corpus of Jewish Inscriptions*; Cesare Colafemmina, "Un'iscrizione venosina", 261–263; Id., "Tre iscrizioni ebraiche", 443–448; Id., "Documenti per la storia", 3–11; Id., "Archeologia ed epigrafia", 199–211; Id., "Tre nuove iscrizioni ebraiche", 201–209; Id., "Gli ebrei in Basilicata", 9–32; Id., "Epigraphica hebraica Venusina", 353–358; Giancarlo Lacerenza, "L'epigrafia ebraica in Basilicata".

Languages

The language used by western European Jews changed during the first millennium. From comprising *circa* 2% of the Jewish inscriptions in Rome (3rd – 5th century), the employment of Hebrew increased significantly reaching almost a 100% of those in Venosa in the 9th century. On the other hand, the Greek language, particularly spread in Roman inscriptions (more than 80%), disappeared from the epigraphical record after the 6th century. Finally, Latin grew from *circa* 20% to more than 70% in Taranto (7th – 8th century) and, after that, disappeared from the register[6]. It is worth mentioning that some of the inscriptions were bilingual and trilingual.

All these figures point to a process called hebraization[7]. It is impossible to determine with precision which language was spoken by the Jews during the period, although Christian sources seem to indicate that they used the common language of each region. The truth is that epitaphs were written by the Jews in Greek or in Latin in the 5th century but, by the 9th century, they had turned to Hebrew. We can thus affirm that the main epigraphical language of the Roman-Jewish community during the 2nd to the 5th century was Greek. This can be seen as something strange but it is important to remember that Greek was, alongside Hebrew, a Jewish sacred tongue in Antiquity. The diffusion of the Septuagint and the dynamism of the Alexandrian community spread Greek deep into Judaism, even in Palestine[8].

Nevertheless, some inscriptions in Rome – less than 10 – introduce words in Hebrew into otherwise Greek epitaphs. The word written is usually שלום (*shalom,* "peace") and שלום על ישראל (*shalom al Israel,* "Peace to Israel"). What does this means? For what reasons certain individuals (or their families) decided to use Hebrew in a Greco-Latin scenario? It is possible that they chose Hebrew against community customs? Perhaps those Jews decided that Hebrew was more important than Greek. Were they acting in opposition to their community? We do not know, but it is plausible. We must stress, however, that their decision may have been influenced by the Palestinian Jewish community, where Hebrew was used alongside Aramaic and Greek. If this idea proves to be correct, the introduction

6 These figures are drawn from my analysis based on the evidence quoted in notes 4 and 5.
7 Some works on *hebraization*: Nicholas De Lange, "The Revival of the Hebrew", 342–358; David Noy, "Writing in Tongues", 300–311. This process needs to be considered side by side with the rabbinization process. See Seth Schwartz, "Rabbinization in the Sixth Century", 55–69.
8 Peter van der Horst and Judith Newman, *Early Jewish Prayers in Greek*; Tessa Rajak, *Translation and Survival*; Nicholas De Lange, Julia Krivoruchko, and Cameron Boyd-Taylor, eds., *Jewish Reception of Greek Bible Versions.*

of Hebrew in ancient Rome was an individual deviation that resulted from the influence of other Jewish communities, in this case, the Palestinian and Babylonian. We may see here, then, that conflicting community patterns – both Jewish – influenced individual decisions.

The choice of an alternative custom can be seen, for example, in II JIWE n. 193. The inscription, found in a grave in Monteverde's catacomb, was written in Greek but contains the following words in Hebrew: שאלום על ישראל . First, the individual (or the persons who ordered the grave) decided to employ Hebrew words against the practice of the majority of the other Roman Jews. If this decision was taken due to a matter of expertise, i.e. knowledge of the language, it is impossible to know. Perhaps most of the Jews in Rome had forgotten Hebrew but not all. It should be remembered that David Noy highlighted that the form of the Hebrew letters in this grave are very similar to those found in Beth She'arim (Palestine)[9]. From this perspective we could argue that the person who ordered the grave was from Palestine and was following another community pattern, not Roman, because of his place of origin. However, there is something in the spelling that points to a different direction: the word "peace" is incorrectly written. The correct spelling is שלום, which makes us wonder, why did the writer employ an *alef*? The same phenomenon can be detected in Venosa one century later[10]. We are here in uncharted waters, but it is possible to see the work of a Greek or Latin-speaking individual who confused the *alef* as an *alfa,* forgetting that Hebrew does not need written vowels. Was this person following an eastern community pattern or was he influenced by his own knowledge, built upon his own Jewish-Roman community? There are no definite answers in this respect, just working hypotheses.

Another example is the word "peace" found in I JIWE n. 72 (Venosa, 5th century). This time it is written in Greek letters: σάλωμ. Surprisingly, almost half of the Jewish inscriptions in Venosa contain, at least, one word in Hebrew. Why *shalom* in Greek? It can be seen here the struggle between a pro-Hebrew and a pro-Greek community compulsion. Although the most important language in Venosa was Greek, Hebrew frequently appeared in isolated words such as *shalom,* and much more than in Rome. It is possible to consider that the hebraization process was beginning in southern Italian cities. In addition, it is plausible that an individual was trying to adopt the new identity markers arrived from Palestine and Babylon but, again, only with knowledge of the sound but not of Hebrew letters. In contrast to the situation in Rome, we see here a balanced conflict within the community given that there were many individuals using Hebrew and, consequently, there was an emerging community pattern.

---

9 JIWE II, n. 156.
10 JIWE I, n. 47 and n. 64.

A Christian source enables us to observe the tensions between pro-Greek and pro-Hebrew tendencies within the Empire. In fact, Justinian's *Novella* 146 established the mandatory use of Greek inside Jewish communities. However, Justinian affirmed that he took the decision due to a request from the Jews. According to the emperor, the Jews were discussing between them the proper language to read the Bible[11]. It is not possible to determine the actions of individuals in this case. What we can perceive – as in the case with epitaphs – is that inside Jewish communities there were different identity markers and patterns (and compulsions) that conditioned individual decisions. Let's turn now to onomastics.

Names

A similar development can be observed with respect to Jewish names. In this case, the use of biblical and Semitic names grew during the first millennium. In fact, more than 85% of names in the Jewish inscriptions of Rome were in Greek or Latin. Almost four centuries later, in Venosa, they represented less than 40%. In opposition, biblical and Semitic names reached 60% in the same period, four times more than the proportion in late ancient Rome[12].

It is even harder to understand why a name is chosen. As we are working with epigraphical evidence, discerning if an individual chose Ἵλαρος because he was thinking in the Hebrew name יצחק or, simply, due to the adoption of a widespread Greek name, is almost impossible.

A second problem is that there is not a clear identity marker in the period. We saw that the employment of biblical and Semitic names grew during the first millennium, but the use of Greco-Latin names survived. Similarly, in the first centuries, although Greco-Latin names were mostly used, almost 15% of Jews employed biblical and Semitic names. It is not possible to ascertain with these figures that the use of a biblical name in Rome during the first centuries of the millennium was an individual deviation[13].

Moreover, sometimes it is possible to see the variation of names within several generations in a family. Surprisingly (or maybe not) there is no continuity in the type of name chosen: someone bearing a Greek name could opt for a bibli-

11 Vittore COLORNI, *L'uso del greco*; Giuseppe VELTRI, "*Die Novelle* 146", 116–130; Leonard RUTGERS, "Justinian's Novella 146", 49–77.

12 These figures come from my analysis based on the evidence quoted in notes 4 and 5.

13 Onomastic studies on ancient Jews: Naomi COHEN, "Jewish Names", 97–128; Gerard MUSSIES, "Jewish Personal Names", 242–251; Margareth WILLIAMS, "The Use of Alternative Names", 307–327.

cal one for his son and, subsequently, this person could choose a Latin name for his offspring. In addition to this, there is no link between a biblical name and the use of Hebrew.

There is only one name in all the Jewish epigraphical record of Rome that broke the pattern: The name *Abundantius*[14] that comes, according to Tal Ilan, from the Persian name *Abandanes*[15]. Is this the decision of someone that infringed the onomastic patterns of his community? It is probable, but it represents only one exception. The freedom which the community of Rome gave its members to choose between Greek, Latin, biblical, or Semitic names was enough to avoid individual deviations. The flexibility guaranteed, in this case, the persistence of community patterns.

Art

There is an open debate about the concept of "Jewish art" in Antiquity and the Early Middle Ages[16]. It is not within the scope of this paper to discuss this topic, but I want to stress that ancient and early medieval Jews had a specific repertoire of icons. Those icons were employed side by side with the common and shared art used also by pagans and Christians. Hence we have, in ancient Jewish art, a combination of Jewish and non-Jewish motifs. I will suggest that the specific Jewish set of icons represented the preferred identity community markers. Individuals could choose between using those icons or those pertaining to other communities, or motifs belonging to pagans or Christians.

Many Jewish graves have specific Jewish icons: *Menorah*[17], *Etrog*[18] and the other *Sukkot* species[19], Torah ark, and *Shofar*[20]. The same motifs were found in the remaining ancient synagogues in Italy, Ostia and Bova Marina[21]. The utilization

---

14 JIWE II, n. 217.

15 Tal ILAN, *Lexicon of Jewish Names*, 623.

16 Rachel HACHLILI, *Ancient Jewish Art*; Jas ELSNER, "Archaeologies and Agendas", 115–128; Zeev WEISS and Lee LEVINE, eds., *From Dura To Sepphoris*; Steven FINE, *Art and Judaism*; Katrin KOGMAN-APPEL and Mati MEYER, eds., *Between Judaism and Christianity.*

17 The menorah is the seven-armed Candelabrum. See Rachel HACHLILI, *The Menorah.*

18 The *Etrog* is a yellow citron used during the feast of *Sukkot.*

19 The four species of *Sukkot* are composed by the *Etrog* (see the previous note), a frond from a palm tree, boughs with leaves from a myrtle tree, and branches with leaves from a willow tree.

20 The *Shofar* is the ritual horn that is sounded during *Rosh Hashanah* (New year) and *Yom Kippur* (Day of Atonement).

21 This is not the place to discuss the archaeological evidence from Ostia and Bova Marina. See Enrico TROMBA, *La sinagoga dei giudei*; Liliana COSTAMAGNA, "La sinagoga di Bova Marina";

of common motifs such as plants, flowers or birds does not imply, automatically, that there was a religious pagan or Christian influence on the Jewish community, nor does it mean that an individual was trying to break the communal tradition. In fact, this kind of art was also Jewish art. Nevertheless, the use of common art demonstrates that other communities had an impact on the Jewish[22]. Therefore it is important to emphasize that the employment of figural images, even though prohibited in some biblical and rabbinical sources, was common in the Jewish communities, not only in the Diaspora but also in Palestine[23].

But there are some cases in which an individual decided to use motifs belonging to other community. For example, an image of the goddess *Fortuna* is found in room I of Vigna Randanini catacomb[24]. Although scholars have discussed whether the room was originally Jew or pagan, the fact is that, even if we accept a pagan origin, it is clear that whoever buried the dead there did not deemed it unsuitable to reutilize the pagan image. This decision is very interesting because those individuals – voluntarily or not – accepted an alternative set of identity markers, challenging – again, voluntary or not – their own community identity.

The same situation can be found in the reutilization of sarcophagi containing pagan images[25]. The best one conserved contains a *Menorah* alongside four figural representations of the seasons[26]. There is a controversy about the person buried here, but most scholars agree that it was chosen by the remaining relatives from the stock available in a Roman stone-cutter's store. Again, we have individuals that, acting against the widespread community pattern, decided to use other identity markers.

Overall, art allows us to see the interaction between different community patterns and the capacity of individuals to choose between them. The majority of the Jews of Rome, as we saw, opted for the specific Jewish repertoire but, some of them – defying the community compulsion – adopted alternative and controversial artistic motifs.

Lee LEVINE, *The Ancient Synagogue*; Birger OLSSON, Dieter MITTERNACHT and Olof BRANDT, *The Synagogue of Ancient Ostia*; Anders RUNESSON, Birger OLSSON and Olof BRANDT, *The Ancient Synagogue*.

22 In fact, many scholars suggest that they shared the same workshops. See Leonard RUTGERS, *The Jews in Late Ancient Rome*, 50–99; JAS ELSNER, "Archaeologies and Agendas", 117–119.

23 See the books quoted in note n. 17.

24 Tessa RAJAK, "Inscription and Context", 226–241.

25 See Adia KONIKOFF, *Sarcophagi from the Jewish Catacombs*.

26 Ibid., 38–41.

Main activity

One last issue that should be addressed is the reference in inscriptions to economic activities or to religious offices. In Rome, almost 20% of the graves mention religious offices[27], but only four inscriptions, less than 1%, allude to economic activities. Therefore, the pattern does not seem to have included references to professions but to synagogue offices.

Nevertheless, there are four individuals who decided to record their economic activities. Perhaps, not surprisingly, one of those individuals was a kind of painter, a ζωγράφος[28]. In addition, the inscription was found in a sarcophagus of Randanini's catacomb, where pagan images were also found and, what is more, it was embellished with animal representations. The most plausible hypothesis seems to be that the Jewish artist decided to stress his profession against the community custom, even though it was a rare profession within the community. From the perspective of the individual, this deviation did not clash with the community, a statement that can be confirmed by the fact that he was buried inside a Jewish catacomb.

The other professions mentioned in Rome were butcher, merchant and medic. These were common activities, inside and outside the Jewish community. In these cases, the deviation consisted in the decision of recording the economic activity rather than the position occupied inside the religious organization.

The situation in the rest of Italy is similar: there are few mentions to economic activities and many references to religious offices. We find, again, an activity apparently rare among Jews: a *pantomimus*[29]. In fact, some scholars suggested that this Ostian Epitaph did not belong to a Jew. The inscription is indeed ambiguous, referring to Judea but not stating explicitly a Jewish character[30]. The interpretation of this inscription shows how our ideas about community patterns and compulsions can modify our perception of individuals and identities.

---

27 Ἄρχων –48 mentions–, γραμματεύς (also in Latin, *Grammateus*) –25 mentions–, γερουσιάρχης –25 mentions–, ἀρχισυνάγωγος –6 mentions–. πατὴρ συναγωγῆς (also in Latin, *Pater*) –9 mentions–. Analysis based on JIWE II.

28 JIWE II, n. 277.

29 JIWE I, n. 15.

30 The inscription says: For Marcus Aurelius Pylades, son of of the Teretine tribe, …. from Scythopolis, the first pantomimus of his time in …. , and approved by the Emperors Valerian and Gallienus …. from the province of Judaea …. after the death of his father Juda. Also a decurion of the cities of Ascalon and Damascus. To him, second, the order of the Augustales not only in memory of his father, but also because of his own consummate skill, with all the citizens demanding it equally… Translation according to David Noy (JIWE I, n. 15).

Concluding remarks: individual deviations and community compulsions in epigraphy

As I stated before, understanding the relationship between community and individual in ancient times is a very difficult task, even more in cases where there is only epigraphic material. However, I hope to have demonstrated that, despite the paucity of evidence, it is possible to detect community patterns, identity markers, compulsions and, on the other hand, individual autonomy and deviations. Language, names, art, and the importance assigned to religious or economic activities are fields in which we can perceive strong community patterns altered by individual decisions.

Paradoxically, in our endeavor to understand the weight of the community in the challenges of life, the words written in a burial context highlight the opportunities and the limitations of individuals to reinforce or weaken communal identity. The Jews of the Italian Peninsula in the first millennium were conditioned by their communities but they were not entirely determined by them. Their decisions, considered within the context of competing communities (Jewish and non-Jewish), contributed to shape the European Jewish identities of the period.

Bibliography

Cassuto, Umberto, "Le iscrizioni ebraiche del secolo IX a Venosa", *Qedem* 2 (1944), 99–120.
Cohen, Naomi, "Jewish Names as Cultural Indicators in Antiquity", *Journal for the Study of Judaism* 7 (1976), 97–128.
Cohen, Shaye and Joshua Schwartz, eds., *Studies in Josephus and the Varieties of Ancient Judaism. Louis H. Feldman Jubilee Volume* (Ancient Judaism and Early Christianity, 67). Leiden/Boston, Brill, 2005.
Colafemmina, Cesare, "Tre iscrizioni ebraiche inedite di Venosa e Potenza", *Vetera Christianorum* 20 (1983), 443–448.
Colafemmina, Cesare, "Archeologia ed epigrafia ebraica nell'Italia meridionale". In *Italia Judaica. Atti del convegno internazionale, Bari 18–22 maggio 1981* (Pubblicazioni degli archivi di stato. Saggi, 2). Rome, Pubblicazioni degli Archivi in Stato, 1983, 199–211.
Colafemmina, Cesare, "Documenti per la storia degli ebrei in Basilicata", *Bollettino della Biblioteca Provinciale di Matera* 6 (1983), 3–11.
Colafemmina, Cesare, "Epigraphica hebraica Venusina", *Vetera Christianorum* 30 (1993), 353–358.
Colafemmina, Cesare, "Gli ebrei in Basilicata", *Bollettino Storico della Basilicata* 7 (1991), 9–32.
Colafemmina, Cesare, "Tre nuove iscrizioni ebraiche a Venosa", *Vetera Christianorum* 24 (1987), 20 –209.
Colafemmina, Cesare, "Un'iscrizione venosina inedita dell'822", *La rassegna mensile d'Israel* 43 (1977), 261–263.

Colorni, Vittore, *L'uso del greco nella liturgia del giudaismo elenistico e la novella 146 di Giustiniano*. Milan, Giuffre, 1964.
Costamagna, Liliana, "La sinagoga di Bova Marina". In *I beni culturali ebraici in Italia: situazione attuale, problemi, prospettive e progetti per il futuro*, edited by Mauro Perani (Le tessere, 6), Ravena, Longo, 2003, 93–118.
De Lange, Nicholas, Julia Krivoruchko, and Cameron Boyd-Taylor, eds., *Jewish Reception of Greek Bible Versions. Studies in their Use in Late Antiquity and the Middle Ages* (Texts and Studies in Medieval and Early Modern Judaism, 23). Tübingen, Mohr Siebeck, 2009.
De Lange, Nicholas, "The Revival of the Hebrew Language in the Third Century CE", *Jewish Studies Quarterly* 3 (1996), 342–358.
Elsner, Jas, "Archaeologies and Agendas: Reflections on Late Ancient Jewish Art and Early Christian Art", *The Journal of Roman Studies* 93 (2003), 115–128.
Feldman, Louis, "Josephus (CE 37 – *c.* 100)". In *The Cambridge History of Judaism*, vol. III, edited by William Horbury, William Davies and John Sturdy. Cambridge, Cambridge University Press, 2006, 901–921.
Fine, Steven, *Art and Judaism in the Greco-Roman World. Toward a New Jewish Archaeology*. Cambridge, Cambridge University Press, 2005.
Frey, Jean Baptiste and Baruch Lifshitz, *Corpus of Jewish Inscriptions. Jewish Inscriptions from the Third Century B.C. to the Seventh Century A.D.*, vol. 1: *Europe*. New York, Ktav Publishing House, 1975.
Hachlili, Rachel, *Ancient Jewish Art and Archaeology in the Diaspora* (Handbuch der Orientalistik, 1/35). Leiden/Boston, Brill, 1998.
Hachlili, Rachel, *The Menorah, the Ancient Seven-Armed Candelabrum. Origin, Form and Significance* (Journal for the Study of Judaism. Supplements, 68). Leiden/Boston/Köln, Brill, 2001.
Ilan, Tal, *Lexicon of Jewish Names in Late Antiquity*, vol. 3: *The Western Diaspora 330 BCE – 650 CE* (Texts and Studies in Ancient Judaism, 126). Tübingen, Mohr Siebeck, 2008.
Kogman-Appel, Katrin and Mati Meyer, eds., *Between Judaism and Christianity. Art Historical Essays in Honor of Elisheva (Elisabeth) Revel-Neher* (The Medieval Mediterranean, 81). Leiden/Boston, Brill, 2009.
Konikoff, Adia, *Sarcophagi from the Jewish Catacombs of Ancient Rome. A catalogue raisonné*. Stuttgart, Franz Steiner Verlag, 1990.
Lacerenza, Giancarlo, "L'epigrafia ebraica in Basilicata e Puglia dal IV secolo all'Alto Medioevo". In *Ketav, Sefer, Miktav, La cultura ebraica scritta tra Basilicata e Puglia*, edited by Mariapina Mascolo. Bari, Di Pagina, 2014, 189–252.
Laham Cohen, Rodrigo, "Los judíos en el *Registrum epistularum* de Gregorio Magno y la epigrafía judía de los siglos VI y VII", *Henoch. Historical and Textual Studies in Ancient and Medieval Judaism and Christianity* 35 (2013), 214–246.
Levine, Lee, *The Ancient Synagogue. The first Thousand Years*, New Haven, Yale University Press, 2005.
Murphy, Frederick J., *Pseudo-Philo. Rewriting the Bible*. New York/Oxford, Oxford University Press, 1993.
Mussies, Gerard, "Jewish Personal Names in Some Non-Literary Sources". In *Studies in Early Jewish Epigraphy*, edited by Jan Willem van Henten and Peter Willem van der Horst (Arbeiten zur Geschichte des antiken Judentums und des Urchristentums, 21). Leiden, Brill, 1994, 242–251.

Noy, David, "Writing in Tongues: The Use of Greek, Latin and Hebrew in Jewish Inscriptions from Roman Italy", *Journal of Jewish Studies* 48 (1997), 300–311.
Noy, David, *Jewish Inscriptions of Western Europe*, vol. 2: *The City of Rome*. Cambridge, Cambridge University Press, 1995.
Noy, David., *Jewish Inscriptions of Western Europe*, vol. 1: *Italy (excluding the City of Rome), Spain and Gaul*. Cambridge, Cambridge University Press, 1993.
Olsson, Birger, Dieter Mitternacht and Olof Brandt, eds., *The Synagogue of Ancient Ostia and the Jews of Rome: Interdisciplinary Studies* (Skrifter utgivna av Svenska Institutet i Rom, Serie in 4°, 57). Stockholm, Åström, 2001.
Rabello, Alfredo, "La datazione della *Collatio legum Mosaicarum et Romanarum* e il problema di una seconda redazione o del suo uso nel corso del quarto secolo". In *"Humana Sapit". Études d'antiquité tardive offertes à Lellia Cracco Ruggini*, edited by Jean-Michel Carrié and Rita Lizzi Testa (Bibliothèque de l'antiquité tardive, 3). Turnhout, Brepols, 2002, 411–422.
Rajak, Tessa, "Inscription and Context: Reading the Jewish Catacombs of Rome". In *Studies in Early Jewish Epigraphy*, edited by Jan Willem van Henten and Peter Willem van der Horst (Arbeiten zur Geschichte des antiken Judentums und des Urchristentums, 21). Brill, 1994, 226–241.
Rajak, Tessa, *Translation and Survival. The Greek bible of the Ancient Jewish Diaspora*. Oxford, Oxford University Press, 2009.
Ramelli, Ilaria, "L'*Epistola Anne ad Senecam de superbia et idolis*. Documento pseudo-epigrafico probabilmente cristiano", *Augustinianum* 44 (2004), 25–50.
Rodgers, Zuleika, *Making History. Josephus and Historical Method* (Supplements to the Journal for the Study of Judaism, 110). Leiden/Boston, Brill, 2007.
Runesson, Anders, Birger Olsson and Olof Brandt, *The Ancient Synagogue from its Origins to 200 C.E. A Source Book* (Ancient Judaism and Early Christianity, 72). Boston, Brill, 2008.
Rutgers, Leonard, "Justinian's Novella 146 between Jews and Christian". In *Making Myths. Jews in Early Christian Identity Formation*, edited by Leonard Rutgers, Leuven, Peeters, 2009, 49–77.
Rutgers, Leonard, *The Jews in Late Ancient Rome. Evidence of Cultural Interaction in the Roman Diaspora* (Religions in the Graeco-Roman World, 126), Leiden/New York/Köln, Brill, 1995.
Schwartz, Seth, "Rabbinization in the Sixth Century". In *The Talmud Yerushalmi and Graeco – Roman Culture*, vol. 3, edited by Peter Schäfer (Texte und Studien zum antiken Judentum, 93). Tübingen, Mohr Siebeck, 2002, 55–69.
Sharf, Andrew, "Shabbetai Donnolo as a Byzantine Jewish Figure". In *Jews and other minorities in Byzantium*, edited by Andrew Sharf. Jerusalem, Bar-Ilan University Press, 1995, 160–177.
Skinner, Patricia, "Conflicting Accounts: Negotiating a Jewish Space in Medieval Southern Italy, c. 800–1150 CE". In *Christian Attitudes toward the Jews in the Middle Ages. A Casebook*, edited by Michael Frassetto (Routledge Medieval Casebooks, 37). New York, Routledge, 2007, 1–15.
Tromba, Enrico, *La sinagoga dei giudei in epoca romana. Presenza ebraica a Reggio Calabria e provincia*. Reggio Calabria, Istar, 2001.
Van der Horst, Peter and Judith Newman, *Early Jewish Prayers in Greek*. Berlin/New York, De Gruyter, 2008.
Veltri, Giuseppe, "*Die Novelle* 146 περί 'Εβραίων. Das Verbot des Targumvortrages in Justinians Politik". In *Die Septuaginta zwischen Judentum und Christentum*, edited by Martin Hengel and Anna Schwemer (Wissenschaftliche Untersuchungen zum Neuen Testament, 72). Tübingen, Mohr Siebeck, 1994, 116–130.

Weiss, Zeev and Lee Levine, eds., *From Dura To Sepphoris: Studies in Jewish Art and Society in Late Antiquity* (Journal of Roman Archaeology. Supplementary Series, 40). Portsmouth, Journal of Roman Archaeology, 2000.
Williams, Margareth, "The Use of Alternative Names by Diaspora Jews in Graeco-Roman Antiquity", *Journal for the Study of Judaism* 38 (2007), 307–327.

Adriana Kemp

Community, Illegality and Belonging

Undocumented Migrant Workers and Anti-deportation Campaigns in Israel

> Suddenly, everyone is talking about the expected 'deportation'. Activists are going around the African churches to warn them about a probable deportation in the summer ...others are getting ready to hide children in their homes... anything that does not concern the deportation, has been put on hold.[1]

On June 20, 2009, the head of the newly established Israeli Population, Immigration and Border Authority (PIBA) announced the beginning of an operation for deporting migrants with no legal status. He made it clear that this time – unlike previous deportation operations – the PIBA and its enforcement unit (aka Oz) would detain and deport migrant workers with children living in Israel: "We'll give them one month to get ready and anyone who's illegal will have to leave. Once we detain adults they will have to take the children with them. The children do not protect the parents." This declaration signaled the beginning of what was to become an unprecedented anti-deportation campaign waged under the banner of "Israeli Children."

Unsurprisingly, the campaign mobilized human rights organizations that routinely advocate for migrant workers' rights. But the large-scale of support it amassed from across a broad range of citizens and activists, many of whom had not previously been involved in political action but had a personal acquaintance with migrants and their children, shined the spotlight on two sets of links between community, illegality and belonging that often go unnoticed in both academic scholarship and public discourses.

First, links between illegality and community. Contrary to popularized images of the undocumented migrants as outlaws shunned from the normative and legal community of citizens, the campaign disclosed the web of associations and often submerged social ties that they forge with a variety of individuals and social groups across ethnic and nationality boundaries[2] and not only with their own insular "ethnic" or "migrant" communities.[3] This raises the question, what are the communities in which migrants struggling with precarious socio-legal

1 Diary notes, fieldwork, May 16, 2009.

2 See also Jane Freedman, "The Reseau Education Sans Frontieres", 613–626; Maya Shapiro, "The Development", 423–441.

3 Jose Moya, "Immigrants and Associations", 833–864.

status participate in their everyday life? And how do they reflect on their modes of integration in the broader society?

Second, links between illegality and belonging. The anti-deportation campaign underscored the implications of deportation for how we understand and conceptualize national membership and belonging, and where do undocumented migrants and their families fit in definitions of the national community. Deportation is usually assumed to be the ultimate act of defining membership by establishing in a powerful and definitive way, who has the unconditional right of residence and who hasn't and by extension, who is normatively fit for membership and who is not.[4] However, the Israeli Children anti-deportation campaign called precisely into question who decides about who is (or can be) "Israeli" and on what grounds, pointing that deportation may also serve to highlight contending views within the national community regarding how citizens conceptualize membership and who has the right to determine membership.

Based on a long-term qualitative research on undocumented migrant workers' communities in Israel[5] and on an ethnography of the anti-deportation campaign in Summer 2009[6], the article seeks to explore these questions by underscoring the contradictory ways in which undocumented migrants and their families can be simultaneously included in the national community through various forms of social integration (also if partially and in contested ways) and excluded from it by the law. To do so, I draw on recent sociological and anthropological studies of illegality and migration that emphasize the multidimensional character of migrants' "illegality" as a juridical status, a socio-legal conditions, and a form of liminal belonging.[7]

These studies posit that binary notions of legality as a juridical status or mode of entry are inadequate for apprehending the gray socio-legal areas in which many migrants in democratic countries are situated as they are caught in

---

4 Bridget ANDERSON, Matthew J. GIBNEY and Emanuela PAOLETTI, "Citizenship, Deportation and the Boundaries of Belonging", 548 sq.

5 See Adriana KEMP, Rebeca RAIJMAN, Julia RESNIK, and Silvina SCHAMMAH-GESSER, "Contesting the Limits of Political Participation", 94–119; Adriana KEMP and Rebeca RAIJMAN, "Christian Zionists in the Holy Land", 295–318; Adriana KEMP and Rebeca RAIJMAN, "'Tel Aviv Is Not Foreign to You'", 26–51; Adriana KEMP and Rebeca RAIJMAN, *Workers and Foreigners*; Adriana KEMP and Rebeca RAIJMAN, "Latinos in the Holy Land".

6 Fieldwork of the 2009 anti-deportation campaign was carried out by Nelly Kfir during her PhD research under my supervision. Part of these materials have been published in Adriana KEMP and Nelly KFIR, "Making Migrants' Rights," 82–116; Adriana KEMP and Nelly KFIR, "Wanted Workers but Unwanted Mothers".

7 Nicholas P. DE GENOVA, "Migrant 'Illegality' and Deportability", 419–447; Sarah WILLEN, "Toward a critical phenomenology", 2–7.

ideological divisions and political stalemate over migration control. The concept of "legal liminality" captures the ambiguity of legality in migration by drawing analytical attention to emergent tensions and gaps between the legal and social personhood of undocumented migrants.[8] More specifically, studies of legal liminality focus on how legal status intersects with specific legal and policy arrangements and socio-economic context that illegalize migrants on the one hand[9], and the migratory projects and social actions of migrants themselves as they engage in i/legalizing moves, on the other.[10] Overall, this line of inquiry has shown that immigration policies at the national and local levels matter a great deal in determining the spaces and practices of migrants' socio-legal incorporation; but so does the social agency of migrant communities and of those advocating on their behalf.[11]

I draw on Menjivar's concept of "liminal legality" to capture underlying contradictions between the practical social integration of undocumented migrant workers and their families in Israel, and their official designation as ilegal. I argue that contradictions surrounding the legal liminality of migrants' families resurface most strongly in extreme situations such as deportation because that is when solidarity ties become manifest and because they elicit contending ethical evaluations of migrants' social inclusion, deservingness and legality. As suggested by Anderson, Gibney and Paoletti,[12] the practice of deportation is uniquely helpful in revealing how the legal and the normative boundaries of membership in national communities are not only constructed but also contested.

The article is organized as follows: After introducing the general background of labor migration in the Israeli context (section 1), I analyse the inconsistencies of Israeli policies that create the liminal position of undocumenteted migrant workers, paying particular attention to the situation of families and children (section 2). Section 3 describes the effects of deportation on migrant communities. Section 4 offers an ethnographic analysis of the anti-deportation campaignsthat took place in 2009 showing how the liminal legality of migrant workers' families stir social mobilization and elicit contending moral evaluations among citizens regarding who can be included in society and why. The concluding section analyzes the possibilites and limitations that mobilization struggles mediated by

8 Cecilia MENJIVAR, "Liminal Legality," 999–1037.

9 Kitty CALATIVA, *Immigrants at the Margins.*

10 Leisy ABREGO, "'I Can't Go to College Because I Don't Have Papers'", 212–231; Susan B. COUTIN, *Legalizing Moves*; Roberto GONZALES, "Learning to be illegal", 602–619.

11 Cecilia MENJIVAR and Susan B. COUTIN, "Challenges of Recognition", 325–331.

12 Bridget ANDERSON, Matthew J. GIBNEY and Emanuela PAOLETTI, "Citizenship, Deportation and the Boundaries of Belonging", 548 sq.

personal and communal ties yield for changing the situation of undocumented migrants and the boundaries of membership in Israeli society.

1 Migrant Workers in Israel: General background

Israel has a sizable number of labor migrants who constitute a structural feature of its labor market.[13] Since the beginning of the 1990s, Israel enacted a managed migration scheme for low-skilled migrant workers from overseas to replace Palestinian commuters from the occupied territories who had been working in the Israeli secondary labor market since 1967. Permission to employ migrant workers is limited to the lowest wage occupations in three sectors: construction (workers mainly from Romania, China, Turkey, and the former Soviet Union), agriculture (mainly from Thailand), and long-term care (LTC) (mainly from the Philippines, but also from Sri Lanka, India and Bulgaria). Live-in care has become the largest and fastest growing sector, comprising more than half of all permits, 80 percent of them are granted to women.[14]

As elsewhere, policy-makers recruited migrant workers as a temporary solution for shortages in the local labor force, and the length of their stay was restricted in order to prevent their settlement. As is also often the case, the official recruitment of labor migration brought about an influx of undocumented migrants. Undocumented migrant workers arrive mainly from Eastern Europe, South Asia, Africa, and South America. While most of them enter through the tourist loophole, there is a margin of overlap between the countries of origin of documented and undocumented migrants either because migrants overstay their visa once it expires or because they leave their original employer to whom they are bound (see below). Undocumented workers work in in the same sectors as documented migrant workers but also in other sectors for which there are no permit policies such as housecleaning and services.

In 2002 the government decided to clamp down on the further recruitment of foreign labor and reduce the numbers of undocumented migrants. This policy succeeded in temporarily reducing the overall number of foreign workers, but by 2011 their proportion in the labor market was again on the rise, comprising an estimated 9 percent of the total labor force.[15] Given these figures, Israel ranks among the industrialized economies that rely most heavily on foreign labor.[16]

13 Adriana Kemp, "Labor Migration in Israel".
14 Gilad Natan, "Employment of Foreign Workers".
15 Adriana Kemp and Nelly Kfir, "The Politics of Reform", 535–572.
16 Central Bureau of Statistics, "Labor Migration: Chosen Data".

Israel's economic dependency on low-skilled migrant labor conflicts with its immigration and settlement policies for non-Jews. Israel's immigration policies have two main features: 1) an ethno-national definition of citizenship based on jus-sanguinis. Jewish immigration, or *alyiah* (Hebrew literally: ascent), is ideologically constructed as a 'return' to the homeland and conceived as a natural right of Jews of which the state is only its trustee.[17] The Law of Return (1950) is the legal embodiment of this idea, creating a legal definition of the right of return for Jews and their relatives (up to the third generation) and granting Israeli citizenship immediately upon immigration. Israel's Citizenship Law (1952) complements the right of return. At the same time, the Israeli regime is highly exclusionary towards non-ethnic immigrants and lacks an institutional framework for their incorporation.

Although Israel has not formulated a comprehensive immigration law to regulate the settlement of non-ethnic migrants, the growing influx of non-ethnic immigrants during the 1990s, either in the form of non-Jewish relatives of immigrants from the former Soviet Union who entered Israel under the Law of Return, family reunification of Palestinians from the occupied territories and Arab citizens, or in the form of labor migrations, resulted in the gradual formulation of internal procedures aimed at regulating the new situation. Applications for legal status for non-ethnic immigrants are conducted via the Inter-Ministerial Committee for Humanitarian Affairs (ICHA). Decisions depend on the Ministry of Interior (MOIN)'s discretion or unpublished directives. Procedures have constantly changed throughout the years, and they often differentiate between categories of non-citizens according to the state's interests and prerogatives.

One of the main aspects of the application for legal status is family reunification. Although Israeli law protects the right to have a family and conduct family life, the extent to which that right has to be realized within Israel has been debated in the courts since the mid-1990s, centering mainly on non-Jews applying for legal status on the grounds of marriage. Concerning migrant workers, specific procedures that aim at preventing claims for permanent status on the basis of family formation evolved with the years, resulting, for example, in restrictions on migrants' length of stay especially in the sector of nursing. Most of the migrants who work in this sector are young women who are presumably more likely to be starting a family. In addition to restrictions on the length of stay, policies that aim to prevent family formation among migrant workers include issuing work visas only to migrants who do not have a first-degree family member working in Israel or requiring that one of the partners of migrant couples leave the country.[18] Up

17 Ayelet SHACHAR, "Whose Republic?", 241.
18 Hanny BEN-ISRAEL and Oded FELLER, *No State for Love.*

until recently, the MOIN's 'pregnant foreign workers directive' also revoked the work permits of migrant women who become pregnant.[19] In 2011, the High Court of Justice ruled against the procedure in its original form.

However, the no-family policies and the limited to nil channels for the acquisition of Israeli citizenship have not prevented the creation of families among migrant workers, nor the state's obligation to recognize their children's basic social rights such as education and healthcare.[20] The total number of migrant workers' children is difficult to estimate due to the lack of official data. According to the municipality of Tel-Aviv, where one in eight residents are foreign nationals, in 2012 1,626 minors with foreign passports were registered in the school system, most in the southern neighborhoods of the city.[21] In 2015, the number of children aged 0–6 (not in the school system) increased to 4,500, due to the more recent influx of asylum seekers, mainly from Sudan and Eritrea since 2006.[22]

2 Legal liminality as a socio-legal position

Being "legal" or "illegal" entails not only a legal status but also a social position and form of belonging affirmed in some aspects and denied in others.[23] Ironically, rules regulating the employment and social life of documented migrants in Israel place them systematically at disadvantage compared to undocumented migrants who enjoy greater mobility within the labor market and possibilities for social integration. Migrants without permits are vulnerable to the abuse of informality and deportation policies.[24] However, illegality in the Israeli context also means that migrants are 'invisible' to governmental control effected through the household employment and agencies. Unlike documented migrants that are "bound" to their employers in ways that restrict their mobility and allow latitude for abuse and exploitation,[25] undocumented migrants have more space to nego-

19 MOIN, "Procedure of Treatment".

20 Adriana KEMP and Rebeca RAIJMAN, *Workers and Foreigners.*

21 Gilad NATAN, *Data on Non-Israeli Children.*

22 Author's interview with the Director of the Center of Assistance and Information for the Foreign Community (MESILA), Tel Aviv, 9 June 2015.

23 Cecilia MENJIVAR, "Legal Liminality".

24 Ami GILL and Yossi DAHAN, "Between Neo-Liberalism", 347–386.

25 The Israeli "binding" system that restricts visa portability and catalyzes the creation of "illegalized" workers was declared by the Israeli High Court of Justice (HCJ) as a violation of "the inherent right of liberty" and a form of "modern slavery." HCJ 4542/02, Kav LaOved et al v. The State of Israel (2006) (Hebrew). Since the ruling, the government introduced "lighter" forms of binding and documented migrant workers are now bound to agencies.

tiate the terms of their employment and may leave their employers if they are not satisfied.[26]

Deportation is the main risk facing undocumented migrant workers. However, vulnerability to deportation varies according to the erratic implementation of deportation policies. Thus, while in 2002–2004 massive deportation campaigns terrorized undocumented migrants and broke up their communities,[27] enforcement since has been considerably more lax and focused on those who overstay their work visas or on criminalized African asylum seekers, both of whom are easier to detect than migrants who entered the country without work permits.[28]

At the institutional level, illegality does not mean full exclusion from public services. Indeed, despite the official "no-family" policies, undocumented migrants are caught between the contradictory public policy logics and bureaucracies most relevant for their family life. For example, undocumented migrant women do not have access to the national health insurance. Yet, they are entitled to reproductive health services such as labor and delivery at public hospitals covered by the National Insurance Institute if registered by an Israeli employer, subsidized prenatal care at municipality centers and full preventive medical services for babies until the age of 1–1.5, regardless of their parents' legal status. These provisions include migrant women and their children within the pro-natalist Israeli welfare system despite the explicit no-family policy.[29] Furthermore, while migrant workers' children are undocumented by virtue of their presence in Israel,[30] they are entitled to medical coverage partly subsidized by the state and to compulsory free public education[31]. Thus, whether driven by public health logics protective of citizens or by residence-based definitions of entitlement, these regulations include migrants lacking legal status in the Israeli social protection system in ways that acknowledge their reproductive life.

Illegality also means that the social position of undocumented migrants is shaped by a wider variety of actors operating at different levels. For example, according to data from a municipal aid center for the migrant communities, the majority of their customers are single mothers from the Philippines, with over two years of tenure in the country, only 11% of them with legal status. These

26 For similar cases, see Pei-Chia LAN, "Legal Servitude", 253–277; Marina DE REGT, "Ways to Come", 237–260.

27 Sarah WILLEN, "Toward a critical phenomenology", 2–7; Adriana KEMP and Rebeca RAIJMAN, *Migrants and Workers*; Galia SABAR, *"We are not here to stay"*.

28 Gilad NATAN, *Data on Non-Israeli Children.*

29 Sarah WILLEN, "Birthing 'Invisible' Children", 55–88.

30 The *jus-soli* rule does not apply to migrant workers' children. Therefore, children born in Israel have the same legal status as their parents and at the time of birth as non-resident foreigners.

31 Adriana KEMP and Rebeca RAIJMAN, "'Tel Aviv is not Foreign to You'", 41.

figures are, among other factors, the result of gendered deportation policies that created a large pool of single mothers with children by expelling men in hopes that women and children would follow.[32] Having children enlarges considerably the circles of association and interaction of undocumented migrants with Israelis beyond the worksite. The main foci of interaction with Israelis are in schools in South Tel-Aviv, where most undocumented families live, and with volunteers working in schools such as Bialik-Rogozin where half of the students are migrant workers' children, or in national youth movements which created special troops for migrant children.

As undocumented migrants do not have to take residence in their employers' homes and are not bound up to a single employer, they have also more chances and incentives for creating their own communities. The lack of state regulation of working and living conditions for undocumented migrant workers has been a powerful catalyst for the development of a rich associational life in religious and self-help associations. In the 1990s, three new ethnic communities developed in Israel among migrant workers from West Africa, Latin America and the Philippines.[33] The mismatch between the social capital accrued by undocumented migrants and their families with Israelis and their political and economic marginalization, has led to the formation of what Shapiro called a "privileged underclass" among Filipino migrant women.[34] Similarly, Kalir found that one of the salient characteristics of undocumented migrants from Latin America in Israel, was their ability to achieve "practical accumulation of belonging," namely a form of cumulative symbolic capital that defies the "either-or" logic of formal citizenship and is recognized as legitimate and legitimating by the dominant national community.[35]

Albeit partial and segmented, these forms of social integration accorded undocumented migrants a social personhood anchored on everyday cooperation and interpersonal networks of solidarity which proved crucial during the anti-deportation campaigns that led to the naturalization of children and their families in 2005 and 2010.[36]

---

32 Mesila Annual Report 2011.

33 By 1996, undocumented Latino and African migrants created the largest communities of undocumented migrants, comprising about 15 and 14 percent of the total undocumented labor migrant population respectively. Adriana KEMP and Rebeca RAIJMAN, *Workers and Foreigners*. The Filipino community includes migrants with papers and those that lost their papers either by leaving their employers or overstaying their visas. Claudia LIEBELT, *Caring for the "Holy Land"*.

34 Maya SHAPIRO, "The Development", 423–441.

35 Barak KALIR, *Latino Migrants in the Jewish State*, 13.

36 Adriana KEMP, "Managing Migration"; Adriana KEMP and Nelly KFIR, "Wanted Workers but Unwanted Mothers".

3 Deportation and its effects on migrant communities

In 2002, the Immigration Police was established with the official aim of enforcing the control over the entrance and stay of foreign residents and reducing the number of undocumented migrants. To accomplish this goal, the government engaged in massive deportations and promoted an agenda that stigmatized foreigners as a social and national threat.[37] In the years following its creation, 40,000 out of the 118,105 foreigners that left Israel, were deported.[38]

The Immigration Police represented a turning point not only in the scope of deportation but also in its target: families and communities. Until then, politicians in office attempted to arouse a "moral panic" about the demographic threat posed by undocumented migrants that create families and settle down in Israel. As the Minister of Interior, who was otherwise responsible for the growing numbers of permits issued to migrant workers, explained, "They [undocumented migrants] have to be deported before they make children". Nevertheless, despite such declarations, deportation campaigns focused on men under the wrong assumption that once they would be deported, the rest of the family would follow. As soon as the Immigration Police was formed, it declared that it would deport families and crack down on whole communities.

Although authorities did not follow through on these threats, they had a significant effect on the communities who now felt under siege. Being "illegal" became a permanent burden affecting migrants' everyday lives, and it constituted a recurrent theme in community gatherings. At the time of fieldwork in the Latino community, many of the members of the community shared with us their consternation regarding their illegal status.

> I live with tension. We never have a moment of peace. We are not free to act and interact as normal people. People take advantage of this situation. Take for instance the questions of renting an apartment. You know that they charge you more just because you are illegal and you cannot say a word. I already left two apartments and my money guarantee was not returned. Again, I couldn't complain. If you need a medicine, you cannot get it because illegal workers don't have medical insurance. Most employers do not want to pay for that (Viviana, Oct. 2003).[39]

---

37 Adriana KEMP and Nelly KFIRK, "The Politics of Reforms".

38 Roni BAR-TZURI, *Foreign workers without permits.*

39 Heretofore, I use pseudonyms.

Suddenly home and street turned into dangerous places. Fearing arrest and deportation, migrants avoid gathering in public places. Police raids targeting the old and new central bus station area in Tel Aviv led many migrants to move out to the surrounding areas with a lower concentration of foreign workers, hoping to reduce the chances of being caught. In other cases, women working as live-out domestics considered the possibility of taking a live-in position as a strategy to avoid being arrested on their way to work. As rising numbers were detained and deported many decided to leave the country of their "own accord."[40]

Deportation campaigns were particularly hard on families. As parents realized that they were not immune to deportation and that integration into Israeli society was not a viable option for non-Jewish migrants, they began thinking about the moment of returning home to their countries of origin. During the campaign, approximately 1,000 families, mainly from Gahna, agreed to leave the country in "voluntary repatriation" campaigns organized by the Ministry of Interior.[41]

In 1999, preparing children for return led to the creation of *La Escuelita,* a Sunday school-like framework operated by Latino migrants and Israeli volunteers. La Escuelita was the initiative of Cristina, a Colombian migrant worker who run an "underground" kindergarten for undocumented Latino children. With the aid of a municipal organization, her project was materialized and it soon became a focus of cooperation between volunteers from the Latino community of undocumented migrants and Israeli Jews of Latin American origins. Its main original objective was to preserve the cultural heritage of Latinos' children until they returned to their (parents') countries. But this objective was to be changed after the first legalization campaign in 2005 that granted legal status to many of these children and their parents. Then, the Escuelita became a center for helping children and adults of the Latino community to assimilate in Israeli culture and society and meet the needs of those that remained.[42]

---

40 See Rebeca RAIJMAN, Silvina SCHAMMAH and Adriana KEMP, "International Migration".

41 Adriana KEMP and Rebeca RAIJMAN, *Workers and Migrants*, 135.

42 Today the target groups of La Escuelita's activities are mainly female migrant workers from Colombia, Ecuador, Peru, Bolivia and Puerto Rico. With respect to children, the emphasis is on offering tutoring as a compliment to the regular school's educational agenda. Unlike their parents, who define themselves by country of origin and take pride in it, the children, many of them Israeli-born, define themselves as Israelis and are less interested in their parents' culture of origin. The staff composition has also changed: presently the majority of volunteers and staff are Israeli-born Jews of Latin American descent; there is no representative of the migrant workers' community. See Yvonne LERNER, "Two Latino NGOS in Israel".

4 Legalization campaigns and the negotiation of membership

Official attempts at deporting families during the 2000s drove NGOs to shift their advocacy strategies from litigation in the courts to lobbying campaigns among policy-makers. The campaign eventually resulted in a government decision in 2005[43] to legalize 562 families of the 862 requests that were filed.[44] According to the decision, children who met carefully drafted criteria were granted permanent residence and citizenship upon enlisting in the Israeli army, and their siblings and parents were granted temporary residence that would eventually turn into permanent residence. The decision was defined as a one-time temporary arrangement that did not change governmental policies. Children who were not eligible remained either deportable or had their case referred to the ICHA in charge of examining humanitarian requests.[45]

Threats to deport children and families recurred in 2009 when the newly established Population, Immigration and Border Authority (PIBA) announced an operation for deporting undocumented migrants. Officials made clear that this time "The children do not protect the parents".[46] This declaration signaled the beginning of an unprecedented public campaign that took on the streets to prevent the deportation of what activists said were "Israeli children."

Drawing on fieldwork at the time of the campaign during summer 2009, I single out three major components of these efforts: constituency, the creation of consensus, and the ethical frames invoked for negotiating the limits of national membership. These components underscore the two main topics of this article: first, the significance of legal liminality as a form of belonging embedded in daily relations and the accumulation of social capital with local communities; second, the ways in which deportation, as an extreme form of social and physical banishment from the national community, generates conflicts amongst citizens and between citizens and the state over the question of who is or can be part of the normative community of members and why.

The anti-deportation campaign crystallized during the summer of 2009 and lasted until December 2010. From the moment the NGOs heard about the plan to deport undocumented migrants as of July 1st, they began organizing against it. NGOs' staff members reached out to migrants in community gatherings, most prominently during religious ceremonies in churches, to inform them about

43 The Government of Israel. Decision # 3807, 27 June, 2005.

44 Gilad NATAN, *Issues Regarding Foreign Workers' Families*, 2.

45 Noa YACHOT, *The Limits of Discretion*.

46 Nurit WURGAFT, "The Ministry of Interior's New Authority Starts Working".

future deportation plans. However, the key actors in the anti-deportation campaign were independent activists who had a prior connection with migrant families.

Activists gradually created a network called Israeli Children, which also included migrants themselves. Inbar, one of its organizers, arrived from long-term work with mothers and children at the municipal Mesila. Neta, who was familiar with migrants' families from her prior volunteer work at Physicians for Human Rights, said that one of her biggest fears in front of the deportation plans was her "post-trauma" after the 2003–2004 massive deportations when people whom she knew "disappeared." She also mentioned that living in south Tel-Aviv she witnessed arrests of her neighborhors and felt she could not be a bystander.

Migrants whose families received legal status in 2005 and could now risk standing beside undocumented fellows were also active participants. Bridget, a Filipino migrant whose child received legal status, said that she was called to join the campaign through her volunteer work with undocumented families in Mesila's Moms' Committee. Although she was afraid to "be exposed and lose her privacy," she explained that she wanted to help other mothers whose situation she knew well.

Initially, supporters included youth leaders and educators working with migrants' children, and students, artists and public figures involved with migrant communities through volunteer projects. They stood alongside migrant mothers and their children in the front lines of the demonstrations holding signs in English, Hebrew and Spanish saying: "Let our children be" or "No child is illegal".[47] They wrote petitions to the Minister of Interior expressing their fervent objection to the deportation of migrant workers and their children whose arrival to the country was "not part of a tourist trip... they were born and raised in the country and want to be citizens of it." Over 15,000 people signed the petition that was distributed in the social media.[48] Dozens of activists from Israeli Children patrolled the streets in South Tel Aviv armed with cameras to deter immigration inspectors from detaining families. Shai and his girlfriend explained that as many of the other volunteers, they were not political activists and they heard about the campaign from friends. "I guess we have to thank Eli Yishay [the former Minister of Interior, A.K.] for making an activist of me," he explained, "until now I was the type that sat at home and just got furious".[49]

Relations anchored in everyday interactions with migrants and sentiments of moral shock and indignation constitute powerful drivers for mobilizing sup-

47 Ofri Ilani, "Hundreds of People Demonstrated in Tel Aviv".
48 See http://www.atzuma.co.il/zarim.
49 See http://www.haaretz.co.il/news/education/1.1223599.

porters regardless of their ideological inclinations.[50] As the campaign developed, their supporters grew numerically – the largest demonstration comprised between 8,000–10,000 people – and in diversity. This depended to a large extent on how the campaign was strategized.

From the outset, activists debated about the wisdom of challenging the principles of the no-family policy or seeking broader public consensus. Ultimately, the latter prevailed. Israeli Children decided to concentrate on children rather than parents and accord them visibility. Previous studies show that focusing on a vulnerable group such as children is crucial for campaigns aiming to garner wider public support.[51] To accomplish this goal, activists hung thousands of colorful posters all over the city, each of them carrying a picture of a child designated for deportation and the word "Deported." Inbar explained the idea behind it:

> A poster of a child seen by every person in this cafe creates a buzz...there's a photo here of a child suddenly showing you their face. "They deport children," that's a very remote image...and that's why, instead of them being just numbers, you get to see their faces and understand what's going to happen to them.

Recruiting the media was crucial for the success of the campaign despite the danger of exposing the children eligible for deportation. At first, the activists were the ones who called the reporters, but as Inbar explained:

> ...at a certain point there was a change. When they came and took pictures of the children at school, suddenly an argument started of who knows more of Bialik's poems [Israel's national poet, Authors]...the reporters were shocked ...and from that moment on, my cell phone bill increased to three thousands NIS per month, just for talking with reporters...

Throughout the campaign the media actively participated in mediating the children's story. Nationwide newspapers followed and published new developments. Reporters maintained close contact with the campaign organizers and were called by activists to cover the raids and arrests of the migrants both as a means of deterring the inspectors and shaming them. Children standing in front of the cameras and telling their personal stories became a common feature of the media coverage, and activists who knew the families personally carefully guided them in speaking publicly. According to Bridget, raising the profile of candidates for deportation helped "Israelis understand the life of migrant workers here and their experience and what it really means to have children born here, deported."

---

50 Jane FREEMAN, "The Reseau Education Sans Frontieres", 613–626.
51 Lindsay O'DELL, "Representations of the 'Damaged' Child'", 383–392.

Winning the support of politicians from various political parties was another way of strategizing the representation of the cause. Given that Israeli Children consisted of activists with a broad spectrum of political identifications, they disagreed about the desirable boundaries of consensus. From Inbar's perspective:

> ...many of our activists, they come back from a demonstration against the Wall [the wall separating Israel from the Palestinian Authority, A.K.] and go to an anti-deportation demonstration. And for me this was the greatest danger to the struggle. From the beginning I made it clear that we are not a radical left-wing organization, that although this is a political question on a policy issue, it is not partisan... the radical left automatically creates antagonism.

Disagreements were generally resolved by strategizing the campaign as non-partisan on behalf of the children. Activists with more radical left-wing identifications did not express their views in official statements but rather individually or in personal blogs and social media. The primary lobbying goal was to recruit politicians in office and public figures to support the struggle publicly. According to Inbar:

> ...I had two people I decided I wanted to reach from the first day of the struggle who are high in consensus...the Minister of Education and the President. I personally turned to the President, which was quite a success, because he wrote a letter calling on the Prime Minister to abort the deportation.

As consensus and media exposure grew around the children, so did the circle of supporters from across the political spectrum. At the same time, NGOs' staff members lobbied ceaselessly in the Knesset [Israeli Parliament]: they called for emergency sessions, filed legal motions with Knesset members, and invited migrants and their children to the Knesset sessions. Since the debate concerned deporting children, the first session took place in the Children's Rights Committee rather than in the usual KCFW. The Committee was headed by a right-wing coalition Knesset member (Likud) who, together with a Knesset member from the left (Meretz), later filed a private motion against detaining children.[52] The proposal was written with the help of lawyers from the human rights NGOs. The Minister of Education at the time (Likud) also filed a motion against deporting children and for creating orderly criteria for granting legal status. Michal, from the Hotline for Migrant Workers, explained that the politicians who joined the struggle understood that their support did not require a more principled political stance on Israel's labor migration policies and that therefore, they will not pay an electoral price for their support. She pointed to the process in which most of the government's ministers took stands against the deportation, which were

52 Michael BENGAL and Michal GRINBERG, "Instead a decision, another committee".

weakly linked to their positions regarding migration and matters of citizenship. "It became the consensus," she explained, "and no one wants to stay outside the circle." However, attempts at achieving consensus across the board also shaped the framing of claims about the legitimacy of the children.

Framing claims reveals the content of moral reasoning – what principles are justifiable and on what moral grounds – and the process, because the framing of claims changes as the struggle progresses.[53] From the beginning Israeli Children stressed the fact that children "were born here and are Israeli," arguing that deportation would drive them into "cultural and social exile". This reasoning was based to a great extent on the manner in which the children themselves presented their desire to remain in Israel. For example, a seven-year-old whose parents came from Colombia said: "All my friends are Israeli and I live in Tel Aviv...I also do not speak any other language, only Hebrew." Similar claims about the center of one's life as grounds for legalization had proved unsuccessful in the past when advanced in courts on behalf of adult caretakers, but they seemed to work in relation to their children.

However, the activists and veteran NGOs who participated in the anti-deportation coalition also tried to establish a broader framing, linking claims about "cultural belonging" with demands for a "just migration policy" and putting an end to the exploitative "revolving door" of recruitment-deportation that creates illegality.[54] Some activists feared that invoking notions of "justice" and "immigration" would backfire and that third parties would change their meaning as they saw fit. This concern was justified.

For example, in advocating for a "'just migration policy" a well-known publicist differentiated between the moral obligation owed to migrants' children and the "immoral" demand of "bogus African refugees" migrating for "economic reasons".[55] Others criticized the activists as hypocrites for opposing the deportation of non-Jews while not objecting in the past to the "deportation of Jews from Gush Katif" (the dismantling of Jewish settlements in the Gaza Strip in 2005).[56] In response, the activists coined the slogan "Jews do not deport children," a paraphrase of Gush Katif's slogan "Jews do not deport Jews."

As the campaign evolved, claims about "justice" slowly dropped away, and the frame of "cultural belonging" prevailed in the wider circles of supporters. On one hand, this framing narrowed the principled demands for overhauling the policy about migration. On the other hand, it allowed for new interpretations

53 Robert D. BENFORD and David A. SNOW, "Framing Processes", 611–639.

54 Israeli Children, "Petition Against Deportation".

55 Eithan HABER, "There is No Free Entrance".

56 Michael FEIGLIN, "I Prefer the children of this country".

of the legitimate grounds for belonging and ultimately, for what it means to be Israeli. For example, President Shimon Peres sent a public letter to the Minister of Interior and the Prime Minister maintaining that the children belonged in Israel and emphasizing the children's "deep connection and love for Israel and their desire [...] to serve in its army".[57] Under these moral arguments, objections to the deportation of children and families turned into a claim for their naturalization not as a right but as a form of recognition based on justificatory rationales widely resonant in Israeli hegemonic discourses.

Conversely, petitions and declarations by educators, physicians, lawyers and social workers' associations emphasized their own Jewish and democratic principles based on the tradition of the humane treatment of foreigners, justice and charity, and the biblical command for Israel to serve as "a light unto the nations" as the basis for caring about the children.[58] The demonstrators also invoked frames related to the history of the Jewish people and their moral obligation to the foreigner, holding placards that read: "Our parents too came here as refugees" or "We are all immigrants." Another common claim was based on the "lessons of the Holocaust," whose meaning in the Israeli context is multivalent. It can be invoked as a particularistic claim to preserve Israel as a refuge of and for the Jewish people, but also as a universal lesson against racism and xenophobia. Activists and sympathizers with the campaign such as the association of Holocaust Survivors in Israel stressed the latter, arguing that the Jewish people must stand up against deportation.[59]

As a result, the campaign spiraled from a campaign to "protect the children" from immigration policies into a tribal debate about belonging, enhancing on one hand internal social divisions over the meaning of Israeliness while simultaneously re-affirming its ethno-national boundaries.

5 Conclusions

> The act of expulsion simultaneously rids the state of an unwanted individual and affirms the political community's idealised view of what membership should (or should not) mean. Deportation thus shows the citizenry not simply as a community of law, but also as a community of value.[60]

57 Ynet, Letter from Shimon Peres.

58 Yael BARNOVSKY, "NGOs to Government Ministers".

59 ID., "Holocaust Survivors Against the Deportation".

60 Bridget ANDERSON, Matthew J. GIBNEY and Emanuela PAOLETTI, "Citizenship, Deportation and the Boundaries of Belonging", 548.

The article showed the contradictory ways in which migrants and their families can be simultaneously integrated in the community of value (also if partially and in fragmentared ways) and excluded from the community of law. Drawing on Menjivar's concept of "legal liminality", I explored the ways in which migrants in tenouos socio-legal status participate in community life. Different from what is usually emphasized by scholarship on migrants' associational life, they do so not only through the creation of their own "ethnic" associations and self-support networks that compensate for their vulnerable status and unfulfilled social, spiritual, and symbolic needs.[61] But also through the personal interactions that they forge with citizens and local communities in a variety of formal and informal settings – school and after school activities, neighborhood, welfare, worksites, and others – as they go on their daily struggles against precariousness. Although asymmetrical and often weak, social ties that reach out across ethnic or nationality boundaries with a variety of local actors contribute to their practical integration in the social fabric of local communities. As the Israeli case shows, they can also prove critical for mobilizing support and solidarity.

I argued that the significance of the practical forms of social belonging and social capital that migrants accrue, comes out strongest in points of collition and times of crisis. The practice of deportation, which has become a standard tool in the practices of states in their dealings with unwanted migrants,[62] is uniquely helpful in revealing not only how the legal and the normative boundaries of membership are constructed but also how they are contested. Contestations are a key feature of the many local anti-deportation campaigns that currently operate in support of individuals and families facing expulsion in liberal democratic states.[63] Indeed, although often used by governmental elites as a way of reaffirming the shared significance of citizenship, deportation accentuates the multidimensional character of migrants' "illegality" as a contradictory form of belonging where the legal (lack of) membership of migrants does not necessarily match their social standing in society.[64]

Nevertheless, as the analysis in the previous sections showed, the politics of membership that emerge during anti deportation campaigns is not unconditional or without limitations. First, appeals to recognize the human lives and needs of migrants' families are deeply influenced by how state policies and regulations shape the socio-legal position of undocumented within society to begin with.

61 Jose MOYA, "Immigrants and Associations", 833–864.

62 Matthew J. GIBNEY, "Asylum and the expansion of deportation", 139–143.

63 Jane FREEMAN, "The Reseau Education Sans Frontieres", 613–626; Barbara LAUBENTHAL, "The Emergence of Pro-Regularization Movements", 101–133.

64 Susan B. COUTIN, *Legalizing Moves.*

As we saw, the tacit acceptance of migrants' presence as cheap and disciplined labor; contradictory governmental decisions and logics of governance operating within the state that render migrants and their children invisible in some cases but not in others; the stricter control over documented migrants that place those that are undocumented in better employment and living conditions – all these, among others, point at the institutional and structural forces that create the legal liminality of migrants. Moreover, they often have a vested interest in maintaining it that way.

Second, the nature of the politics of membership in anti-deportation campaigns depends to a great extent on the ways in which claims are strategized. The slippage from "adults" to "children" that took place throughout the campaign, is not accidental, for several reasons.

First, it depends on which moral claims are perceived by activists as justifiable and which are not. Calling for the legalization of a limited number of children who are not held responsible for their parents' deeds in Israel, was found defendable, meaning, consensual, whereas demanding a just immigration policy that does not create undocumented families, was not. This type of strategizing relies on the idea that children are "ideal" vulnerable subjects and therefore emotional and highly personalized calls for protecting them are likely to garner wider public support for humane solutions.[65] Yet, while personalized and consensus-based mobilizations may prevent particular acts of deportation, like they did in Israel, they often result in the depolitization of the factors leading to them through their "humanitarianization"[66] or through a politics of consensus that silences challenging voices and bolster hegemonic policies.[67]

Third, the politics of membership underscored by anti-deportation campaigns generates, perhaps unwittingly, hierarchies of moral deservedness and inclusion in the national community. Raising the profile of the children and framing demands in terms of their being Israeli created public pressure that resulted in a legalization arrangement, on the one hand. But it also considerably narrowed the meaning of Israeliness to hegemonic understandings of it, on the other. Based on "humanitarian" and "Zionist" arguments, and demanding migrants to prove their cultural and social assimilation in the host society by speaking Hebrew, serving in the army, and looking and doing "Israeli things," on August 1, 2010, the government announced a one-time decision granting legal status to children who meet several criteria.[68] Since this decision, 636 applications were filed by

65 Nando SIGONA and Vanessa HUGHES, *No Way Out, No Way In.*

66 Jane FREEDMAN, "The Reseau Education Sans Frontieres", 613–626.

67 Peter NYERS, "The Politics of Protection", 1069–1093.

68 The same as in 2005. See The Government of Israel. Decision # 2183, 5 August, 2010.

families; as of 2013, 259 were accepted, 60 rejected and the rest are still in processing.[69] Children who did not meet the criteria became deportable.[70] For the activists in Israeli Children, the arrangement not only excluded those who did not meet the criteria but its justifications also reinforced the perception of migrants' children as humanitarian exceptions from the non-immigration regime. Meanwhile, Israeli Children formalized as an NGO and continues advocating for the needs of children whose situation has not been regularized and for new deportable children. As Alon, an active staff member of a human rights NGO, succinctly summarized: "As long as there are migrants, there would be illegal children."[71]

Bibliography

Abrego, Leisy, "'I Can't Go to College Because I Don't Have Papers': Incorporation Patterns of Latino Undocumented Youth", *Latino Studies* 4 (2006), 212–231.
ACRI, "The High Court Failed to Uphold Basic Human Rights" January 12, 2014. http://www.acri.org.il/en/2012/01/12/citizenship-law-petitions-rejected/ (last accessed on June 21, 2014).
Anderson, Bridget, Matthew J. Gibney, and Emanuela Paoletti, "Citizenship, Deportation and the Boundaries of Belonging", *Citizenship Studies* 15 (2011), 547–563.
Bar-Tzuri, Roni, *Foreign workers without permits that were deported in 2008*. Jerusalem, Research and Economy Department, Ministry of Industry, Trade and Labor, 2009 [Hebrew].
Barnovsky, Yael, "NGOs to Government Ministers: It is Inhumane to Deport Children," Ynet, July 30, 2009a. http://www.ynet.co.il/articles/1,7340,L-3754270,00.html (accessed June 10, 2014).
Barnovsky, Yael, "Holocaust Survivors Against the Deportation of Migrant Workers' Children: This is the Lesson from the Past," Ynet, October 14, 2009b. http://www.ynet.co.il/articles/0,7340,L-3789996,00.html (accessed June 10, 2014).
Ben-Israel, Hanny and Oded Feller, *No State for Love*. Tel Aviv, ACRI, 2006. http://www.acri.org.il/he/wp-content/uploads/2011/03/NoStateForLoveEng.pdf (accessed June 10, 2014).
Benford, Robert D. and David A. Snow,"Framing Processes and Social Movements: An Overview and Assessment", *Annual Review of Sociology* 26 (2000), 611–639.
Bengal, Michael and Michal Grinberg, "Instead a decision, another committee", Ma'ariv, November 1, 2009.
Calavita, Kitty, *Immigrants at the Margins: Law, Race and Exclusion in Southern Europe*. Cambridge, Cambridge University Press, 2005.
Central Bureau of Statistics, "Labor migration: Chosen data" (2012).
http://www.cbs.gov.il/www/publications/alia/t2.pdf (accessed 22 October, 2014).

69 ACRI, "The High Court Failed to Uphold Basic Human Rights".
70 According to Israeli Children's estimations over 300 families who did not meet the criteria were deported since 2011. Personal correspondence with Neve and Inbar, December 2013.
71 Interview, ACRI, Tel-Aviv, December 25, 2011.

Coutin, Susan B., *Legalizing Moves: Salvadoran Immigrants' Struggle for U.S. Residency.* Michigan, The University of Michigan Press, 2000.
De Genova, Nicholas P., "Migrant 'Illegality' and Deportability in Everyday Life", *Annual Review of Anthropology* 31 (2002), 419–447.
De Genova, Nicholas P., "The Deportation Regime: Sovereignty, Space and the Freedom of Movement". In *The Deportation Regime: Sovereignty, Space and the Freedom of Movement*, edited by Nicholas De Genova and Nathalie Peutz. Durham, Duke University Press, 2010.
De Regt, Marina, "Ways to Come, Ways to Leave: Gender, Mobility, and Il/legality among Ethiopian Domestic Workers in Yemen", *Gender & Society* 24 (2010), 237–260.
Feiglin, Michael, "I Prefer the children of this country, the children of refugees and their parents should be deported", NRG, August 2, 2010 http://www.nrg.co.il/online/1/ART1/924/659.html (accessed 5 May, 2010).
Freedman, Jane, "The Reseau Education Sans Frontieres: Reframing the Campaign Against the Deportation of Migrants", *Citizenship Studies* 15 (2011), 613–626.
Gibney, Matthew, J., "Asylum and the expansion of deportation in the United Kingdom", *Government and opposition* 43 (2008), 139–143.
Gill, Ami and Yossi Dahan, "Between Neo-Liberalism and Ethno-Nationalism: Theory, Policy, and Law in the Deportation of Migrant Workers in Israel", *Law and Government in Israel* 10 (2006), 347–386 [Hebrew].
Gonzales, Roberto G., "Learning to be illegal: undocumented youth and shifting legal contexts in the transition to adulthood", *American Sociological Review* 76 (2011), 602–619.
Haber, Eithan, "There is no free entrance," Yediot Achronot, August 2, 2009.
Ilani, Ofri, "Hundreds of people demonstrated in Tel Aviv against the new Immigration Authority", Ha'aretz, July 4, 2009 http://www.haaretz.co.il/hasite/spages/1097691.html.
Israeli Children, "Petition against the deportation," Ha'aretz, July 28, 2009.
Kalir, Barak, *Latino Migrants in the Jewish State. Undocumented Lives in Israel.* Indiana, Indiana University Press, 2010.
Kemp, Adriana, "Managing Migration, reprioritizing National Citizenship: Undocumented Labor Migrants' Children and Policy Reforms in Israel", *Theoretical Inquiries in Law* 8 (2007), 663–691.
Kemp, Adriana, "Labor Migration in Israel: Overview", *Social Policy, Employment and Migration* 103 (2010).
Kemp, Adriana and Nelly Kfir, "The Politics of Reform and the Construction of a Social Problem: Labor Migration Trends in Israel During the 2000s", *Megamot* 54 (2012), 535–572 [Hebrew].
Kemp, Adriana and Nelly Kfir, "Making Migrants' Rights in 'Non-Immigration' Regimes: Ethnography of Labor Migrants' Rights Activism in Israel and Singapore", *Law and Society Review* 50 (2016), 82–116.
Kemp, Adriana and Nelly Kfir, "Wanted Workers but Unwanted Mothers Mobilizing Moral Claims on Migrant Care-Workers' Families in Israel", *Social Problems* (forthcoming).
Kemp, Adriana, Rebeca Raijman, Julia Resnik, and Silvina Schammah-Gesser, "Contesting the Limits of Political Participation: Latinos and Black African Migrant Workers in Israel", *Ethnic and Racial Studies* 23 (2000), 94–119.
Kemp, Adriana and Rebeca Raijman, "Christian Zionists in the Holy Land Evangelical Churches, Labor Migrants and the Jewish State", *Identities: Global Studies in Culture and Power* 10 (2003), 295–318.

Kemp, Adriana and Rebeca Raijman, "'Tel Aviv Is Not Foreign to You': Urban Incorporation Policy on Labor Migrants in Israel", *International Migration Review* 38 (2004), 26–51.
Kemp, Adriana and Rebeca Raijman, *Workers and Foreigners: The Political Economy of Labor Migration in Israel.* Tel Aviv, Van-Leer Institute of Jerusalem and Kibbutz Hamehuhad, 2008 [Hebrew].
Kemp, Adriana and Rebeca Raijman, "Latinos in the Holy Land: The Making and Unmaking of an Undocumented Community". In *Global Latinos: Latin American Diasporas and Regional Migrations*, edited by Mark Overmyer-Velázquez and Enrique Sepulveda. Oxford University Press, forthcoming.
Lan, Pei-Chia, "Legal Servitude, Free Illegality: Migrant 'Guest' Workers in Taiwan". In *Asian Diasporas: New Conceptions, New Frameworks,* edited by Rhacel Parrenas and Lok Siu. Stanford, Stanford University Press, 2007, 253–277.
Laubenthal, Barbara, "The Emergence of Pro-Regularization Movements in Western Europe", *International Migration* 45 (2007), 101–133.
Lerner, Ivonne, *Two Latino NGOs in Israel: Languages and Identities*. Unpublished manuscript, Tel Aviv University, 2014.
Liebelt, Claudia, *Caring for the "Holy Land": Filipina domestic workers in Israel.* New York, Berghahn Books, 2011.
Menjivar, Cecilia, "Liminal Legality: Salvadoran and Guatemalan Immigrants' Lives in the United States", *American Journal of Sociology* 111 (2006), 999–1037.
Menjivar, Cecilia and Susan B. Coutin, "Challenges of Recognition, Participation, and Representation for the Legally Liminal: A Comment". In *Migration, Gender and Social Justice Perspectives on Human Insecurity*, edited by Thanh-Dam Truong, Des Gasper, Jeff Handmaker and Sylvia I. Bergh. Berlin, Springer, 2014, 325–331.
Mesila Annual Report 2011. Tel-Aviv-Yafo Municipality.
MOIN (Ministry of Interior), Procedure of Treatment of Pregnant Foreign Worker that Give Birth in Israel, 2009. http://www.moin.gov.il/Apps/PubWebSite/publications.nsf/All/B96025D3EC6D1EA9422570AD00463773/$FILE/Publications.3.0023-1.8.2009.pdf?OpenElement (accessed on 2009).
Moya, Jose C. "Immigrants and Associations: A Global and Historical Perspective", *Journal of Ethnic and Migration Studies* 31 (2005), 833–864.
Natan, Gilad, *Employment of Foreign Workers in the Caregiving Sector Through Corporations.* Jerusalem, Knesset Research and Information Center, 2010.
Natan, Gilad, *Data on Non-Israeli Children in the Tel Aviv-Jaffa School System*. Jerusalem, Knesset Committee on the Problem of Foreign Workers, June 28, 2012a.
Natan, Gilad. *Issues Regarding Foreign Workers' Families awaiting Legal Status through their Children*. Jerusalem, Knesset Committee on the Problem of Foreign Workers, January, 2012b.
Nyers, Peter, "The Politics of Protection in the Anti-Deportation movement", *Third World Quarterly* 26 (2003), 1069–1093.
O'Dell, Lindsay, "Representations of the 'Damaged' Child: 'Child Saving' in a British Children's Charity Ad Campaign", *Children & Society* 22 (2008), 383–392.
Raijman, Rebeca, Silvina Schammah and Adriana Kemp, "International Migration, Domestic and Care Work: Undocumented Latina Migrant Women in Israel", *Gender & Society* 17 (2003), 727–749.
Sabar, Galia, *"We are not here to stay." African migrant workers in Israel and back in Africa.* Tel-Aviv, Tel-Aviv University Press, 2008 [Hebrew].

Shachar, Ayelet, "Whose Republic? Citizenship and Membership in the Israeli Polity", *Georgetown Immigration Law Journal* 13 (1999), 233–272.
Shapiro, Maya, "The Development of a 'Privileged Underclass': Locating Undocumented Migrant Women and Their Children in The Political Economy of Tel-Aviv, Israel", *Dialectical Anthropology* 37 (2013), 423–441.
Sigona, Nando and Vanessa Hughes, *No Way Out, No Way In: Irregular Migrant Children and Families in the UK.* Oxford, COMPAS, 2012.
The Government of Israel. Decision # 3807, 27 June, 2005.
The Government of Israel. Decision # 2183, 5 August, 20
Willen, Sarah, "Birthing 'Invisible' Children: State Power, NGO Activism, and Reproductive Health among Undocumented Migrant Workers in Tel-Aviv, Israel", *Journal of Middle East Women's Studies* 1 (2005), 55–88.
Willen, Sarah, "Toward a critical phenomenology of 'illegality': State power, criminalization, and abjectivity among undocumented migrant workers in Tel-Aviv, Israel", *International Migration* 45 (2007), 2–7.
Wurgaft, Nurit, "The Ministry of Interior's New Authority Starts Working", *Ha'aretz*,
June 19, 2009 http://192.118.73.5/hasite/spages/1093960.html?more=1 (accessed
June 10, 2014).
Yachot, Noa, *The Limits of Discretion: Humanitarian Immigration in the Jewish State.* M.A. thesis, Department of Sociology and Anthropology, Tel-Aviv University, 2011.
Ynet, Letter from Shimon Peres to Eli Yishai, 30 July, 2009 http://my.ynet.co.il/pic/news/30072009/1.pdf (accessed August, 2009).

Omar Ribeiro Thomaz

The Passage of Time and the Permanence of Fear

Long-Lasting Tensions in Contexts of Conflict in Mozambique

I

In recent years, the possibility of war has been haunting Mozambique. More than two decades after the peace agreement signed between FRELIMO and RENAMO to put an end to a violent civil war,[1] hostilities between armed groups and inflammatory political speeches are creating once again an atmosphere of constant fear.[2] The memories of past conflicts and the imminence of a new war in the country

1 The conventional chronology assumes that the civil war – rarely mentioned in these terms – began in 1977 and ended with the General Peace Agreement signed between the two belligerent forces in October 1992 in Rome. Hence the recent popularization, largely promoted by the government, around the term "16-year war". In fact, there has always been a great deal of resistance by the FRELIMO, the ruling party since the independence, to refer to the conflict as a "civil war". Thus, in the early years of its rule, the name favored by the ruling party was "war of destabilization", emphasizing the aggression waged by white minority regimes, initially Rhodesia and then South Africa during the apartheid, and therefore the external causes of the war. Given the clear territorialization and generalization of the conflict in the 1980s, FRELIMO made a concerted effort to obfuscate the consolidation of a warring force, with which it could negotiate any sort of settlement, and preferred to apply to their contenders the designation of "armed bandits". In the final years of the conflict, the acronym RENAMO was more widely adopted and its leader was eventually accepted as the counterpart for peace negotiations. Between late 1970s and early 1990s, a debate on the causes of the war among anthropologists, sociologists, political scientists and journalists polarized those who blamed it entirely on the hostile action of white minority governments of neighboring countries and those who perceived the war as a result of the failure of policies adopted by the FRELIMO party in the wake and in the name of the revolution. Cf. Bertil EGERÖ, *Mozambique: a dream undone*; Paul FAUVET, "Roots of counter-revolution"; William FINNEGAN, *A Complicated War*; Christian GEFFRAY, *La cause des armes*; Joseph HANLON, *Mozambique: the revolution under fire*; ID., *Mozambique: who calls the shots?*; Lina MAGAIA, *Dumba Nengue*; Willian MINTER, *Apartheid's Contras*; Alex VINES, *Renamo: terrorism in Mozambique.*

2 Nearly as a harbinger of history repeating itself, recent conflicts have been heralded on the official press and on social media linked to the government as mere "hostilities", while I had the opportunity to witness that, in suburbs of big cities like Maputo and Beira and among peasants of the central and southern regions of the country, there is already some currency to the notion of an outright "war". At least since 2012, the reorganization of RENAMO troops has become evident, particularly in the central region of the country, and reports have been mounting up regarding attacks waged against vehicles traveling along the National Road #1 and situations of recrudes-

reveal the frailty of bonds that rely on any given idea of community. But what is a community in Mozambique? In this essay, based on ethnographic work carried out in the last 15 years between the southern and central regions of the country, I argue that perceptions of the passage of time and the constant fear of disorder are intertwined in stories and rumors connecting different moments in Mozambican history. It is through its very opposite – the war or fear of war – that the ideal of community, always fragile, always unfinished, finds its voice.

Carrying out field research for such a long time in the same region offers many advantages. Not only can one get a grasp on objective shifts among a given population – the end of a civil war (which has reportedly caused more than a million casualties), the return of refugees and displaced persons, reconstruction efforts, the alternation between periods of scarcity and abundance – but, above all, it offers the opportunity to assess in dynamic terms the shift in worldviews and to follow a prolific debate that tends to take place on its own terms and in its own pace over long stretches of time. Time here also implies a transformation of the anthropologist himself: along with his informers, he comes to see the world – his own and the one corresponding to this research interest – in a completely different way.

I returned to Mozambique in July and August 2015, remaining most of the time in the city of Beira and making incursions to its suburbs, to Chimoio and the border with Zimbabwe. It surprised me more than ever to witness again something I had already observed in my previous field research back in 2013: rumors[3] stressing the evidence of a looming war, stories of mass killing occurring all across the national territory, fear of powerful wizards operating from across the border – in Zimbabwe, Zululand or Tanzania[4] – and references to previous historical periods, mentioning wars fought in the nineteenth century, like those resulting from the Nguni expansion and from the pacification promoted by the Portuguese troops,

---

cent violence between government troops and peasant populations, which has unleashed a new wave of internally displaced persons and refugees fleeing particularly towards Malawi.

3 Rumors are not limited to the usual murmured lines that spread through more or less clandestine conversations, but they also appear in the press and on the radio in the shape of news reports accounting for the death of dozens of people in one single day, either in skirmishes between FRELIMO and RENAMO or in massacres carried out by one of the sides, but without presenting any figment of journalistic evidence and without any continuity in the coverage, i.e., headlines about episodes of bloodletting on such a jaw-dropping scale appear on the frontpage of a major Mozambican daily paper on one day and, on the next day, nobody even makes the slightest mention of it or of what happened next.

4 In the short time I was there, there were countless rumors of Zimbabwean sorcerers who ordered human heads to be harvested on the Mozambican side of the border, as well as daily (and anguished) news reports about the widespread killing of albinos in the northern provinces of the country so that their bodies could be used by Tanzanians sorcerers.

or the war of national independence and the ensuing internecine wars within the nationalist camp between 1964 and 1974, as well as references to the recent civil war that scarred the entire country between 1977 and 1992, causing a huge number of deaths and producing among peasant populations an enormous wave of refugees and displaced persons.

The recurrent stories were not circumscribed to massacres and battles: there was constant talk of the forced displacement of men and women. Indeed, over the period stretching from the late nineteenth century to the present, the war stories are overridden with narratives about the transit to and from the mining jobs in South Africa,[5] deportations of Mozambican peasants to serve sentences of forced labor in São Tomé[6] or still of forced labor in large Portuguese companies operating in Mozambique itself.[7] The forced labor and displacement, typical traits of the colonial period, are thus perceived and portrayed as a continuation of the kidnapping and slavery that existed in the region until the late nineteenth century, or of the abductions that became widespread during the recent wars. Both in the war of independence and in the civil war, the armies were largely composed by men, women and children who were deprived of their communal ties and saw themselves thrown into the only remaining figment of community, the one formed by the soldiers themselves and the world around them.[8] And just as those who kidnap people in order to man their ranks and form an army, the independent state itself emerges as a powerful agent capable of depriving men and women of their freedom of movement or forcing them to engage in forms of labor analogous to the forced labor typical of the colonial period or ancient forms of slavery.

II

Recurring representations of the revolutionary idea in the second half of the twentieth century, particularly with reference to images of the Cuban revolution (but not only), often tend to imbricate the revolutionary process with enthusiasm and joy, at least on the part of previously silenced and suppressed majorities that are suddenly released to embrace the future. Revolution is also associated with a spirit of youthful strength: long-haired youngsters entering a town that welcomes them in jubilation. The long march that Samora Machel, the leader of the Mozam-

5 Cf. Ruth FIRST, *Black Gold*; Centro de Estudos Africanos, *O mineiro moçambicano*.

6 Cf. Augusto NASCIMENTO, *Desterro e contrato*.

7 Cf. Allen F. ISAACMAN, *Cotton is the mother of poverty*; Allen F. ISAACMAN and Barbara ISAACMAN, *Mozambique: from colonialism to revolution*.

8 Christian GEFFRAY, *La cause des armes*; Michel CAHEN, *Les bandits*.

bican revolution, undertook from the North to the South of the country was often described as enthusiastic, with huge crowds flocking to hear his speeches, while Portuguese troops fraternized with FRELIMO guerrillas suddenly turned into national heroes. The memory of the revolution that is actually conveyed by the contemporaries' narratives, albeit punctuated by short bursts of joy, is radically different: the Mozambican one was a rather sobering revolution.

The FRELIMO soldiers' entrance in Lourenço Marques, as Maputo was then called, in September 1974, amid that "huge confusion", was marked by a climate of suspicion, ignorance and fear.[9] The soldiers were afraid of those city dwellers they were coming in contact with for the first time, while the inhabitants of Lourenço Marques were afraid of the soldiers, who often spoke languages from the northern regions of the country. The tension was great and little had been done to build up trust among them. It was in the midst of this confusion that the overwhelming majority of whites left the country.[10]

FRELIMO elites imparted not only an ideal for the future embodied in the idea of a New Man,[11] but also a military experience shaped by the distrust of enemies, who, as such, ought to be eliminated either physically or spiritually. To this end, a whole repressive and correctional apparatus was created, finding its structural expression as "camps" – re-education camps, prison camps and collective labor camps (also known as communal farms). Gradually, an encompassing judicial apparatus spreads its reach to the point of completely restricting the movement of individuals (through the restrictive issuance of internal passports called *guias de marcha* or "travel protocols"), legalizing the death penalty for those considered traitors and reviving the colonial practice of flogging (Thomaz, 2008). Individuals were ranked according to a number of categories, and the ensuing climate of mutual accusations and suspicions nurtured the conviction that the enemy could be anywhere: over women hovered the ceaseless and obsessive suspicion of prostitution; those who had shunned the battlefronts faced the constant threat of being identified as someone committed to the old colonial system; any individual seen as *foreign* to his immediate surroundings could be accused of vagrancy and accordingly charged and prosecuted; local businessmen and petty traders, particularly those of Indian and European background and without patent con-

9 Isabella Oliveira, years later, recalls the memory of the deep enthusiasm felt when the guerrillas entered Lourenço Marques. Her account, however, must be set against the backdrop of a profusion of narratives that tend to circumscribe the euphoric outbursts to specific moments and specific sectors of the society. Cf. Isabella Oliveira, *M.& U. Companhia Limitada*; Omar Ribeiro Thomaz, "Duas meninas brancas".

10 Omar Ribeiro Thomaz and Sebastião Nascimento, "Nem Rodésia, nem Congo".

11 Christian Geffray, *La cause des armes*.

nections to the higher circles of power were unremittingly suspected of crimes against the economy, as potential *hoarders* of basic goods; onto those who dared to present themselves as an alternative in the middle of this process, the dissidents, was hurled the taint of traitors and reactionaries.[12]

The revolution was intrinsically associated with the figure of the charismatic leader – Samora Machel –, with promises of a bright future and with a reality deeply marked by civil strife[13] and a repressive structure created by a state that was neither less nor more violent than the former colonial state, but that was above everything else and from the point of view of many contemporaries infinitely more unpredictable.

The end of one-party rule, the 1992 peace agreements and the multiparty elections held in 1994 announced a period of relative peace and reconstruction, which, increasingly since 2000,[14] has been marked by economic growth, the establishment of big companies and by an increasing concentration of wealth, side by side with endless rumors and with an increasingly patent fear of war.

In my listening efforts, I could perceive more strongly than ever the fear of this looming war, explained by reference to a myriad of past wars and conflicts. The war is mainly explained because there is a shared conviction that the rules that sustain the *good community* have been violated. In fact, it seems that the community is always at risk: its very existence implies the risk of its suspension. The daily need to affirm it implies the suspicion that it, in fact, does not exist.

III

But what is a community? Both in rural and urban contexts, community implies an array of ties associated with rather fluid ideas of consanguinity and alliance

---

12 These categories have been mobilized over the years and are remembered as typical of the revolutionary period, often referred to as "Samora period", which stretches roughly from the transition and independence period (between September 1974 and June 1975) to the death of Samora Machel (1986).

13 Although the death of Samora Machel and the "end of the revolution", which became evident for the population after the adoption of the *PRE* (Economic Rehabilitation Programme) in 1987, did not mean an immediate end of the war – in some regions, the conflicts were even intensified and became even more violent), the start of the war and its territorial spread are undeniably associated with the revolution.

14 On the crisis during the late 2000, see Michel CAHEN, "Mozambique: l'instabilité comme gouvernance".

– in the sense proposed by the already classic text written by David Webster in regard to Southern Mozambique.[15]

Ancestry is absolutely crucial, without which there is no possibility of performing the appropriate rituals dedicated to the ancestors. The non-ancestry or improper ritual performance – forgetting the dead, for instance, or lacking access to the lands where they were buried, or still the negligence or abandonment of the bodies – result in the emergence of angry spirits, which demand vengeance and may be responsible for triggering new wars.[16] One war may foster another, for every armed conflict produces its inevitable share of abandoned bodies, which shall never return home or be properly buried. Their return may occur after many years, maybe decades, and the longer it takes, all the more forgetful of the abandoned bodies the descendants become. That is when the spirits get increadingly annoyed and then decide to return, imbued with fury, to scold and lambaste the living for their forgetfulness and to exact their due retribution – after all, they belong to the community just as well, and towards them every single member of the community has his or her obligations. Now, all of them are returning at once: those who fell victim of the Nguni massacres during the final decades of the nineteenth century and those who have been forgotten in the last 16-year war, and they return speaking other languages, which demands the intervention of powerful wizards and healers and ends up provoking widespread disorder.

Progeny requires alliances, through marriage, established by the *lobolo* (the progeny price), which does not imply the acquisition of the bride, but represents a guarantee of control over the progeny: in Southern Mozambique, the correct payment of the *lobolo* means that the progeny of any given union shall belong to the father's lineage and shall not result in the wife's disconnection from her own lineage. The commitments between both groups are thus sustained, unless one is confronted with the father's or the mother's infertility, which requires the intervention of healers and enfolds this alliance relationship (the *lobolo*) in the clutch of rumors, accusations, gossip and, of course, all kinds of suspicion. In such cases, adultery, usually surrounded by gossip as well, turns from problem into a possible solution.

The alliance, however, is not limited to marriage: it also concerns nomination rituals. Naming someone implies the establishment of a relationship between the lineage of the individual named at birth and the lineage of the name's master. Between the namesakes, an identification arises, as well as obligations, and a classificatory kinship between both lineages is created: a baby is thus defined as father of his namesake's children, who now have obligations towards their

15 David Webster, *A Sociedade Chope.*

16 Alcinda Manuel Honwana, *Espíritos vivos, tradições modernas.*

new parent, which evidently involve the spirit world. When growing up, the child must spend part of his life in his namesake's house and work for him. The occasional breach of these obligation is wrapped in gossip and accusations, giving rise to a perpetual imbalance in the communal universe.

Following the paths opened by Webster[17], we realize that alliances are not strictly limited to the spheres of marriage and naming rituals. They also encompass friends and neighbors. Neighborhood relationships are just as strong as those based on consanguinity: people know their close neighbors and tend to nurture a deep affection for them, which also translates into everyday commitments. Similarly, friends require formalities according to their age, place of residence, social status or gender. We find ourselves immersed into a world with absolutely no room for indifference, and failure to abide by the rules governing the relationship between neighbors and friends, or even a mere hint of non-compliance, may culminate in suspicions and accusations of witchcraft.

IV

The fact that there is no place for indifference has powerful consequences in a world marked by the incessant dislocation of individuals and groups across the territory and by considerably strong tectonic perceptions. The claims regarding the legitimate ownership of the land has laid roots and has been deepened during the colonial period. The land occupation by the Portuguese produced an Indigenous Code that established the terms for the territorialization of rural communities through the figure of a traditional authority imbued of full recognition by the colonial power. Every indigenous individual should belong to a specific plot of land, and should accordingly speak the corresponding language and be submitted to an authority whose legitimacy would derive concomitantly from lineage recognition and colonial power.[18]

If it is true that things are not so rigidly established, a temporal idea is nevertheless consolidated, attaching the "owners of the land" to territories marked by an explicit heterogeneity, by migration and by intense territorial shifts. Categories are thus created to deal with those who are perceived either as foreigners or as guests, and who, although often established for generations in a given territory, must live under constant threat. In all villages, one can witness the strong

17 David WEBSTER, *A Sociedade Chope.*
18 For a better understanding of this model for organizing the colonial state power, see the work of Mahmood MAMDANI, *Citizen and Subject.*

presence of Indian families and individuals who are considered as *vientes* (newcomers), strangers kept under constant surveillance and upon whom any sort of harm or injury can be blamed.

A brief note on the Indians is due here: are they part of the community? As well from the perspective of local communities in Mozambique as from the point of view of those who claim to speak on behalf of the nation, we are faced with extraordinary ambiguity. If there is no doubt regarding their foreign origin, as opposed to the *originários* (the "autochthonous", a term that has been gaining strength and causes serious discomfort to Mozambican Indians, whites and mulattoes), nor is there any doubt about their commitment to the country during difficult times – unlike the overwhelming majority of whites, who fled the country after the independence and during the war, many among the Indian families remained in Mozambique. It is said that, during the most difficult years under the rule of Samora (colloquially referred as *tempo Samora* or "Samora period"), Indians were sent to the camps and, in regard to the war, stories abound of Indians who were abducted and murdered. Furthermore, regarding witchcraft, Indians are reputed to be powerful wizards. No less important is to remember that there is a good deal of awareness of the important role played by the Indians in local trade networks, enabling and shaping the flow of peasant goods and offering these same farmers access to extremely valuable products – textiles, soap, oil, kerosene, salt and sugar. In other words, while portrayed as non-autochthonous and even less entitled to ownership of the land, they may still be assigned the category of guests. As such, though, they must behave properly at all times, otherwise even this assigned status may be suppressed, resulting in the risk of a more or less violent expulsion from the country, just as happened under different circumstances in other regions of East Africa.

In this universe of relationships, there is no place for someone who is indifferent: either you are regarded as a relative, neighbor or friend, or else you are considered a potential enemy. Ultimately, everybody is a potential enemy, because everybody is subject to constant suspicion of seeking for the help of wizards and witches or turning to witchcraft themselves.

V

More than two decades after the peace agreement between FRELIMO and RENAMO that put an end to the violent civil war, recrudescent hostilities between armed groups and the inflammatory political discourse from all sides recreate an atmosphere of fear and make the possibility of open war palpable again. In daily life, the perception of the passage of time and the constant fear of disorder intertwine in stories and rumors that connect different moments in the country's history.

It is on the basis of this everyday ritualization of communal ties that the imminent disorder is expected to be contained or prevented. In the suburbs of the capital Maputo or in a small district in the country's South, men and women strengthen their alliances with neighbors, friends and namesakes, negotiate the terms that should regulate a future wedding (the *lobolo*) and strive to keep the necessary balance in relations concerning the progeny, particularly those involving the ancestors. Within the community there is no room for strangers. An individual is either a relative or a friend, with whom relations based on joy and affection are held (but from whom suspicions may be held just as well), or a guest whose behavior must be constantly monitored, or else a stranger (*viente*) and, ultimately, an enemy – who can be held responsible for casting spell and promoting disorder in the worlds of both the living and the dead, capable even of kidnapping children, taking them to the bush and turning them into child soldiers. Between both extremes, the reverse side of the community is heralded is its inner side, among its very members. Ritualization of alliance and progeny takes place amid constant suspicions that, at any moment, can turn anyone into an enemy.

The everyday challenges to the concept of nation in cities like Maputo, Beira, Quelimane or Nampula are directly connected with the perception of the community at the local level. Disputes over who should be accepted as truly Mozambicans (the autochthonous and, within the realm of the state, those who are black) and who are to be seen as lesser ones (Mozambican Indians, whites and mulattoes), over who were the ones responsible for past wars and over the real intentions and identity of the nation's founding fathers are all devised to detect or rule out the eventual presence of mere opponents, since disagreement on such issues imply the presence not of an opponent, but of an enemy, with whom no dialogue is possible. The world of state politics is perceived as the reverse side of an unfinished nation, because in it what is accomplished is the translation of local conflicts into an all-encompassing fear of war, of another nationwide conflagration looming in the horizon.

The definition of politics as a foreign or undesirable world is recurrent. Politics belongs is what sets the individual apart from the community, in which one may attain power, but only drawing it from an authority whose legitimacy must have a traditional basis. The politics of the socialist revolution, with its promises to overcome all the suffering and distress, in the end only deepened both suffering and distress through policies that gave rise to a world peopled by unknown elements – the socialist New Man should have neither ancestors, nor relatives or friends. The war was thus regarded as intrinsically linked to the necessary destruction of that unsettled and unsettling New Man, whose isolation and atomic nature was deemed unbearable. The only possible way to overcome it would be an effort to destroy its very foundation: the sphere of the city, of the state and, above all else, of the camps, the re-education and labor camps characterized by a rigid discipline that held off any ties to neighbors, relatives, friends and namesakes.

The post conflict period in Mozambique after 1992 was actually a mere suspension. The return of the war is announced not only by quite palpable references to hostilities between those groups that in recent years have compete with increasing violence for state power and for access to the mineral riches recently discovered in the country. In city neighborhoods and rural districts, the inevitability of war is expressed in the perception of time, in the stories told of past wars that left unburied a countless number of dead, who are now returning to confront the living with their unsettled accounts. There is talk of wars fought to resist the Portuguese colonizers, of the many massacres perpetrated during the colonial period, of the war of independence and the civil war that succeeded it. After much wandering, the ancestor spirits return to settle the debts of their own offspring and to remind the living of the connection between all the previous wars and the one about to come.

Attempts to placate the fury of the ancestors demands the resort to the dangerous and magical world of witchcraft. The 16-year war was neither overcome nor forgotten by means of purification rituals and the violence of the "times of fury" does not amount to any sort of zero-sum game in which the contending poles of FRELIMO and RENAMO would account for all the actors engaged in the conflict. A complex and multilayered conflict, whose terms and episodes were usually punctuated by systematic kidnappings and forced marches, indiscriminate murder and torture, widespread use of child soldiers among the RENAMO "armed bandits" and not fewer children soldiering among the FRELIMO troops. All these elements are out there, making their unrelenting presence felt in stories that tell of immense anxiety and reveal an almost paralyzing fear.

But fear of what? It would be easy to say that it is the fear of a state apparatus that, in reality, still represents a complete alien structure to the overwhelming majority of the population, at least in regard to its promises. For this reason, much of the fear is driven by this alienated state, but not only by it. Here, the dead of each family are not stored in their respective cabinet drawers or hidden in chests under the bed or behind the door. Those who did not receive a proper burial, which would ensure a satisfactory post-mortem life, are now resurfacing to demand their due and reclaim the dignity they were denied.

Notwithstanding all the broken commitments, the new world that emerged from the ashes of the war was also permeated by political promises: the secret ballot, as a basic requirement for democracy, represented the possibility of a new order where political participation would guarantee justice and wealth for all.

Now, in the different languages of the country, the word "secret" has become synonymous with witchcraft. Then how could anything like a secret ballot be good, after all? The secrecy surrounding the polls and politics in general are, thus, what explains the sudden wealth acquired by powerful figures of the democratic and capitalist world. Witchcraft would be the underlying reason for the wealth amassed by those invested in the state apparatus and settled in the cities. Theirs is a suspicious and illegitimate wealth. Challenging this democratic and modern world may be seen as a means to restore and reaffirm the community, through the death not only of witches and wizards, but also of all those regarded as political figures.

Bibliography

Cahen, Michel, *Mozambique: la révolution implosée. Études sur 12 ans d'indépendance (1975–1987)*. Paris, L'Harmattan, 1987.

Cahen, Michel, "Mozambique: l'instabilité comme gouvernance", *Politique Africaine* 80 (2000), 111–135.

Cahen, Michel, *Les bandits. Un historien au Mozambique, 1994*. Paris, Centre Culturel Calouste Gulbenkian, 2002.

Centro de Estudos Africanos, *O mineiro moçambicano*. Maputo, Universidade Eduardo Mondlane, 1988.

Egerö, Bertil, *Mozambique: a dream undone. The political economy of democracy, 1975–1984*. Uppsala, Scandinavian Institute of African Studies, 1987.

Fauvet, Paul, "Roots of counter-revolution: the Mozambican National Resistence", *Review of African Political Economy* 29 (1984), 108–121.

Figueiredo, Isabela, *Caderno de memórias coloniais*. Coimbra, Angelus Novus, [2]2009.

Finnegan, William , *A Complicated War: the harrowing of Mozambique* (Perspectives on Southern Africa, 47). Berkeley, University of California Press, 1992.

First, Ruth, *Black Gold: the Mozambican miner, proletarian and peasant*. Brighton, Harvester Press, 1983.

Geffray, Christian, *La cause des armes au Mozambique: anthropologie d'une guerre civile*. Paris, Karthala, 1990.

Hanlon, Joseph, *Mozambique: the revolution under fire*. London, Zed Books, 1984.

Hanlon, Joseph, *Mozambique: who calls the shots?* London, James Currey, 1991.

Isaacman, Allen F., *Cotton is the mother of poverty: peasants, work, and rural struggle in colonial Mozambique, 1938–1961*. Oxford, James Currey, 1996.

Isaacman, Allen F., Isaacman, Barbara. *Mozambique: from colonialism to revolution, 1900–1982*. Boulder, Westview Press, 1983.

Honwana, Alcinda Manuel, *Espíritos vivos, tradições modernas: possessão de espíritos e reintegração social pós-guerra no Sul de Moçambique*. Maputo, Promédia, 2002.

Magaia, Lina, *Dumba Nengue: Histórias trágicas do banditismo*. Tempo: Maputo, 1987.

Mamdani, Mahmood, *Citizen and Subject: contemporary Africa and the legacy of late colonialism*. Princeton, Princeton University Press, 1996.

Minter, Willian, *Apartheid's Contras: an inquiry into the roots of war in Angola and Mozambique*. London, Zed Books, 1994.
Nascimento, Augusto, *Desterro e contrato. Moçambicanos em São Tomé e Príncipe, 1940–1960*. Maputo, Arquivo Histórico de Moçambique, 2003.
Oliveira, Isabella, *M. & U. Companhia Limitada*. Porto, Afrontamento, 2002.
Saul, John, *The State and Revolution in Eastern Africa*. London/New York, Monthly Review Press, 1979.
Thomaz, Omar Ribeiro, "Escravos sem dono: a experiência social dos campos de trabalho em Moçambique no período socialista", *Revista de Antropologia* 51 (2008), 177–214.
Thomaz, Omar Ribeiro, "Duas meninas brancas". In: *Itinerâncias, Percursos e Representações da Pós-colonialidade / Journeys, Postcolonial Trajectories and Representations*, edited by Elena Brugioni, Joana Passos, Andreia Sarabando, and Marie-Manuelle Silva. Ribeirão (Portugal), Edições Húmus, 2012, 405–427.
Thomaz, Omar Ribeiro and Sebastião Nascimento, "Nem Rodésia, nem Congo: Moçambique e os dias do fim das comunidades de origem europeia e asiática". In *Os outros da colonização. Ensaios sobre colonialismo tardio em Moçambique*, edited by Cláudia Castelo, Omar Ribeiro Thomaz, Sebastião Nascimento, and Teresa Cruz e Silva. Lisboa, Imprensa de Ciências Sociais, 2012, 315–340.
Vines, Alex, *Renamo: terrorism in Mozambique*. London, James Currey, 1991.
Webster, David, *A Sociedade Chope: indivíduo e aliança no Sul de Moçambique 1969–1976*. Lisboa, Imprensa de Ciências Sociais, 2006.

South American Conditions

Diana Braceras

The Contemporaneity of 'The Savage Mind' In the Andean Communities

Introduction

As from the logic based on the principles of identity and non-contradiction, the opposing pair individual/community branches off into movements that go from alienation to segregation, spanning such a variety of shades and combinations which allows the dualism that fosters reflexive thought. From Masses to Individualism, the weakening of social bonds expresses the ways of contemporary subjectivity.

I propose to think about a different logic: the one implied in the practice of native people of our continent, particularly the Andean culture, whose matrix of "wild thought" such as defined by Lévi-Strauss, becomes now contemporary and archaic at the same time.

I will particularly refer to the culture of the *Qheschwa* people, and their language, the Runasimi, that I encountered in my early childhood, in the provinces of Catamarca and Santiago del Estero, in Argentina. Even now, in rural areas of the northern parts of the country, they still speak the indigenous language, originally spread by the *Inkas*, and later by the Catholic Church, as lingua franca. During the last decade, in some local schools a bilingual program has been established, to include Spanish.

The home-school where I spent my early childhood, served a scattered population near Las Salinas Grandes, within a schedule starting when the torrid northern summer allowed it, until the spring, when new generation of goats was born. From then on, this *increase*, as they call it, demanded all hands of family members, along extended wild areas. In this context, the teaching of Spanish as a second language was taught in public schools and partially learnt by the local children, through considering their mother tongue a disadvantage for future opportunities, in a mixed society that still considers even today their native origin as shameful.

The acquisition of the Spanish language and its corresponding literacy become a means of social promotion, fostering at the same time the oblivion of a diverse cultural matrix, different from the one currently dominating in the nation-states. An approximation to the traditional configuration of the called "Original Population" today is possible through the study of their languages, their practice, and their iconography, by which we can recreate a field of millenary experience containing contemporary expressions, in spite of the colonization and modernization processes suffered by our native societies.

In general these people from the Argentinean Norwest are called *Qolla*, communities organized in a lax sense as extended families, whose historical backgrounds are the traditional *Andean ayllus*. Towards the North the relative ethnic unity is recognized in certain practices and symbols currently more present in Bolivia and Peru than in Argentina. The ceaseless immigration did not recognize national borders, normally erasing the matter of origins and making the *Qolla* and their relationships –symbolic or bloodline- the only motherland.

Peerage as unity

The actual composition of these people with their authorities – confirmed by the names of the original heroes[1], - stems from a minimal unit, which is not the individual but a duality. The conjugation of opposing elements of the most diverse dimensions within a scale that spans from the intimacy of the male/female in a couple, to the attributions of male/female of species of plants, minerals, landscape entities, atmospheric phenomena, ceremonies, food, musical instruments, utilitarian or artistic activities, architectural placements, etc.

The "individual" then, would not be a category of composition for this "community", but rather a logic belonging to a collective nature or entity, in proximity with fractal geometry and complex systems rather than to the two dimensions of the plane or the Euclidian elements. Even the word with which we translate as the most akin to an individual, a person or a human being: *runa*, is the result of a noun with the suffix '*na*', of a semantic unit, the prefix '*ru*' which is added to expressions of "lacking", useless, silly, deaf, half empty, untidy, decrepit, unfocused. The incompleteness or the deficiency of the isolated human being is conducive in this culture to conceive the union with another as something necessary, not contingent or accessory. Therefore, the propitiatory inclination towards relationships and the celebration of the bonds between individuals has led to the prejudice of seeing these communities as "licentious" due to practices understood as of a greater tolerance of sexual pleasures[2] and "idleness", as well as for

1 The heroes of the rebellions during colonial times are remembered by their brave stand, as well as of their women: *Tupac Amaru* (José Gabriel Condorcanqui) y Micaela Bastidas Puyucahua; *Tupac Katari* (Julián Apaza) and Bartolina Sisa. See the classical study of Boleslao LEWIN, *La rebelión de Túpac Amaru*, 1967, 427–443.

2 The institution called *sirinaku* for example, was banned as from colonial times: *the marriage by trial*, in which, by the lapse of a year each couple could try their compatibility before deciding

the profuse dedication to festive activities, rituals and socializations within and without the community. The *runa-ayllu* would be the closest expression for community as we understand them from western thought, and as we will see the *ayllu* is not reduced to a plurality of persons.

This conception of common bonds, considered to be a vestige from primitive societies heading towards extinction from the point of view of civilizing progress of humanity, does it matters today from an horizon of expectations threatened by the supremacy of a capitalist social model?

The hypothesis of this paper stems from the ontological principle of understanding the social realm as a discourse space, producing sense and signification. Therefore, there is also a difference in the consequences of human actions beyond the conscious rationality, the subjective intentions and wills, including the instrumental power during different epochs of the historic flux.

I propose to think the collective logic coded in the traditional institution of *Ayllu* – basic unit of the Andean communities- that is organized through the practice of the *Ayni* – principle of reciprocity of all existence-, and that is represented by the symbol of the *Chakana* – cosmological ideogram known as 'the Andean cross'. It is only now that we find a correspondence between this ancestral knowledge, the new paradigms of western thought known as the "Emerging Science"[3] and psychoanalysis. In this direction it is possible to think on subjects not ascribed to a dominating logic, but to a know-how directed towards life, that is to say engaged in the preservation of the social bond based on the gift of love, the increase of social interchange, and the respect of limits.

The Sciences and the "Wild thought"

Already towards the middle of the last century, Lévi-Strauss posed new problems to Anthropology, in the task of elucidating the logical operations implied in the "wild thought":

on a more permanent union. Virginity was not a value to be preserved for the future alliance. *Carnival*, now associated with the Christian liturgy, continues engaging native communities on a yearly period of sexual liberty and a joyful celebration with permissiveness towards that what in daily life would be considered excess. Exceptions, as we know, help to preserve the rules.

3 From this perspective and against the reductionism of the classic sciences, physics laws are rules of collective behavior. Emergent endeavors to decipher the behavior of the whole, not of the parts. The emerging phenomena imply an organizational phenomenon. Operating under such a principle, it becomes independent from the concept of material property. Robert B. LAUGHLIN, *Un universo diferente*.

> This wild thought is not, strictly speaking, the thought of 'primitive society' but thought in a wild state, that is in a non domesticated state, not subjected to goals of efficiency. In this way, it tries to escape from the evolutionist work-frame that keeps worrying anthropology, proposing structures of a thought developing its logical possibilities in a non-lineal manner.[4]

This "complex" and "emerging" conception of the "Wild thought", is more in harmony with the latest developments of the western hard sciences than the common sense which percolates from the modern vision of the world, still ruling the social sciences.

> (...) There are two different modes of scientific thought, both of which are a function, not of unequal stages in the development of the human spirit, but of the two strategic levels in which nature allows to be addressed by scientific thought: one of them adjusted to perception and imagination, and the other, displaced; as if the necessary relationships composing the object of any science – be it Neolithic or modern – could be reached by two different ways: one of them close to sensitive intuition, and the other further apart. (...) This science of concrete objects had to be, in essence, limited to results other than the ones promised to the natural exact sciences. But was not less scientific, and its results were not less real. Reached ten thousand years before the other ones, they keep representing the underlying base of our civilization.[5]

The work of Levi-Strauss destroys the supposition of a hierarchical difference between "scientific thinking" and "primitive thinking". So much so that he points out in his rituals and beliefs to veritable anticipations of science itself, and to methods or results that science belatedly includes in the systematization of its territory.

It is the very category of fundamentals, of progress and evolution, which will justify the "wilderness" of the century. Moving against the historicist imaginary, certain ways of life born by concrete practices of a community sense which manages to survive, not without some difficulties, to the rupture of social bonds and solidarities imposed by the narcissistic regime amplified by capitalism.

Within this promiscuous XXIst century of modern and postmodern layers, the political experience of equality, of the common, and of justice, has something to say about the contemporary network of societies not identified with the neoliberal subject, but rather with a counter-experience.

Emerging from the borders of the hegemonic system the so-called "organic societies" are being considered archaic remains of humanity's past. The forms of organization, beliefs systems and practices of those contemporary people

4 Fréderic Keck, *Lévi-Strauss y el pensamiento salvaje*, 9.
5 Claude Lévi-Strauss, *El pensamiento salvaje*, 33–35.

are anthropologically described with particular categories, therefore taken to a level of ideological generalization. The disqualified "primitive magical thought" of these animist, totemic and "analogical societies", even shares an ontological epistemic shelf compatible with the recent Theory of Complexity[6] though.

A place in the world: the *Chakana*

In truth we are witnessing a fundamental transformation of our way of looking at the world, according to which the objective of understanding natural phenomena disassembling them into their smaller component parts is being replaced by the purpose of understanding how nature is collectively organized.

Instead of addressing the search of the ultimate causes of phenomena, now the paradigm is that on the "organization". Its consequences are lethal to the scientific omnipotence: The impossibility of predicting the qualitative changes caused by minor events implies the impossibility of controlling everything. At the end, this is not about a scientific discussion but a controversy on what is our place in the world. Science has passed from the Era of Reductionism to the Era of Emergence:

> The principles of collective organization are not just a curious aspect of reality, but rather constitute the whole: the true source of all laws of physics. The level of precision we have reached when measuring them enables us to say that the search of a unique ultimate truth has reached its end, and at the same time it has failed, for nature now reveals itself to us as a vast tower of truths (...).[7]

Human bonds, proportions, relationships, organization, more than the elements, are the protagonists of the new orientation of this emerging thought as represented by the symbol of the cross of *Tiawanaku*, a philosophical geometrical concept thousands of years old: the *Chakana*[8], in the *runasimi* language. Thus, humans would be only one case in this emerging result of relationships and collective organization of life.

---

6 The Theory of Complexity refers to a branch of mathematics created in the 1970s. It deals with caos, fractals.

7 Robert B. LAUGHLIN, *Un universo diferente*, 255.

8 *Chakana*: from the verb *chakay* to cross, turned into a noun by the suffix '*na*'. It means crossing, bridge, transition, link and rotation. It is represented in the shape of a cross, point of encounter of four quadrants, an element of a relational connection whose center/origin, *cozco*, navel, is a void.

The *Chakana* is one of the principal symbols in the sacred iconography of the American continent. Being a structure of quadrupleness of institutions based on the complementary, correspondence, and reciprocity among the opposites, not antagonist, it was addressed profusely in the pre-Columbian symbolism and nowadays popularized with the concealed name of *Pampa Cross* in Argentina (due to its inclusion in the fabric of *mapuche* or *Araucanean* materials, people who have been influenced by the *Inka* culture, which reached southward down to Neuquén on the Patagonian Andes).

Perhaps the thesis that Jean-Claude Milner proposes in relation to the centrality and persistence of the quaternary structure in the Jewish culture may also be extensible to the native *Indian*[9] from the Americas, explained as at the reason of their survival in a chronic context of persecution:

> To explain with a single word what is at stake, I will seek inspiration in the translators of Heidegger and coin the term *Quadrupleness.*
>
> The quadrupleness masculine/feminine/parents/son, this is what is expressed by both the calm expression "from one generation to the next", as with the disturbing question, "What will I tell to my son?" It may be said that groups of speaking beings encounter the quadrupleness. All the names that these groups give themselves, or they refuse, rest on this quality (...). They could do it for a long time, and could still do it if need be. Finally, persistence has no other material base than that which makes it possible. The study supposes this base, indeed, the rites suppose it; it is supposed, in the end, the simple startle the name Jew sometimes produces in most European Jews.[10]

Also in the *Bask* people, one of the principal symbols, inherited from the Celts, is, in the *Euskadi* language, the one called *Lauburu* (four heads), quaternary structure, common to the most ancient people on the planet, and effectively persistent in the safeguard of their identity.

For the Andean thinking the actual human being is *Chakana*. In its geometric structure, the male/female (right/left) side is condensed, as well as the succession of generations: ancestors and descendants (below/above), and at the same time serves as a calendar of the annual planetary cycle in the succession of the seasons, and therefore, of the activities related to the earth, its festivities and celebrations, the musical instruments prescribed for each agrarian time, the ritual practices dedicate to the four elements: water, air, earth, and fire.[11]

---

9 The vindication o the name Indian for the originals of America, apart from the mistaken origin, was established by the "indigenism" current founded by the Bolivian writer Fausto Reinaga, *La revolución india* (1970); *Tesis India* (1971); *El pensamiento indio* (1991).

10 Jean-Claude Milner, *Las inclinaciones criminales*, 118.

11 Jorge Miranda Luizaga, Javier Medina *et al.*, *Aportes al Diálogo*.

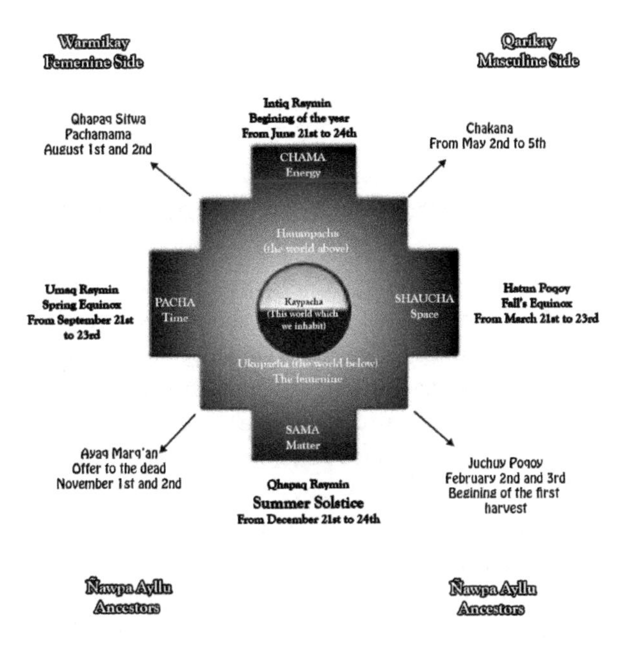

Contributions to the dialogue

The four terms related to the ritual structure of the *Chakana* mark the sexual difference and the affiliation, their own marks of the language, inasmuch as it prints the symbolic order produced by the language: it represents "castration" in the speaking being. That is to say his impossibility of self-generation to become eternal, one, full, complete and consistent, without necessary bonds with other beings.

On the contrary, the *Chakana* implies a structure of relationships with mortal and sexual beings; collectively in transit through life in a network within the cycles of nature, beginning from a void: the navel of the origin. Plurality inhabits the beings and knits bonds originating in the mouth (*simi*), iconographic seat of the language (*runa-simi*), through the reproductions of the image and through the protagonism of the musical wind instruments.

The *Chakana* condenses the very structure of the speaking being: mortal and sexed. Precisely that which is threatened by the ruling of the homogeneous logic of the "Capitalist Discourse"[12]: The destitution of sense -effect of differences-, an

12 Discourse: different modes of social bonds, that is of the relationship of the Subject with Others, implying different ways of inhabiting the language. Historically psychoanalysis distin-

extension of the 'concentration-camp logic', ceaseless increases new forms of generalized segregation, cruelty and indolence. The condition of possibility of such anti-social decomposition is to divert the other from the place of his fellow being, that is, to flay him of all fraternal features in the sense of affiliation.

The *Ayllu*, multiplicity of the Andean organization

The *Ayllu* condenses the form and unit of the community's organization, based on a participating co-responsibility of conservation, not of the individual but of the collective balance and the continuity of life. The *Ayllu* is composed of three community dimensions: *runa* ('people'), *salka* ('wilderness'), and *waka* ('sacred'), within a time-space conviviality. It is about an inhabitable scale of the *Pacha*[13] whose intersection itself manifests and becomes concrete in family work, the *Chajra*. Not only the unity of vegetable garden or physical space of production and work but of the "collective breeding", in the direct experience of the care, growth and the distribution of the resulting benefits of the "doing together", thus recreating the dynamics of an active construction of the social-communitarian order. Its relatedness is constitutive while the individuality is secondary, derived and transitory.

A multiple and simultaneous functioning of the three dimensions implies logical properties of a necessary and contingent knot, articulated around the "doing the *chajra*", similarly to the three dimensions of the structure of human experience: Real, Symbolic and Imaginary.[14]

The word *Salka*, the savage, not ruled, the level of 'nature', according to our western dichotomy which antagonizes with the 'cultural' – what is not an intentional product of human activity-, but also what is unexpected, what is threatening in the *aymara* language (mother of *runasimi*) has a specific name: *khä pacha*, an approximation to the dimension of "the Real".

---

guishes four different matrixes: discourse of the Master, of the hysteria, of the University and of psychoanalysis. The discourse of capitalism is proposed by Lacan as a variant of the Master, thus originating a perverse form of the discourse tending towards the dissolution of the social bonds. Jacques Lacan, *El reverso del psicoanálisis*.

13 *Pacha*: Cosmos, space/time, conceive as moving always. A dual concept in which *pa* means 'two', and *cha* stands for 'energy'. There the opposition operates as in quantum physics, the symmetries of space-time and the interaction between particles of matter and antimatter. '*taqikunas panipuniw akapachanxa*': 'everything in the universe is paired' (old *Aymara* proverb).

14 Jorge Alemán, *En la frontera*, 25–26.

> The world of the yet- not- known, of the unknown, is born as a sort of potentiality permanently moving, facing a perpetual disjunction: anything can be ruined and human actions may end in a catastrophe, or may redeem the world of what it exists and turn it into an act of liberation and completion.[15]

The community of the *Wacas*, relates to the 'sacred', realm of the Symbolic, articulated in a narrative, the organization of time and space, the articulation with places which guarantee the succession of generations, organizes the difference in sex, the relationship with death and the word. A legal order of repetitions and sequences in accordance with conventions or laws, which establish exclusions and anticipations that, if not observed, would imply a departure from the "Good Living" or Sumakawsay: the good walk along life[16]. Both the knowledge of astronomy and the organization of the Inka calendar in Cuzco, sufficiently studied as to confirm this analogy with the symbolic order, is related to their own social organization, the regency of the *Wacas* (sacred places/entities) assigned to each group, together with the distribution of the physical spaces of the *ayllu-panaca*, (formed by the family of direct descendants of each Inka), which, twenty in number, made up the permanent population of the capital city of the Tawantinsuyu.[17]

And we also include within the structure of the Ayllu, the Runa community, the humans, participants of the social bonds, with the consistency of the Imaginary prevailing in the relationships among equals, inasmuch as they link with the other two dominions, as a projection surface of everything that sutures what is incomprehensible or unknown, with the dynamics always equivocal of the interpersonal relationships, the intentions and senses that is mirrored, isolated as "anthropocentrism" by the classic anthropology. Present in this dimension, we find the domestic breeding animals, what is cultivated as crops, the world of affections and learning, solidarities and mutual care, and also harm, as in all human interactions.

The graph[18] shows the written strength of the structure which poses the Andean relational concept of an "individual" linked in a knot to "another", incarnated in the community/ies.

Though the image is a contemporary graphic interpretation resorting to a formalization to show the intersections of the dimensions spanning the conceptual

---

15 Silvia RIVERA CUSICANQUI, *Sociología de la imagen*, 212.

16 Fernando HUANACUNI MAMANI, *Buen Vivir / Vivir Bien*, 10.

17 Dick Edgar IBARRA GRASSO, in his work *Ciencia Astronómica y sociología incaica*, returns to the Works of Huamán Poma, Leo Pucher, Lehmann-Nitsche, José Imbelloni, R.T Zudeima, Nathan Wachtel, etc. About the distribution of places according to the calendar/astronomic organization and the respective *wacas*.

18 Andean conception of the *ayllu*. Carlos MILLA VILLENA, *Ayni*, 37.

fabric condensing the Andean concept of Ayllu, we must also bear in mind the particular semantics of the Andean language. The words we find with the chart, are there to ad explanations to the graphic, and should not be taken in the ideal and finished sense suggested by the Spanish language, but must be interpreted within the context of Quechua/Aymara significations, in which the nouns indicate the tendency that is followed, the movement towards where it is going, corresponding to the verbal time of "to be/being". That is a key to the interpretation of the *ethos* of these people who consider change and the transit -temporality- as the natural state of life in all its dimensions.[19]

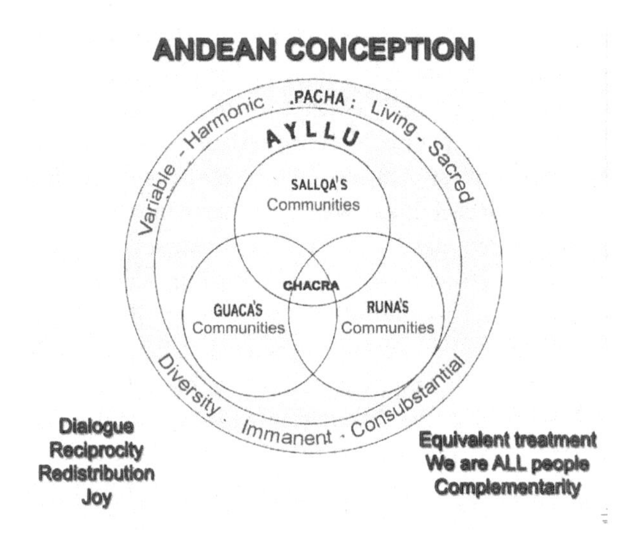

Andean conception of the Ayllu

The places of being are places to be at. Each language, with its grammatical specifications designs the entrance door of each one to his or her culture, that is to each social bond that makes him human. That open structure we come to, is, besides the living memory of times, including us in the generational chains beyond the bodies and stories. The "past" is active in the speaker, and such a

19 Socio-linguistics provides us with important elements to avoid the attribution of identical sense that sometimes produces mistakes in traditional literature interpreting sources translated from native languages. In *Variación y significado. Y Discurso* (2014), the 'principle of reinterpretation' formulated by linguist Beatriz Lavandera, establishes a difference between stylistic meaning and situational and social significance. We resort to them to underline the variations in the significance.

condition enables to operate devices trying to include what is not said, what is not known, and even what is impossible to express in the social fabric.

The approximation to other ways of being in the world with languages without the essential being substantiated, but rather an existing being organically territorialized along the social bonds in the collective belonging and following the cycles of planetary life.

Within the dimension of thought, permeated by modern anthropocentrism, we are unable to think of bodies, symptoms, current pathologies, displaced from territory, crossed by jargons, reflected only upon a surface of insignificance, unwoven from the social fabric, lacking the love of others, reduced to the menace of a persecuting rival, and not only for being an "other", but for disputing a unique absolute point of reference: the possession of that sense that will make us free of social, political, religious and even biological determinations. Being one, without the Other acting as limit.

The Law of Reciprocity: *Ayni*

The dynamics of the *Ayllu*, is the effect of a legality of a specific circulation of gifts: The *Ayni*, included now as a practice with a constitutional level at the Pluricultural States.[20] It is a base for "Good living" or *Sumakawsay*: the Law of reciprocity for the community or *common conviviality*.

The cohesion and recognition of a brotherhood among the members of the *Ayllu* that form part of each community[21] are interwoven with the interchanges and solidarities of those participating of the *Ayni*, in the provision of resources necessary for leading their lives and for the celebration of life which, as mentioned, it is an important and permanent activity with a place in the common calendar of the *Chakana*, apart from spontaneous celebrations for particular reasons or festive gatherings during moments of free time.

20 "Ancestral people, in their permanent reflection and deliberation from their communities pose a structural change, understanding that the pluri-nationality expresses the existences of the different cultures, they therefore say that all should be equally respected. After some four decades of having presented this proposal, they were voted in Ecuador (2008) and Bolivia (2009) Constitutions in which they declare themselves as Plurinational States". Huanacuni Mamani, *Buen Vivir / Vivir Bien*, 10.

21 The word "brother" for the other, is not reserved to the bloodline relation. The resonance with the Latin *frater germanus*, implies a personal sense of true, authentic bond based on a symbolic common ancestry, constituting themselves under the same law marking the historic memory of each community.

It is curious that in the phonetic composition of this word condensing all practices of collaboration, assistance, and joint efforts of the community without personalized payments, the knotted point of the different existential dimensions of the *Ayllu*, because it articulates all practices related to *Runa*, *Sallqa*, and the *Guaca*, we once again encounter Mortality and Language:

> Ay-a: corpse, dead.
> *Ni:* Verbal suffix, first person singular
> *Ni-y* Verb to say (the ending in "y" is valid for all verbal infinitive modes

The verb articulating experience and practices in the Andean communities is the verb *Qoy* (to give), plus a suffix from *Runasimi*, which is *naku*: "reciprocity". *Qonakuy*: is to give in reciprocity. This is the backbone of life in the Andes, not related to any law nor to a social contract nor a mirror pact of the "I give you, you give back" type. It is expressed in absolutely everything: with the other, with the animals, with the earth, among communities. The *Qonakuy* is constantly practiced, while gives and receives the benefit of building a community together with the pride of feeling useful and doing things "right". It is fundamental to "Living well", or the "Good living".[22]

Many established practices with different variations are akin to this communitarian conception: *mink'a; ayni; yupanaki; mitima; kumpa*, have very specific names and legalities, and still currently in use. To *live well* or the *good living* is to participate in this: *qonakuy*.

The symbolic death, being the exclusion from the *Ayni* and its various forms of *qonakuy*, will be the consensual punishment by the communal justice: to sever the transgressor's social ties. It is a living death. To become a *guacho*. From this word the native rural population transformed that word into another that names the destitute peasant we find in the Argentinean literature such as the famous characters Martin Fierro and Don Segundo Sombra. From the *guacho* Indians came our *gauchos*, whose translation from *Runasimi* would mean: "the ones without mother nor father", but not in the biological sense.

The filiation in this logic is not fundamentally carnal. It is interlaced to the territory: it is the *Ayllu*, the community work at the *Chacra* where the "upbringing" is collective and reciprocal among *Guacas*, *Runa* and *Sallqa*, in conversation, talking, or interpreting the languages of the different community dimensions.

22 Grimaldo Rengifo Vásquez, *Los caminos Andinos de las Semillas*.

The "Crossed hands" symbol of Ayni

The ones exiled from their Ayllu become aimless wanderers on earth, pressed into military service they became soldiers, or cowhands in large farms. Originally they were the guachos, the ones who do not belong to the qonakuy.

The *Ayni* exceeds the "mutual help" of "person to person", it is about legality, it does not necessarily give to the same person from whom he or she received something. The ties are on a community basis, so reciprocity is satisfied beyond personal bonds. Giving circulates. The reciprocity for what is received, may be destined to the deceased, to the *Pacha*, to the *Guaca*, to the animals, to the brotherhood of life, all members of the *Ayllu* in the common task of raising the greater family. Controlling through a complex network of male and female authorities, the respect of the "Law of the Ayni" from which stems the measure of what is 'fair' preserves community cohesion, limiting and dissolving conflicts among "brothers", or repairing the damage and administrating punishments for not complying or because harm caused to others.

This functioning explains the resistance to trust Justice in independent organs from National States. Justice in the Ayllu is communitarian, answering to its own logic.

In the iconography of the Andean people the Law of the Ayni has been symbolized by the "Crossed hands"[23]: one in a position of giving, the other of receiving. The image is repeated, carved in the geography, on stone and clay objects, as well as in drawing and colonial paintings - a mould of millennia on the long memory of our people.

In conclusion

Each culture organizes ways of social bonds and enforces them by the "Law" constituting them: not necessarily written, manifest and recognized by the community, but active in the style of life they transmit along generations.

Particular to the Andean world with its organization in *Ayllus*, based in the *Ayni,* the Law of the Andean Communality, symbolized by the "Crossed hands", naturalizes a well known spirit of constitutional hospitality of these nations. Its identifying logics and filiations not based on ideals of purity and common bloodline resulted alien to the dictum of totality, which requires exclusion and the elimination of the Other and its difference. A lethal condition before a drive for domination and power as the one inaugurated by the European invasion in the Americas.

Linguistically let us say that such a policy is based on figures of reciprocal recognizing, as a difference from other human groups which to determine themselves and the others they produce contrary asymmetric concepts[24] whose effects produce segregation, exclusion, and the negation of the other.

The structure of the language is inscribed in the social reality, in the unconscious, streamlining the social bonds and presiding over the enjoyment of the bodies, the interpersonal relations, and therefore, the dominant affection in a given epoch.

23 Archeological findings of the representation of *Ayni*, are dated as being 5000 years old (*Wak'a Kotosh*), found in ceremonial sites. Its stylization in the crossing of diagonals, a typical Andean iconography, is studied by paleo-semantics. Image: engraved in Stone of Sechín, Carlos MILLA VILLENA, *Ayni*, 10.

24 Reinhart KOSELLECK, *Futuro pasado*, 205–250.

The style of the bonds is thus transmitted through an identification with the social collective, by means of the symbolic structure of the language, and of the enjoyment it fosters, which are a product of, a trace of that processing that appears in the form of affections, desires, ideals and also sufferings and anguish. In that way resulting in an architecture, a general or predominant form of social bond, characterizing different communities or cultures. Due to this, sometimes they are difficult to translate and to accept the reasons put forward to judge people or compare them in respect to the approximation or distance from a singular "civilizing" model.

Obviously, a "cultural *ethos*" doesn´t dissolve the tensions between the universal and the singular, between the necessary and the contingent, inherent to life.

To bring forth here today an originating and contemporary conception of our Andean communities, that logic of the collective is different from the built-in matrix of the social ties volatized by the techno-capitalist circuit, and it has the intention of presenting[25] a civilizing invention, a supplement composed by social bonds weaving ties over the inconsistency and the incompleteness marking any symbolic order. By the same token it is not about vindicating an ancestral identity, complete and homogeneous -as a substitute of another historical subject teleologically constituted-, it is rather the consideration of making common bonds through another logic, diverse to the one historically used by the colonization.

It is based on practices and procedures of non cumulative circulation of gifts, in structures conceived according to a logic of the multiple articulated from the minimum parity of opposing but not antagonist: the quaternary structure of the *Chakana* as witnessing in its own way the human experience, the world we live in and where we learn and experience the policies of life,[26] or also where extermination is planned.

As from the immanent "lacking" of the "individual" and the necessary bonds with others for the caring and the reproduction of life, stemming from the contemporaneity of an historical legacy of our plural heritage of mixed races, we are surprised by their interpretations of contemporary urgencies, through their "savage" wisdom not domesticated by the imperatives of the yield and abolition of the symbolic, own universal features of being human. Western science is cur-

25 To produce a testimony: *Kawsaywillakuy:* in the language of the *Qheshwa* it means not only news, communication, or message, it participates of the *Ayni*, to give and receive. it is conformed by the verbs *Kawsay*: to live, to exist, and *willa*: to tell, to let it be known. and the suffix *Kuy*: mutually, reciprocating. The function of the testimony is ethical. It implies having an experience with others, it has a dialogical character, it constitutes the other interlocutor.

26 Jean-Claude Milner, *Por una política.*

rently reaching through other roads a conception of how life emerged and the ethical possibility of conserving it, in terms we consider harmonious with original/native visions of the cosmos.

The difficulty of resorting to written sources forces on us a keener look towards research and conceptualizations from diverse fields, supplying us with thinkable forms of signification, contrasts and productive similarities.

Qhipnayra[27] implications of the collective memory allows recuperating the cognitive statutes of human experience and systematizing it, in particular those implied in practices, languages and iconographies of our people. That invites us to a dialog with past narratives and at the same time with the current problems, its hermeneutical limits and amplified horizons towards a radical perspective of the future.

Convoking to think about the Community as a challenge of life is an implicit invitation to think about politics, always implied in the logics of the collective. It is to ask ourselves about the possibilities of survival of the human species, that one which reduced to an individual level results in a formal and homogenized abstraction; or if taken as a mass, can only be an object of utilitarian calculations or a surplus to be eliminated.

Author's note

The epistemological potential of the oral history as implemented by the method of the Seminar on Andean Oral History (THOA) directed by the *aymara* sociologist Silvia Rivera Cusicanqui in La Paz, Bolivia, and conformed by native intellectuals, have produced important works for research on forms of knowledge *quechumara* (qheschua/aymara), avoiding the academic colonialism in the practices of the Social Sciences: Carlos MAMANI CONDORI, *Los aymaras frente a la historia. Dos ensayos metodológicos*, La Paz, Chikiyawu, 1992; Domingo LLANQUE CHANA, *La cultura aymará. Desestructuración o afirmación de identidad*, La Paz, Tarea, 1990; Denise Y. ARNOLD, Domingo JIMÉNEZ A y Juan DE DIOS YAPITA, *Hacia un orden andino de las cosas*. La Paz, Hisbol, 1992; Roberto Choque Canqui, *Educación indígena, ciudadanía o colonización?* La Paz, Aruwiyiri, 1992. *The reports of* the THOA, Ayllu: *Pasado y futuro de los pueblos originarios*. La Paz, Ediciones del THOA, 1995 y María Eugenia CHOQUE, *La reconstitución del ayllu y los derechos de*

27 *Qhipnayra*. Future/Past. In the *Aymara* language: *Qhip* means at the same time the Back and the Future; *Nayra*, means Eyes and the Past. It is about a conception of Time inherent to human historicity, consistent with the works of Reinhart Koselleck.

los pueblos indígenas, THOA, Mimeo, 2000. Are added to the tradition of the historical work of Ramiro Condarco MORALES, *Zarate, el temible Willka. Historia de la rebelión indígena*, La Paz, Editorial el País, 1966; *Orígenes de la nación boliviana: Interpretación Histórico sociológica de la Fundación de la República*, La Paz, Instituto Boliviano de Cultura, 1977; *Historia del saber y la ciencia en Bolivia*, La Paz, Academia Nacional de Ciencias de Bolivia,1981; Xavier ALBÓ y Franz BARRIOS, *Por una Bolivia plurinacional e intercultural con autonomías*, La Paz, , Informe Nacional de Desarrollo Humano, 2006; Xavier ALBÓ, *Pueblos indios en la política*, La Paz, CIPCA, 2002 y *Movimientos y poder indígena en Bolivia*, Ecuador y Perú, La Paz, PNUD- CIPCA, 2008; *Interculturalidad en el desarrollo rural sostenible. El caso de Bolivia: pistas conceptuales y metodológicas*, La Paz, CIPCA, 2010. All of them are participants of *amuytáña*: to think, conceptualize and imagine a plural gesture of actualization and recreation of the collective memory, of which I participate with my writings on the field of des-colonial studies.

For the references in *Runasimi* language for the grammatical, orthographic and semiotic corrections I consulted Tayta Carmelo Sardinas Ullpu, member of *Mink'akuy Tawantinsuyupaq* and the Council of Mallkus of Potosí, Cochabamba, Oruro and Chuquisaca.

Besides, I consulted a variety of language publications:

1. *Revista Huaico Lazo Americano*, Jujuy ediciones Huaico, (2004).
2. Kusi KILLA, *Irpa*, Instituto Qheschwa Jujuymanta, Colección Kunanpacha, Jujuy, Ed. Aylluyachaywasi, 1ra. Ed., 2008.
3. Wanka WILLKA, *Aylluymanta*, Instituto Qheschwa Jujuymanta, Colección Kunanpacha, Jujuy, Ed. Aylluyachaywasi, 5ta. Ed., 2008.
4. Wanka WILLKA, *Tallmay*, Instituto Qheschwa Jujuymanta, Colección Kunanpacha, Jujuy, Ed. Aylluyachaywasi, 3ta. Ed. 2008.

Notice. I would like to point out that the categories and concepts stated in this paper has implied the use of a double translation of key words and expressions in the *Runasimi* and *Aymara* language, into Spanish, and then into English. The subsequent difficulty in the precise transmission of primordial significations is a risk that I decided to take.

Bibliography

Alemán, Jorge and Sergio Larriera, *Existencia y sujeto*. Buenos Aires, Grama Ediciones, 2007.
Alemán, Jorge, *Jacques Lacan y el debate posmoderno*. Buenos Aires, Ediciones del Seminario, 2013.
Alemán, Jorge, *En la frontera. Sujeto y Capitalismo. El malestar en el presente neoliberal*. Buenos Aires, Gedisa, 2014.
Colombres, Adolfo, *Celebración del lenguaje. Hacia una teoría intercultural de la literatura*. Buenos Aires, Ediciones del Sol, 1997.
Colombres, Adolfo, Juan Acha, and Ticio Escobar, *Hacia una teoría americana del arte*. Buenos Aires, Ediciones del Sol, 2004.
Freud, Sigmund, *Obras Completas*. Madrid, Biblioteca Nueva, 1973.
Huanacuni Mamani, Fernando, *Buen Vivir / Vivir Bien. Filosofía, políticas, estrategias y experiencias regionales andinas*. Perú, Coordinadora Andina de Organizaciones Indígenas CAOI, 2010.
Ibarra Grasso, Dick Edgar, *Ciencia Astronómica y sociología incaica*. La Paz, Cochabamba, Bolivia, Editorial Los amigos del Libro, 1982.
Keck, Fréderic, *Lévi-Strauss y el pensamiento salvaje*. Buenos Aires, Nueva Visión, 2005.
Koselleck, Reinhart, *Futuro pasado. Para una semántica de los tiempos históricos*. Buenos Aires, Paidós, 1993.
Lacan, Jacques, *Escritos. 1.décima edición en español*. México, siglo XXI, 1971.
Lacan, Jacques, *El reverso del psicoanálisis. Seminario XVII*. Buenos Aires, Paidós, 1992.
Lajo, Javier, *Qhapaq ñan. La ruta inka de sabiduría*. Perú, Amaro Runa Ediciones, 2005.
Laughlin, Robert B., *Un universo diferente. La reinvención de la física en la edad de la emergencia*. Buenos Aires, Katz, 2007.
Lavandera, Beatriz R., *Variación y significado. Y Discurso*. Buenos Aires, Paidós, 2014.
Lévi-Strauss, Claude, *El pensamiento salvaje*. México, Fondo de Cultura Económica, 1964.
Lewin, Boleslao, *La rebelión de Túpac Amaru y los orígenes de la independencia de Hispano América* (1ra. ed. 1943). Buenos Aires, Sociedad Editora Latino Americana, 1967.
Milla Villena, Carlos, *Ayni. Semiótica andina de los espacios sagrados*. Lima, Perú, Ediciones Amaru Wayra, 2005.
Milla Villena, Carlos, *Génesis de la cultura andina*. Lima, Perú, Auspiciada por la Asociación de investigación y comunicación cultural andina Amaru Wayra, 2006.
Milner, Jean-Claude, *Las inclinaciones criminales de la Europa democrática*. Buenos Aires, Manantial, 2007.
Milner, Jean-Claude, *Por una política de los seres hablantes. Breve tratado político 2*, Olivos, Pvcia. de Buenos Aires, Grama, 2011.
Miranda Luizaga, Jorge, Jorge Medina, Francisco Tamayo *et al.*, *Aportes al Diálogo sobre Cultura y Filosofía Andina*. La Paz, Siwa, 1985.
Prigogine, Ilya, *¿Tan sólo una ilusión? Una exploración del caos al orden*. Barcelona, Tusquets Editores, 2004.
Rivera Cusicanqui, Silvia, *Sociología de la imagen. Miradas ch'ixi desde la historia andina*. Buenos Aires, Tinta Limón, 2015.
Vásquez, Grimaldo Rengifo, *Los caminos Andinos de las Semillas*. Lima, PRATEC, 1997.

Axel C. Lazzari

Home is not Enough:

(Dis)connecting a Rankülche Indigenous Collective, a Territory, and a Community in La Pampa, Argentina[1]

For Julia, homing out

"We are bounded by the same need" was Curunao's response to the question about the causes of mobilization of a small indigenous collective aimed at securing a territory of its own. Curunao is *lonko* ("head", in ranküldungun) of "indigenous community" Epumer, one among the several groupings in which Rankülche indigenous people are organized in the province of La Pampa, Argentina.

Which is the foundation of this "same need" that stimulates a union among individuals? At first sight it is about social causes: to defend themselves from threats of eviction of their plots. These menaces refer to an ongoing land conflict confronting Epumer with non-indigenous persons in which the State plays as arbiter and regulator. Likewise, the "need that binds us all" might be explained bringing in the mutual dynamics of power relations and discursive strategies of legitimation by these actors.

In this article I develop a supplementary approach.[2] I maintain that this conflict cannot be reduced only to actors conceived of as human. Agency is distributed among forces, human and non-human forming networks or assemblages more or less effective according to volume and extension. I propose to explore the argument that in order to attain its ends in the struggle for the land, Epumer collective must constantly establish specific links with distinct sets of super-non-human forces. Specifically, I am interested in analyzing the crossings between Epumer's actions and the multiple agencies of "territory" and "community".

In what sense are territory and community non-human forces? In the discursive register of "Our land", "Home" or "Indian Territory", the territory metonymically refers to a multitude of non-human actors/forces such as cattle, pas-

1 I am deeply grateful to Laura Carugati for her support during the writing process and to Anne Gustavsson, Belén Hirose and Gabriela Leighton for helping to translate the Spanish version into English.

2 The "supplement" is an addition, allegedly secondary, to a unit deemed "natural" or "pristine". Unlike the complement it does not close a totality maintaining with it a deconstructive tension. In our case, what is considered "primary" is the sociocultural/natural domain which actor-network theory is the supplement of. See Jacques Derrida, *Of Grammatology*, 269 sq.

tures, water flows, weather, and equipment in general (houses, production and communication facilities and artifacts, etc.). Conversely, the idea of community points to the super-non-human domain, featuring the labour of spirits and souls transcending the human individual. Both non-human and super-non-human modes of agencies converge in networks of different size with with human actors. So my first argument is that Curunao's "need that binds us all" requires us to delve into causes other than socio-cultural and natural. Instead of limiting our inquiry to how human communities are constructed and reconstructed, articulating social powers and cultural knowledge in conflictive milieus, I attempt to grasp the different entanglements between human actors, territory, and community in order to weigh their relative strength and autonomy in the conflict about land. I am obviously nodding to ecological trends in contemporary anthropology, particularly to the Action Network Theory (ANT) developed by Bruno Latour and other authors.[3]

Territorying, communing, linking, dissolving

Anthropocentrism looms in every time the territory is conceived of as a resource-object, a tool, passible of a subjective semio-cultural inscription and political use. Far from being born as an inert fact which is enlivened as "territoriality" or retroacted as discursive power effect the territory is an agency among others, playing in mutual translations along a rhizomatic contact-space ontologically prior to any "entification" that would break it up between an object and an object signified by a subject. Needless to say, the dualism nature-culture and its modern boosting are underscored by this operation.[4]

3 The actor-network is an assemblage of forces (actors, actants) that performs actions. The notion of network refers to the association of irreductible human, non-human and super-non-human monads. Derivate notions such as hybrid, quasi-object, quasi-subject and factish name the fragile entities stabilized by those associations, alignments, compositions. These specifications clearly set Latour apart from the most familiar "network analysis" deployed within an interactionist and formal sociological traditions which start from the ontological principle of reducibility of entities to one another. See, among other works, Bruno LATOUR and Steve WOOLGAR *La vida en el laboratorio*; Bruno LATOUR, *The Pasteurization of France*; ID, *Ciencia en Acción*; ID, *Nunca fuimos modernos*; ID, *La esperanza de Pandora*; ID, *Aramis, or the love of technology*; ID, *La clef de Berlín*; ID, *Politiques de la nature*; ID., *Sur le culte moderne des dieux faitiches*; ID., *Reensamblar lo social*; ID, *Investigación sobre los modos de existencia*.

4 Philippe DESCOLA, "Más allá de la naturaleza y la cultura'"; Bruno LATOUR, *Nunca fuimos modernos*.

"Territory", thus, names the actions of non-human forces such as animals, plants, minerals, weather and artifacts each of them infinitively demultipliable in other forces– which frequently correspond with human practices but occasionally –when they cease being "at hand" or break down- come across the path of humans accustomed to use and signify them, as if they had been dwelling, even haunting from the rear.[5] Therefore we attach to the common phrasing of "social groupings disputing land" –connoting an economic, political and cultural articulation[6]- the cautionary remarks "with territory", "in territory", "in 'political' accord or disaccord with the non-human actants". We thus speak of a "territorying" or becoming-territory as the entangling of these forces. This move will allows us to defocus human intentions and struggles, and projecting them in a wider field of affectation with a multiplicity of non-human things and animated beings. A too human dispute about residence, resources and reproduction must acknowledge the exposition to territorying.[7]

Taking this cue, isn't community the *locus classicus* of a discourse that describes the overcoming of the individual or the divided group by a divine force or a secular spiritual one (social or cultural)? Community always entails a super-non-human power more or less transcendent like a God, a Geist or a fetish that bring people together, despite their socio-symbolic differentiation. Insofar as these super-non-human forces work for the social organization we can speak of communalization, nation-building, etc., all different modes of attaining collective identifications by means of putting things and persons in and out the common. Yet we should take this ghostly character of community literally in order to grasp not only its social means and effects –a national holiday or a witchcraft accusation in order to revitalize the "us"- but also its apparition amidst human collectivities. We then speak of "communing", a limit case of super-non-human agency which in linking (individual or collective) nodes and leaving others aside dissolves all and every entities in con-fusion. Communing is the overflowing of "society", "community" *and* the networks of human and non-human actors. It is the occasion of an ontological event. In a way, Latour indicates this limit by the

5 This Heideggerian tone is reluctantly acknowledged by Latour, who chastises the philosopher for seeking Being just in the wandering trails of the forest – Holzwege –, while Latour deems it expanding everywhere, namely through technological hybrids. Ibid.

6 See Rogério HAESBAERT, *El mito de la desterritorialización.*

7 The opening up of the Pandora's box of the object raises the question of whether a thing exhaust itself in a relational effect. This problem is thoroughly rejected by Latour who is interested in things as actors. However, some commentators of his oeuvre consider it legitimate. See, for example, Graham HARMAN, *Hacia el realismo especulativo.*

notion of “plasma”.[8] Briefly put, while in territorying things, animated beings, minerals and weather go on unnoticed and sometimes may come up to the fore, it is the unbounded effect of spiritual forces which manifest in communing.[9]

In this perspective, the territory is not the object of dispute between human individuals and collectives; the controversy is between different networks of humans-territories. The so-called “real lords of the land” and the “legal owners”[10] are, in the best of cases, the queen bees of non-human swarmings. Besides, they both may turn into “victims” of “spirits” (communing), looming in with an event.

Insofar as things, plants, animals, weather, etc., conform “polities” in spaces non-correlative to secular politics, and given that the plasma “measures the extension of our ignorance” (my translation),[11] the question of the political subject, apart from considerations of power and legitimation, calls for the recognition of its hybrid pragmatic consistence of nature, culture and the spiritual.[12] The author envisages this project as a cosmopolitics distinguishing it from cosmopolitanism “from above” –the universable human subject- and cosmopolitics “from bellow” –the particular-universable cultural subject.[13]

As a methodological corollary we must place the “need that binds us all” within the frame of territorying and communing thrusts affecting Epumer, the “legal owners” and the State technical officials as well as look into the actions deployed by the latter ones in order get along with them or defend themselves from the unexpected occasions of their emergence. That is why we argue that “home” –that is, Epumer- cannot suffice itself, being constantly “disemboweled” by different forces. Yet, home insists in surfacing but it is something else, perhaps more fragile (and promising) that the usual indigenous subject.

In order to probe into Epumer networking with non-humans and super-non-humans we set up a chronological schema to recover some key historical incidents, and gradually zoom in a micro-process as the concrete object of analysis. The questions are always the same: how do human actions assemble with

---

8 Latour introduces the concept of “plasma” as that which overflows the links and nodes of an actor-network. The Latourian general equation of reality with an effective assemblage, however, needs “something else” (a supplement?) which he sketches briefly in *Reensamblar lo social*. The plasma is a reserve, the space of the unformatted, some sort of a virtual. At this point the author comes closer to Deleuze and Guattari. See Gilles DELEUZE and Félix GUATTARI, *Mil mesetas*.

9 Territorying may well become communing in the guise of the *genius loci* but for the sake of simplicity we keep our distinction between things and animated beings, and spirits.

10 Self-referential category authorized by the State but not by Epumer members who refer to them as “so-called owners”.

11 Bruno LATOUR, “Paris, ville invisible: le plasma”.

12 ID., *Politiques de la nature*.

13 ID., “Whose cosmos?”

non-human territorying? Can we point to events of emergence (communing)? How do the membranes of Epumer perform a home?

The winding road to Epumer: from Colonia Mitre to "Agrupación Aborigen Ranquel"

Our history begins around 1900 when more than a hundred indigenous families settled down in Colonia Mitre. Most of these families were related to Rankülche "indios amigos" who helped the army in 1878 and 1879 during the military campaign which culminated in the dismantlement of the sovereign indigenous societies in Pampa and Patagonia.[14]

Colonia Mitre was originally a spatial dispositive designed by the national government to "civilize" the indigenous people through agriculture and shepherding. It was located in a territory deemed inhabitable by the Rankülche which they called "Hualichu Mapu" ("Land of the Demon") given its semiarid characteristics including bad pasture, sand dunes and extreme climate.[15]

Families spread out over this area according to flexible patrilineal houses led by the lonkos that had established political dealings with the white world. Four clusters were formed, one of which maintained a greater degree of cultural continuity, something which can be verified, for example, by a higher level of group endogamy (patrilateral cross-cousin marriage), the conservation of the language and also certain ritual practices. This was due to random factors, the most important being that this northwest area of the Colonia remained free from the direct civilizational influxes brought in by the concrete nodes of a peace court, a police station, a country store, a school and a post office. In time this distant-from-civilization place became known as "El Pueblito" (little village), expressing certain ethnic limits within the Colonia itself which, on the other hand, was progressively being whitened either due to the arrival of non-indigenous inhabitants or because the indigenous themselves were looking to become de-indianized.[16] Ethnicity

---

14 For archaeological and historical reconstructions from 18th to late 19th centuries see, among others, Martha Bechis, *Piezas de etnohistoria*; Rafael Curtoni, *Archaeological approach*; Jorge Fernández, *Historia de los ranqueles*; Axel Lazzari, "'¡Vivan los indios argentinos!'"; Marcela Tamagnini and Graciana Pérez Zavala, *El fondo de la tierra*; Alicia Tapia, *Arqueología histórica*; Daniel Villar and Juan Francisco Jiménez, "Acerca de los ranqueles".

15 Axel Lazzari, *Autonomy in Apparitions*; Claudia Salomón Tarquini, *Largas noches en La Pampa*.

16 For this period see, for instance, Ead., "Redes sociales"; Ead., "Estrategias de acceso y conservación"; Ead., "Los vínculos con los ranqueles", all of them studies in social networks.

was strengthened during the 1950´s with the so-called "arrival of the Evangelio" whereby a Pentecostal Church was formed, consolidating bonds which allowed for a "controlled self-civilization".

During the 1960's, after the provincialization of the ex Territorio Nacional de La Pampa, there was a developmentalist wave which had a direct impact on Colonia Mitre. It involved promoting the entry of "rational producers" to diminish the area's backwardness. Those individuals started to displace indigenous occupants with the support from the Colonia's own police force and judicial apparatus. Thus an indigenous mobilization was organized and managed to revert the situation thanks to judicial maneuvers and a successful campaign in the provincial and national press.[17]

It was the time of the de facto military rule with nationalistic and developmentalist overtones. The case of the "last Rankülche" who resisted the ones coming "from outside" the Colonia provided the opportunity to launch a development program from the provincial government which included public works, entitlements, as well as economic, sanitary and educational improvements, and a complete re-territorialization of the space. This plan, known as the "Operativo Colonia Mitre", was initiated in 1969 and faded away around 1974. The crucial moment came in 1972 with a new distribution of the plots well as the issuing of individual entitlements deeds.[18]

The years comprising to the so called military "Proceso" (1976–1983) registered the almost definitive whitening of Colonia Mitre since most of the indigenous inhabitants which had become owners in 1972 ended up selling at a loss to "people from outside" or to white inhabitants from within. With the return of the democratic regime in the middle of the 1980's, the indigenous problem was gradually visibilized at a national and provincial level culminating in the organization of a provincial Rankülche indigenous movement at the end of this decade. We witness the passage from a dispositive aiming to "integrate the Indian to the nation" to a multicultural one based on the recognition of autonomous "indigenous peoples". In the case of La Pampa this occurrence acquired the form of a transition from the imaginary of the "Phantom Indian", whereby the Rankülche were considered to be at the brink of disappearance according in national mainstream, to another imaginary which accepts Rankülche indigeneity but remains suspicious of its authenticity.[19] In this context, at the beginning of the 1990's the

17 See Axel Lazzari, *Autonomy in Apparitions*; José Ignacio Roca, "Agentividad indígena".
18 See Axel Lazzari, *Autonomy in Apparitions;* José Ignacio Roca and Anabela Abonna, "El 'Operativo Mitre'".
19 See Axel Lazzari, *Autonomy in Apparitions;* Id., "Aboriginal Recognition"; Id., "Autenticidad, sospecha y autonomía".

"Agrupación Aborigen Ranquel Epumer" was created upon the base of families belonging to the evangelical church in "El Pueblito" and to the group mobilized between 1963 and 1975.

The struggle of Epumer and its official recognition as "indigenous community"

After the national debt crisis in 2001 and the consequent devaluation the soya boom finally finished taking hold. Lands, previously destined to ranching, were absorbed into the soya bean frenzy, and marginal areas such as Colonia Mitre started feeling the expansionist pressure. According to some accounts, "people from outside" showed up in the Colonia and began to occupy land for the purpose of introducing cattle. At the same time, other inhabitants were threatened with eviction. At this point the "Agrupación Aborigen" underwent changes in its membership by accepting people with generic indigenous ancestry or non-indigenous "puesteros".[20] This practice is an unconscious syncretism between the traditional *lakutun* whereby the Rankülche and Mapuche of the 18th and 19th centuries used to incorporate "strangers" to their groups, and the more common popular custom of adopting orphans or raise others' children as their own. But the new members weren't real strangers since they were already a part of the evangelical church. Only one of the new candidates wasn't related to the church but did belong to a family who had "forever lived in that place". Besides, and this is fundamental, they were all rural neighbors and, as the lonko said, share the "same need" of defending their plots. All of these people would contribute to the association with their practical knowledge concerning legal procedures in matters of land litigation.

A series of incidents followed in. Defensive actions against evictions, legal complaints, press releases and getting support from provincial indigenous movement and the National Institute of Indigenous Affairs (INAI by its Spanish acronym) were among the most prominent. The collective action transformed previous allegiances based on church membership and neighborhood and gradually an "us", a home, grounded in the common experience of rights enactment, began to settle in.

Meanwhile, Epumer would proceed to claim before the INAI the legal personality as "indigenous community", a recognition finally granted in 2007. The

20 "Puestero" is the caretaker of a "puesto" or rural post, a set of modest houses and ranching facilities. All members of Epumer are puesteros, but not all puesteros are indigenous.

category must not be thought of as entailing the verification of actual communal practices. It is foremost a legal instrument whereby the State makes multiplicity legible as "indigeneity", conditioning the scope of its organization, action, and identity, precisely as a "community" within a plural political environment. From Epumer's standpoint, "indigenous community" is a much valued political tool which serves to authorize it before government and public opinion based on the laws and the Constitution (reformed in 1994) which indigenous citizens with special rights.[21]

Epumer and the process of recognition of an "indigenous territory"

We can now start our ethnographic reconnaissance of territorying and communing of Epumer.[22] How do actor-networks perform during the technical survey of an "indigenous territory"?

The same debt crisis and commodities boom that triggered expansion on marginal lands obliged the State to take pacification and regulative measures of old and new rural frictions. It was in this context that a law was issued in 2006, declaring the "emergency in matters concerning possession and property of lands traditionally belonging to the country's indigenous populations" (my translation), and calling off eviction processes (until September 2017).[23] The law commissioned the INAI a "technical-legal-registral survey of tenancy situation in lands occupied by indigenous communities" (my translation).[24] This gave birth to the program of Territorial Survey of Indigenous Communities (ReTeCI by its Spanish acronym). In a language framed in pluralist terms the program pleads for a multidimensional territorial approach. Therefore, territory is analyzed in political, economic, social, cultural and religious "components" in order to make some room for aboriginal models of it. We will soon be able to assess the practical consequences of this conception.[25]

We must not lose sight of the fact that the ReTeCI is the latest variety of a long history of territorial and demographic policies –entailing conflicts for territories-

21 On the contrary, the figure "Agrupación Aborigen" is restricted to the Civil Code and a provincial jurisdiction.

22 See Axel LAZZARI, "El Relevamiento Territorial", for a more detailed description.

23 Law 26.160.

24 Ibid.

25 Instituto Nacional de Asuntos Indígenas, Programa Nacional de Relevamiento.

that traces back to the 18th century when the "Indian Frontier" was established in the area. The land and people INAI's officials would meet on the ground are but the transformed remains left behind by the constant encroachment of the Rankülche dwelling area prior to the conquest in 1879. The ReTeCI, in line with its unacknowledged military and developmentalist precedents, reveals a governmental rationality whose objective is the building up of security mechanisms in order to protect life and foster productivity in the name of the nation.[26] Its novelty is the framing of the indigenous territorial problem in terms of a respect for the (cultural) autonomy of the governed and the management of risks derived that characterizes a neoliberal mindset.[27] As it will be shown, these socio-structural features of the ReTeCI are not contradictory with our argument. In the same vein, let us recap the human social actors involved and their strategies in the land conflict.

a) Epumer community. Its strategy aims at securing a land for each one of the members at risk. This is not self-evident because Epumer also comprises persons who hold legal titles so that the ReTeCI gets them to decide whether or not the plot they own should be included in a future community title. While some members (mainly the lonko) are "in need" of the communal title, others are not equally prone to turn into potential land "commoners".
b) "Legal owners". We do not have enough information about their discourses and practices but a few comments suffice for our purposes. They include a private corporation, a joint venture society, and in some cases, relatives and neighbors, but unlike Epumer there is no common front among them. It should be distinguished between owners that seek legal eviction, from those whose purpose is, as we will soon show, to reclaim a plot. In general, their strategies consist in litigate to "get rid of a problem". Most of them just want to win a lawsuit, sell the property and "get outta of that place". What they share, however, is the State's authorization to claim a legal individual property. Not exactly a need but rather a "natural right" granted by fiat. Title deeds, of all the para-legal "papers" that hover around land conflicts, would be the key non-human actors on legal owners' side. But how do they act? In the case we will analyze it is the phantasmagoric influx of this paper fetish which is duly recognized.
c) INAI's technical officials. Their fundamental objective is to produce a "traditional territory" according to the unequivocal verification of a human occupation as "present, traditional, and public" (my translation).[28] In order to

---

26 Michel Foucault, *Seguridad, territorio, población.*
27 Nikolas Rose, *Powers of Freedom.*
28 Law 26.160, Art. 2.

> accomplish this, the ReTeCI follows a protocol of participation and consent of the interested party, namely indigenous peoples. The action is determined by legal and administrative provisions and precedents as well as by political negotiations with indigenous movements and between federal and provincial jurisdictions.[29]

The ReTeCI spanned out from mid-2009 to the end of 2010. The task were necessarily performed *in situ*. This obvious point must be stressed because, technicalities aside, the State needs legitimating itself on the ground by the beneficiaries/people. Yet it is this very requisite which exposes the whole enterprise to territorying and communing. The same predicament applies to "participation" and "consent". If it is true that these procedures are tendentially channeled and neutralized, they are also the occasion of super-non-human interference.

Directions were accomplished: croquises of land plots were drawn by each occupant, narratives about community were recorded, and last but not the least some areas of the territory were walked along, capturing with GPS technology, the coordinates of plots' corners, buildings, memorial places such as cemeteries, and remnants of past habitation. In this fashion an important amount of data was collected aimed at producing an "indigenous territory" according to State criteria, namely "geo-referenced" maps.

Upon officials' return some months later in order to get the community's consent of the provisional cartography, some of Epumer associates disagreed with the fact that a plot that had been duly drawn up in the sheets and named in the narratives as "La Reserva" had been snatched from the map. The ReTeCI file itself reads the lonko words referring to this plot as "usurped (...) some 10 years ago (...) my grandfather and my father lived here but from 10 years now it is occupied by a puestero employed by someone from Buenos Aires; now there is a wired fence that cut us off" (my translation).[30] The referential ambiguity of "La Reserva" –a reserve of pastures and an Indian reservation- confirms that this zone was customarily used as a grazing field by dwellers from "El Pueblito". Despite these proofs, the area was left out of the survey.

---

29 There are other sets of human actors which we are unable to follow in this opportunity, namely, the media and educated public opinion, the provincial Rankülche movement and the social scientists, all of them fashioning different ideas about the case. Other important actors are the judiciary, government officials and political representatives.

30 Instituto Nacional de Asuntos Indígenas, File 50277, f. 52.

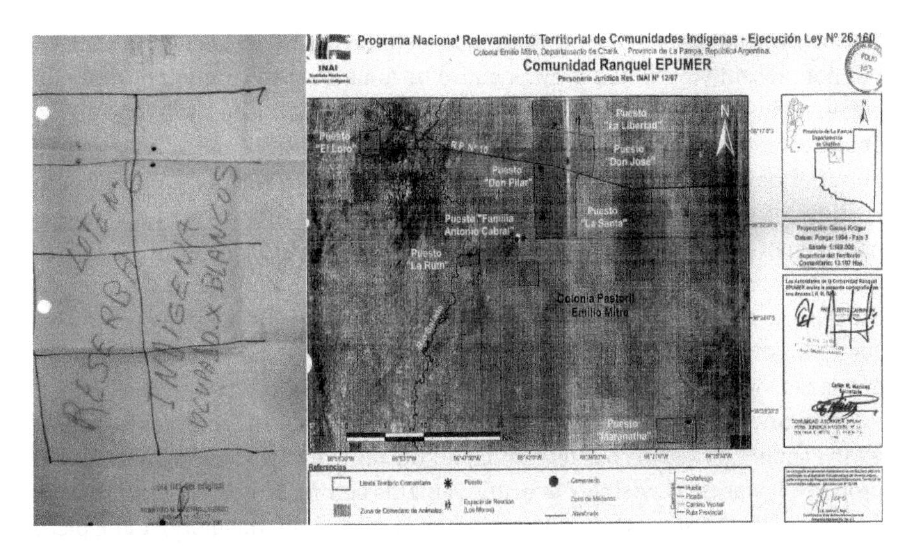

Map 1: "La Reserva", present in Epumer's croquis, absent in provisional map. Archive of Epumer (access authorized).

The report establishes that none of the members of Epumer occupied "in the present" those fields so that "in agreement with Law 26.160 this sector has not been geo-referenced" (my translation).[31] In fact, the sector was "referenced" but not "*geod*", that is, people's sayings did not coincide with what the technical eye was trained to see. It seems an undisputable fact that Curunao's people were claiming a land they were not occupying "at the moment" of State's officer obturating his eye. Everyone knew that this land was being occupied by the "so-called owner". But wait! Was he actually there at the moment of the officer scanning the spot? Definitively not. If the officer could only see an "empty space", what did he fill it with? Let us dwell on this critical conjuncture and the ontological and epistemological problems involved.

First of all, the survey does record territories as culturally mediated. But this recognition is the result of splitting it in "components".[32] So a territory is a combination of a layer of nature (space, resources) and a layer of culture (memory, artifacts, built environment).[33] "Occupation" means for technical reason to live

31 Ibid, 52.

32 Instituto Nacional de Asuntos Indígenas, Programa Nacional de Relevamiento, 27.

33 This conception guides and is reinforced by techniques of map design and visualization, which allow for "freely" superimpose layers of information (culture) on a represented (physical) space. It is no coincidence that truth claims by the ReTeCI program fully depend on the justification of the precision and flexibility of the technical instrument deployed. The description of

"on" a physical space which is derivatively cultured. If "La Reserva" is claimed but not presently occupied it turns out to be a mere representation of desire or worse a cheating. No matter what "strange beliefs" about territory and indigenous community may have, if the "believers" do not live "above" the surface object of belief, their claim must not proceed. In other words, "cultural" conceptions of territories as involving networks of human, non-human and super-non-human powers are only acknowledged as long as they "fit in" in the space actually controlled by that population. We have found the limit of the cultural conception of territory endorsed by ReTeCI. Yet, things get complicated in this case, for as we mentioned the land claimed was being populated by a non-indigenous person. Curunao's "traditional", affective argument –the land as evoking the memories of his dead father and grandfather- had been acceptable if there wasn't a more powerful counter-argument: the law, specially, the right of private property. It was this invisible force, this veritable spiritual force what the officer's eye saw in the "empty space". Between Curunao's culture (memory) and a piece of nature (land), law as a super-non-human phantom appeared. This powerful actor-network that combines nature and culture under the overarching sacred being of private property and the violence is the State. Against the fetish of the State nothing can be done except out-fetishing it.[34]

On a closer look the fact of not accepting Curunao's claim shows the traces of mis recognizing the clash between an indigenous and State territorying. Officials' habitus is oriented to record and objectify "things" in/as maps and photographs, and "words" in/as stories, statements and opinions. In so doing it prevents non-human agencies of the land (cattle, rain, soil, equipment, etc.) to surface in otherwise than "tools" and means. It also hides its own condition as derived from State fetishism. Yet there was a circumstance in which an actant appeared at a crossroad, precisely at the moment the lonko, in company of a State official, approached the gate (*tranquera*) at the border of "La Reserva". In pronouncing the words we have already quoted, the lonko bodily indexed the territory being the object of discourse. Not only did he get himself in contiguity with the place, but he also made the tranquera appeared as a non-human in full power. Two actor-networks were at stake in deciding whether a gate open or remained closed. The lonko meant to open it and walk into "La Reserva" but the State official pre-

---

"Jaguar" a digital GIS (SIG, by its Spanish acronym) amounts to 60 % space in ReTeCI's official booklet. For problems arising from "participatory mapping" in Latin America, see Carlos SALAMANCA and Rosario ESPINA, *Mapas y derechos*.

34 This is the argument developed about Rankülche "struggles for recognition". See Axel LAZZARI, *Autonomy in Apparitions*. Also Michael TAUSSIG, "Maleficium: State Fetishism".

ferred to keep it closed, again, with the unnoticed aid of law as ghost. It was then that he asked if they presently occupied the land "on the other side" of the gate.

Not including "La Reserva" during the first survey obliged Epumer to take an action at its own risk. The response was twofold: recovering this tract of land and, once settled in, presenting a counter-argument to the INAI. Asked about the first incident the protagonists respond with fear, joy, shame and neglect. These mixed sentiments put us on the clue of an event with foundational undertones.

Epumer retook "La Salvación", the puesto of "La Reserva", in January 2010 "counter-evicting" the ones that had "got into" the land in 2003. Interestingly, these people were initially tolerated as long as their cattle shared the grazing fields with Epumer members'. I recall one of them saying "What can we do?" It was not until the uninvited guests cut the cattle's pass off that these people became intruders "coming from outside". In any case, the "so-called owners" were fully aware of their fragile position as they had attempted to sell the land in a judicial auction without success.

One night, taking advantage of the absence of the caretaker, a small group comprising two old persons and a boy slid in "La Reserva" laying their "reales" by the house.[35] The next day the puestero arriving in from town got startled. In the clash of gazes and presences the group declared their intentions to stay. The man went back to town and returned once again, this time with the "so-called owner" and two police car-patrols. Their purpose was evicting the counter-evictors. It did not happen. The deputy, in a conciliatory tone, attempted to take Curunao to the police station in order to solve the trouble, as if he had intuitively noticed the works of territorying against him. In fact, the police car-patrols had been intercepted by a community member's vehicle in the middle of the road. Besides the policemen were being the target of an insidious and bold camera-eye, which would keep the "so-called owner" trapped inside the police vehicle.[36]

35 The "real" or "rial" is a temporary tent. Its name potently evokes reality as an act of dwelling. The "real" also connects with the "toldo", the traditional indigenous abode before the conquest.
36 The videos allow us to reconstruct details of the situation that escape actors' retrospective accounts.

Image 2: "Trapped inside". Video capture. Epumer community, 2010

-Are you authorized to get in here? – demanded the lonko's daughter
-And you? Do you own this property? – replied the deputy.
-I am owner...I am living here – she responded.
-Everybody, everybody, all this belongs to the community – intervened the lonko.

The police, cornered by rhetoric and material deployments, had to retreat to town. After the confrontation, the lonko showed up in the police station, and filed legal complaints against the deputy and the "so-called owners" for intruding indigenous property. Since then "La Salvación", the heart of "La Reserva", is in the hold of a flexible assemblage of human and non-human agents coordinated by Epumer.

The retreat of the policemen was experienced as a great victory. Men, women, children were all excited. Their bodies intermingling with ruins, film cameras pointing at horses staring back at them, cars with their doors opened, the wind leaving its trace on a microphone...It was the closest point to the overflowing of an actor-network by plasma.

Image 3: "Toward communing?" Video capture. Epumer community, 2010 (Image retouched)

After recovering the puesto, some persons went to Buenos Aires in order to convince the INAI of the fact that "La Reserva" was now "presently" occupied by humans. With the puestero not there anymore the Rankülche were trying to conjure up an alliance nature, culture and the spirit of law. They were trying to out-fetish the State. Curunao's argument of a dwelling of the place by old generations was heard but again did not sound convincing. Then came the community's *secretary,* who argued that this "present occupation" was, in reality, part of a dwelling which had been continually performed through the intermediation of cattle until the intruders closed the land off with a wire fence. Thus, the idea of cattle-men assemblage was introduced but, as we have shown, the ReTeCI recognizes these notions as part of a "culture of theirs" and as long as they fit in within the perimeter of their dwellings. So this was not a decisive reason either. However, the INAI decided to send a new team to survey the place. The only reason may seem to be "political". In effect, it was political and in this calculus the critical move had already been made and it was that the spot under dispute was being inhabited "in the present" by Epumer people. In other words, there was a territorying going on as though pushing the Rankülche to recruit the State as ally. The INAI as part of the State, a spiritual-material entity whose office is not only to enforce the law but also to legalize force, recognized the *fait-accompli* of occupation and opted for the latter.[37]

A month later a "complementary" mapping was completed and finally "La Reserva" (and its puesto) was included as a part of the indigenous territory. Yet there was another surprise. The official report reads that "according to the position of this Institute the recovery of the puesto 'La Salvación' is considered as recent (...) since the technical officials were unable to verify in that occasion (during the first survey) that a present, traditional and public occupation of the mentioned sector proceeded" (my translation).[38] The plot was thus incorporated to the indigenous territory with the marks of the exceptional. Through this disclaimer the ReTeCI was insisting in having the last word introducing with the adjective "recent" the suspicion of irregular proceedings. This adjective opens a new set of problems, not addressed in the ReTeCI. By "presently the INAI pragmatically intends not an instant but a portion of passing time, but when does this interval begin and end? The markers of time are people, things, animals and their change of position or identity signals a commencement or a closing. Perhaps Epumer's occupation was still too hot to attain the status of "presently" as if people were still moving in. But we should ask the reversed question. Why are not the occupation of "La Reserva" by "people coming from outside" in 2003 or, even more, the so-called "conquest of the desert" in 1879 deemed as "recent"? The reason lies again in the operation of a human and non-human network guided by

37 We have reconstructed this situation from the memories by Rankülche actors.

38 Instituto Nacional de Asuntos Indígenas, File 50277, f. 112, 113.

the fetish of the State, a powerful hegemonic network that in recognizing Indigenous peoples' rights can only really acknowledge them as exceptions.[39]

Be it as it may be the "sociocommunitary" questionaire (CUESCI by its Spanish acronym) attached to the ReTeCI yielded the following profile of Epumer. It is a community of twelve families comprising about seventy people. They have dual residences, commuting between rural puestos and house in the town of Santa Isabel. Family members mostly live on small scale cattle-raising, doing some textile crafts, getting occasionally hired for rural tasks, and, in a few cases, being public employees (nurses, municipal workers). Educational, sanitary and habitation conditions are acceptable. In sum, Epumer is far from presenting the poverty-stricken image of some indigenous communities in Argentina. Perhaps the only important "non-satisfied basic need" is the provision of water for human consumption, but this applies also to regional population.[40]

Such living conditions and the fact of slightly presenting the diacritical marks of aboriginal difference (vernacular language, esoteric cosmologies, "communal" organization, "non-white race") make Epumer community and the Rankülche in general suspects of inauthenticity. However, what is at stake is not cultural authenticity –a habitus ingrained by State "cartographies of alterity"-[41] but their autonomous practices as indigenous, taking this term literally as "from the place".

Image 4: "La Salvación" in the fields of "La Reserva". Photo by the author, 2016.

39 Our observations do not point to specific individuals but to State bureaucracy which, although traversed by rifts, still are vertical organizations. Nor do we lose sight of the fact that notwithstanding its shortcomings, the ReTeCI opened specific possibilities to actor-networks guided by indigenous interests.

40 Beatriz DILLON, "Modelos de desarrollo".

41 Claudia BRIONES, ed., *Cartografías*.

As of December 2010 the provincial government handed Epumer and other Rankülche communities the so-called "folders", the documents (maps, narratives, legal resolutions) resulting from the survey. Perhaps driven by the impression that the province of La Pampa seem to be one of the few "stories of success" of the ReTeCI, national and provincial authorities decided to announce the expropriation of "La Reserva" in order to definitively grant a community property title. It's been six years since then and nothing has changed. But the Rankülche are allied with two powerful non-human actors: the "folders" and the puesto. Official documentation unambiguously states that Epumer is an indigenous community comprising more than 15,000 hectares recognized by the State. This fuels on expectations about land expropriation. On the other hand, dwelling in the puesto announces the promise of "the common".

After the survey: taking care of the homestead, sharing the land?

The puesto has undergone deep changes. Its inhabitants, the lonko and his wife, say that they just dedicate themselves to "take care" of the homestead. At the moment of recovering the land, the place was in a state of semi-abandonment. One building was a pile of rubble and the other was deteriorated (see previous image). Only the corrals, the water-tank and the windmill were in operation. The causes were not that the puestero lived with his family in town for the new inhabitants also commute every week from the puesto to their urban residence. The reason had to do with a certain "practical-man" turn of mind that transformed the place into an instrument of capital accumulation.

A livable and vital home was born out of the efforts of the lonko's family with the aid of community's hands. Not only cattle "that care for itself" if there is enough pasture but, appropriately, farmyard birds that need a constant attention generally paid by women and children. During my stays in the puesto hens, chickens, cocks, geese and even a *charito* (the offspring of the American ostrich) would move around until dust, when they would take shelter in an abandoned truck turned into a hen house. Simultaneously, a bedroom, a kitchen and, above all, an *enramada* (a patio thatched with dried foliage) were raised from the ground. Notably, people did not tear down any of the previous buildings. A bricolleur libido seemed to be invested in repairing, redistributing, and reorganizing the remains left behind by the puestero. Besides, he was never really considered as an "enemy". Not only were respectful manners kept during clashes, but also we witness the worries of some people for the puestero, a "poor thing like us". It seems that the various ways of acknowledging the traces of the "usurpers" are constitutive of the taking care of the place.

Image 5: The "truck that was left behind" reused as hen house. Photo by the author, 2013.

Cell-phone communications, automobiles coming and going from town, carrying people, water, food, animals, and information also contributed to exponentially refurbish this actor-network. "La Salvación" is now growing as another home for the lonko's family and becoming the home of the community as well, since it is at this node that a subtle and patient territorying of Epumer as a whole goes on. But this is a task that requires practical and character skills. Perseverance is always in need to negotiate arrangements with this shade-less, water-scarce landscape of extreme temperatures. Fires, in particular, are deemed the most untamable of "elements", burning pastures, animals, puesto. One of the members of the Epumer was severely injured by a fire years ago. Despite all this the lonko would stroll his gaze around the puesto and say to me: "over there, the cows", implying that this is all they need to live on. Quite an understatement of human autonomy in an actor-network without claiming any property whatsoever!

Image 6: A living landscape with cattle, fires and horse-rider. Photo by the author, 2016.

What of the promise of "putting things in common"? We refer with this phrase to the level of reciprocities regardless their character as balanced, negative, or generalized. It is the field of debts accountable to subjects whereby the "us" expands and diminishes its perimeter. This resembles, in Esposito's vocabulary, a sort of "communitas of immune beings", an arrangement of the "us" on the basis of an ontological limit imposed by death.[42] Likewise, "communitas" without immunity is, according to the author, the sharing of nothing, a given-ness without devolution, "nothing in common", all of them tropes akin to our idea of communing.[43]

Let's explore now the problem of what and what not, according to people, pertain to "the common". How do Epumer's members react to a probable future of a territory as common property? The figure of indigenous community property is presently undergoing debate but, in principle, it means that the territory becomes an unalienable and inembargable thing.

Let us recap the land tenancy situation among Epumer members. On one hand, there are those occupying plots under litigation who would benefit from a common land property. On the other hand, those holding legal titles would eventually lose their individual property. In these circumstances an agreement was made whereby the latter would not be obliged to donate their property to the common. If the condition of a community member does not entail participation in communal land property, then an indigenous community like Epumer may

42 Roberto Esposito, *Communitas.*

43 A possible confusion may arise out of homonymy. Indeed "communitas" is the influential formulation by British anthropologist Victor Turner to conceptualize ritual. However, the key distinction between Turner's and Esposito's "communitas" is that while the former grasps a situation of human brotherhood beyond social statuses (anti-structure), the latter points to a post-humanist, "in the open", direction. See Victor Turner, *The Ritual Process.*

include individual title holders along with a common land title. Thus regarding the members with legal titles, Curunao's "same need" refers to the possibility of belonging to a community without "putting land in common".

Another problem may arise in future. Many of the litigant members may well gain their lawsuits before the actual granting of indigenous community property to Epumer. In that scenario, would they "take their land plots out" of the territory, thus ceasing the "same need" to "bind us all"? This probable moral test already haunts some of them but so far the intention predominates that eventually they would "donate" their plots to "the common". Interestingly enough, a person told me that his reason was not really Epumer community's but one going deeper, binding the spirits of the dead and the living. By surrendering his title to the indigenous community, excluding land from the market and securing it for his descendants, he would make up for moral, legal, physical, psychological strives made by his forefathers to keep the land in the family. In this way, the "same need" may well entail a sacrificial attitude in each member, the necessity of acknowledging non-human powers, here the dead, in order to keep on living.

On another respect, "use in common" is fairly restricted. For instance, cattle may graze in other member's plot but always under permission and some sort of reciprocity. If water sources may be occasionally shared, houses, tools, vehicles, and above all cattle are unambiguously private property and it comes to each member to decide in what circuits of reciprocity include them. Leasing to third-parties "coming from outside" has raised specific concerns. Can members decide this type of leasing without consulting the rest? The problem is not framed as a limitation to ownership but as a cautionary remark to unilateral permission given to a stranger that would not be bound by the implicit pact of "the same need". Insofar as this person rents the land but does not possess it, he or she would not be one of "us".

"Home", "us" is a situated learning of how to assemble heterogeneous forces. No doubt that "community", the "sacred values of community", is a discourse legitimating the partial interests of some human actors but we must look at this strategy as entangled in Latourian networks. The non-human agencies of territory are regularly recruited into the service of "the common" based on a schema of negotiable reciprocities among members of Epumer. Examples are the typical mutual aid in economic activities and the "non-economic" fiestas. In these occasions, non-humans circulate orderly in the guise of the "social life of things".[44]

When affections of things, animals, and spirits do emerge, they are generally conceived of by humans as accidents (good and bad luck) or plainly catastrophic. For instance, the drought, the omnipresent "natural disaster" in the region, man-

44 Arjun Appadurai, ed., *The Social Life of Things*.

ifests as water that "tastes saltier", pastures that "get themselves burnt in fire", cattle that "lets itself die". It is this middle-voice idea of "action" that characterizes the non-human. Neither passive nor active, non-human intentionality calls for some sort of divination.

If catastrophe becomes the norm, a high test is put to the moral of reciprocity. It is not really "a fact" that people in need have each other. There are perceived degrees of necessity in which the potency of some actants plays a role. In the case of Epumer we have observed some risk situations (lack of water or pasture for animals) in which transactions in money are preferred to "favors" as tough money would perform as a fetishist barrier against natural elements. On the other hand, limit situations gather people together. Fire-combats are one of them and perhaps land recoveries too. These instances reveal that holding a territory "in common" is a complex equation of reciprocities, transactions and having nothing to lose because there is "nothing in common".

Rewinding

We have made some efforts to decipher what Curunao intended with "the same need that binds us all". We have found that that expression evoked the workings of a network of human and non-human actors. Generally, a particular assemblage grows rhythmically in the backstage (latent territorying) but, at certain conjunctures, non-human actions may cut loose from the network being dysfunctional to or interrupting some human actors involved or all of them.[45] We have shown this in the very act of land recovery (emergent territorying). We have also inferred in this situation the inkling of the event of communing. Both situations challenged a cosmopolitical statu quo.

If our central argument expanded the question of how an indigenous collective is affected by territorying and communing, then "what happens to the entity in territorying and communing?" Do we "fall back" into the problems of hegemonic articulation of subjects in power and truth?

We cannot follow Bruno Latour's injunction of "generalized symmetry" as the application to any assemblage of the same rule of "following the action". If action is in and by itself proliferous, how do we identify "limits" in order to attribute causes and values. In short, a human or humanized entity must be postu-

45 See Michel CALLON, "Some elements of a sociology of translation".

lated as starting or ending point.[46] Latour suggests that this bounded actor aspect of the actor-network is produced by other actor-networks such as "panoramas" and "oligoptics".[47] But what are the latter the effects of? Panoramas and oligoptics do not just produce actors of different scales; they are the products of some actor-networks stabilized as "modes of existence". These modes *look like* our more familiar institutions arranged along the nature/culture divide. So it seems we have come full-spiraled.[48] Perhaps we should bring in the idea of a trans-immanent condition whereby inside and outside, up and down, fore and rear, before and after are reciprocal effects of difference along and across "modes of existence" and "plasma". The challenge is to keep on thinking a topological reality and this requires a difficult compatibilization between ontologies of positivity and negativity, between the actor-network and the (phantom) subject, between cosmopolitics and secular politics.[49]

In essaying a supplementary approach to sociocultural constructionism it has not been our purpose to obviate the latter's pertinence. The critique of anthropocentrism does not erase the *anthropos* but only its centrality derived from a postulated interiority. Highlighting non-human powers does not require dispossessing humans from theirs, nor does it curtails their patterning effects. Still, we must draw our attention to multiple and mutual connections of exteriors.

The Epumer collective owes its existence neither to a cosmopolitical network nor to a habitus engrained by a political economy, a biopolitical matrix or the

---

46 Some critics of actor-network theory consider it a superfluous endeavor. After all this "non-human" chatter it gets to "humans" to interpret, decide, etc. This misses the point because there is no "after all" and the unfinished-ness of any process is due to humans not being ontologically alone in their love and hate relations. Interpretation, decision must take into account the cultured other and the Other of culture/nature as well. However we must confess some discomfort with the expression "non-human". It evinces a negative consideration of the "rest" of beings (within an affirmative ontology!) and a too positive one of human, for there are quasi-humans or less-than-humans, namely those mapped by familiar critical theories as "subaltern", "popular", "lower classes", "Indians", etc. It is in this sense that the catch phrase human and non-human network may betray a conservative solution to anthropocentrism if it does not address asymmetry within symetry. See Axel LAZZARI, "Más acá del 'más allá' de naturaleza y cultura'".

47 Panoramas and oligoptics are always localized actor-networks. The first ones produce "big" subject-effects such as the "city of Paris", "universal history" or "Latin America" while the second ones create the correlative "small" ones such as a proton or the history of a land recovery in this very article. See Bruno LATOUR, *Reensamblar lo social.*

48 In his last works Latour fully acknowledges this problem and is currently conducting a research into the "modes of existence", a rephrase of the institutions of the moderns he had previously "actor-networked". The revision of modernity values which is at stake here necessarily presupposes the *sub-jectum* of the actor. See Bruno LATOUR, *Investigación sobre los modos de existencia.*

49 For an ethnographic approach with this question in mind, see Gastón GORDILLO, *Rubble.*

coloniality of power. The home of the "two foxes" (Epu-gner in ranküldungun) -a felicitous cypher of double trickstery- must necessarily remain a riddle.

Bibliography

Appadurai, Arjun, ed., *The social life of things. Commodities in cultural perspective*. Cambridge, Cambridge University Press, 1988.
Bechis, Martha, *Piezas de etnohistoria y de antropología histórica*. Buenos Aires, Sociedad Argentina de Antropología, 2010.
Briones, Claudia, ed., *Cartografías argentinas: políticas indigenistas y formaciones provinciales de alteridad*. Buenos Aires, Editorial Antropofagia, 2005.
Callon, Michel, "Some elements of a sociology of translations: domestication of the scallops and the fishermen of St. Brieuc Bay". In *Power, action and belief: a new sociology of knowledge?*, edited by John Law (The Sociological Review. Monograph, 32). London, Routledge, 1986, 196–223.
Curtoni, Rafael, *Archaeological approach to the perception of landscape and ethnicity in the west Pampean region, Argentina*. Masters dissertation, Institute of Archaeology, University College London, London.
Deleuze, Gilles and Felix Guattari, *Mil mesetas. Capitalismo y esquizofrenia*. Valencia, Pre-Textos, 1988.
Derrida, Jacques, *Of Grammatology*. Baltimore, Johns Hopkins University Press, 1976.
Descola, Philippe, "Más allá de la naturaleza y la cultura", *Etnografías contemporáneas* 1 (2005), 93–114.
Dillon, Beatriz, "Modelos de desarrollo y su impacto en la población Oesteña: labilidad ambiental e implicancias sociodemográficas". In *Territorialidades en tensión en el Oeste de La Pampa. Sujetos, modelos, conflictos*, edited by Beatriz Dillon and María Eugenia Comerci. Santa Rosa, EdUNLPam, 2015, 27–55.
Esposito, Roberto, *Communitas. Origen y destino de la comunidad*. Argentina, Amorrortu Editores, 2003.
Fernández, Jorge, *Historia de los indios ranqueles. Orígenes, elevación y caída del cacicazgo ranquelino en la pampa central (siglos XVIII y XIX)*. Buenos Aires, Instituto Nacional de Antropología y Pensamiento Latinoamericano, 1988.
Foucault, Michel, *Seguridad, territorio, población. Curso en el Collège de France: 1977–1978*. Buenos Aires, Fondo de Cultura Económica, 2006.
Gordillo, Gastón, *Rubble. The Afterlife of Destruction*. Durham/London, Duke University Press, 2014.
Haesbaert, Rogério, *El mito de la desterritorialización*. Córdoba, Editorial de la Universidad Nacional de Córdoba, 2004.
Harman, Graham, *Hacia el realismo especulativo. Ensayos y conferencias*. Buenos Aires, Caja Negra Editora, 2015.
Instituto Nacional de Asuntos Indígenas, *Programa Nacional de Relevamiento Territorial de Comunidades Indígenas*. Buenos Aires, Argentina, 2007.
Latour, Bruno, *Ciencia en acción. Cómo seguir a los científicos e ingenieros a través de la sociedad*. Barcelona, Labor, 1992.

Latour, Bruno, *La clef de Berlín*. Paris: La Decouverte, 1993.
Latour, Bruno, *The Pasteurization of France*, Cambridge, Harvard University Press, 1993.
Latour, Bruno, *Politiques de la nature. Comment faire entrer les sciences en démocratie*. Paris, La Découverte, 1999.
Latour, Bruno, *Aramis, or the love of technology*. Cambridge, Mass: Harvard University Press, 1996.
Latour, Bruno, *La esperanza de Pandora. Ensayos sobre la realidad de los estudios de la ciencia*. Barcelona, Gedisa, 2001.
Latour, Bruno "Whose Cosmos? Which Cosmopolitics? A Commentary on Ulrich Beck's Peace Proposal", *Common Knowledge* 10 (2004), 450–462.
Latour, Bruno, *Nunca fuimos modernos. Ensayo de antropología simétrica*. Buenos Aires, Siglo XXI, 2007.
Latour, Bruno, "Paris, ville invisible: le plasma". In *Airs de Paris. Exposition présentée au Centre Pompidou, Galerie 1, du 25 avril au 16 août 2007*, edited by Christine Macel, Daniel Birnbaum and Valérie Guillaume. Paris, Centre Pompidou, 2007, 1–7.
Latour, Bruno, *Reensamblar lo social: Una introducción a la teoría del actor-red*. Buenos Aires, Manantial, 2008.
Latour, Bruno, *Sur le culte moderne des dieux faitiches*. Paris, La Découverte, 2009.
Latour, Bruno, *Investigación sobre los modos de existencia. Una antropología de los modernos*. Buenos Aires/Barcelona/México, Paidós, 2011.
Latour, Bruno and Steve Woolgar, *La vida en el laboratorio. La construcción de los hechos científicos*. Madrid: Alianza, 1995.
Lazzari, Axel, "Aboriginal Recognition, Freedom, and Phantoms: the Vanishing of the Ranquel and the Return of the Rankülche in La Pampa", *The Journal of Latin American Anthropology* 8 (2003), 59–83.
Lazzari, Axel "La repatriación de los restos de Mariano Rosas: identificación fetichista en las políticas de reconocimiento de los ranqueles", *Estudios en antropología social* 1 (2008), 35–64.
Lazzari, Axel, *Autonomy in Apparitions: Phantom Indian, Selves, and Freedom (on the Rankülche in Argentina)*. Unpublished Ph. D. Thesis. Anthropology Department, Columbia University, 2011 (http://udini.proquest.com/view/autonomy-in-apparitions-phantom-goid:864737978 [consulted July 8, 2014]).]
Lazzari, Axel, "Autenticidad, sospecha y autonomía: la recuperación de la lengua y el reconocimiento del pueblo rankülche en La Pampa". In *Luchas indígenas e identidades en disputa en Argentina: historias de invisibilización y emergencia*, edited by Gastón Gordillo and Silvia Hirsch. Buenos Aires, Editorial La Crujía, 2011, 147–172.
Lazzari, Axel, "Emergence in Re-emergent Indians: Para-history of The Re-Emergent Rankülche Indigenous People". In Lazzari, Marisa (coord.), "El pasado-presente como espacio social vivido: identidades y materialidades en Sudamérica y más allá", *Nuevo Mundo Mundos Nuevos [En línea], Cuestiones del tiempo presente* (2012) (http://nuevomundo.revues.org/64024 [consulted Feb 18, 2016]).
Lazzari, Axel, "'¡Vivan los indios argentinos!'" Etnización discursiva de los ranqueles en la frontera de guerra del siglo XIX", *CORPUS. Archivos virtuales de la Alteridad Americana* 2 (2012) (http://ppct.caicyt.gov.ar/index.php/corpus/issue/view/65 [consulted Feb. 20, 2016]).
Lazzari, Axel, "El Relevamiento Territorial de Comunidades Indígenas (ReTeCi) de la Comunidad Epumer: pasado, presente y futuro". In *Investigaciones acerca de y con el pueblo ranquel:*

pasado, presente y perspectivas. Actas de las Jornadas en Homenaje a Germán Canuhé, edited by Claudia Salomón Tarquini and José Ignacio Roca. Santa Rosa, EdUNLPam, 2015, 247–267.

Roca, José Ignacio and Anabela Abonna, "El 'Operativo Mitre': Desarrollismo y pueblos indígenas en la provincia de La Pampa durante la dictadura de Onganía", *Atek Na* 3 (2013), 167–207.

Roca, José Ignacio, "Agentividad indígena y discursos hegemónicos en una disputa por tierras de Colonia Emilio Mitre (1966–1972). Situando los comienzos de la militancia Rankulche en La Pampa". In *Debates y Perspectivas de la Investigación en Ciencias Sociales y Humanas*, edited by Marisa Eugenia Elizalde. Santa Rosa, Universidad Nacional de La Pampa, Facultad de Ciencias Humanas, 2013, 1–16.

Rose, Nikolas, *Powers of Freedom. Reframing Political Thought*. Cambridge, Cambridge University Press, 1999.

Salamanca, Carlos and Rosario Espina, *Mapas y derechos. Experiencias y aprendizajes en América Latina*. Rosario, Editorial de la Universidad Nacional de Rosario, Argentina, 2012.

Salomón Tarquini, Claudia, "Redes sociales y campos de negociación en una colonia pastoril indígena (Emilio Mitre, La Pampa, principios del siglo XX)", *Estudios Digital* 3 (2010), 1–19.

Salomón Tarquini, Claudia, "Estrategias de acceso y conservación de la tierra entre los ranqueles (Colonia Emilio Mitre, La Pampa, primera mitad del siglo XX)", *Mundo Agrario* 11 (2010), 1–28.

Salomón Tarquini, Claudia, *Largas noches en La Pampa. Itinerarios y resistencias de la población indígena (1878–1976)*. Buenos Aires, Prometeo, 2010.

Salomón Tarquini, Claudia, "Los vínculos con los ranqueles de Emilio Mitre, 1914–1915" In *Un Quijote en la Pampa: los escritos de Manuel Lorenzo Jarrín (1883–1942)*, edited by Claudia Salomón Tarquini and María de los Ángeles Lanzillota. Santa Rosa, Fondo Editorial Pampeano, 2011, 41–64.

Tamagnini, Marcela and Graciana Pérez Zavala, *El fondo de la tierra. Destinos errantes en la frontera sur*. Río Cuarto, Córdoba, Editorial de la Universidad Nacional de Río Cuarto, 2010.

Tapia, Alicia, *Arqueología histórica de los cacicazgos ranqueles (siglos XVIII y XIX)*. Buenos Aires, Caracol, 2014.

Taussig, Michael, "Maleficium: State Fetishism" In Id., *The Nervous System*. London/New York, Routledge, 1992, 111–140.

Turner, Victor, *The Ritual Process. Structure and Anti-Structure*. Ithaca/New York, Cornell University Press, 1991.

Villar, Daniel and Juan Francisco Jimenez, "Acerca de los ranqueles. Los indígenas del Mamil Mapu y del Leu Mapu (1750–1840)". In *Primer Encuentro entre Investigadores y Pueblos Originarios del Centro de Argentina*. Santa Rosa, Universidad Nacional de La Pampa, 2006. M.S.

Walter Delrio

Indigenous communalizations in Patagonia in Post-genocidal Contexts (1885–1950)

Introduction

In Argentina there exists nowadays a National Register of Indigenous Communities which is, at the same time, copied by provincial registers.[1] In order to become part of these records, the government establishes a series of requirements that each community must follow to achieve official acknowledgment. The communities are expected to fill forms, choose authorities, make a council of elders and submit a portfolio with the "history" of the community and the foundations that probe that they belong to an indigenous People. Then, Government agencies determine the validity of each request seeking to certify the accuracy of the information submitted. These agencies tend to stress, first, the legitimacy of the indigenous people based in the continuance of territorial occupation, second, the cultural practices over the territory and, finally, the ties of affiliation of the members in a historical dimension. These three issues allow the group to officially ascribe as part of a recognizable and recognized people/tribe.

This process started since the constitutional reform of 1994 and many things have changed since then on. A variety of different demands have argued against homogenizing State's policies which often act in these processes of State recognition of indigenous peoples. However, the processes of indigenous community building have always been paradoxically related to the State. This paradox reveals sometimes as a political tension, others as an opportunity for politics and some others also as obstacle or limit. On the one hand, the (national and/or provincial) State requires the cultural and social continuity of the indigenous communities when, at the same time, we can identify a century of State policies aimed at the

1 In the present, 925 communities have been recognized Nationwide. The National Register of Indigenous Communities (RENACI) is carried out by the National Institute of Indigenous Affairs (INAI), an institution of the national State has the responsibility to create intercultural channels for the implementation of the rights of indigenous peoples enshrined in the Constitution (Art. 75, Inc. 17). The Register was created by the national law 23.302, in September 1985 as a decentralized body, with indigenous participation and regulated by Decree No. 155 in February 1989. It depends directly to the Ministry of Social Development. More recently the National Law 26,160 declares the "Emergency regarding possession and ownership of the land" and enables the "Territorial Survey of Indigenous Communities" with participation of INAI. It then creates the National Program of Territorial Survey of Indigenous Communities (RETECI).

extinction of indigenous communities and peoples. On the other hand, from the indigenous peoples' point of view, been part of a community – as a marked population – has enabled, both, the possibility of expropriation and the denial of some rights, and at the same time, the official recognition of membership has enhanced other rights and it has become practically the only way to be acknowledged as members of an indigenous people.

These essentialist requests of social and cultural continuity of the indigenous community are another dimension[2] of the process of subaltern status construction of the indigenous peoples as an internal-other within the nation-State-territory matrix. Our aim here is to analyze different ways of understanding indigenous communalizations[3] in relation to that process. Especially considering post-genocidal contexts in northern Patagonia between 1880–1950s.

This temporal and spatial selection allows us to study the particular and complex relationships among three dimensions articulated by the concept of indigenous community in this context: first, the affiliation with a broader socio-political group (people / nation); second, the identification with certain socio-cultural life and an imagined unity; and, third, the relationship between the individual and the collective.

Genocide and normalizing exception

By the last decades of the nineteenth century the Chilean and Argentinean States performed military campaigns of occupation in order to subjugate and incorporate the indigenous population of Pampa and North Patagonia. They produced the violent dismantling of these societies. Both States agreed to occupy and divide these territorial spaces. In the Argentinean case, during the campaigns the "indios" were divided in two types: those who voluntarily submitted to the armed forces and those who were forced to submit. Though the Act 215 of 1867 which ordered the military advance of the southern border to the Negro river (northern Patagonia) and already made this distinction between the indigenous population, it was not until 1878–1879 that this advance could be executed. However the classification of the indigenous people in the discourse of State agencies remained, through different terms, this basic differentiation between rebel-wild and subject-presented persisted in time.[4]

2 There are also demands linked to an indigenous ecological stereotype.
3 About the concept of communalization see James Brow, "Notes on Community", 1–6.
4 The use of different terms in the historical documentation to refer to this dichotomous classification was worked on Walter Delrio, *Memorias de expropiación*.

Thus, the varied distinctions and ways of naming indigenous societies in terms of Peoples, chiefdoms or greater socio-political groupings that had characterized until then the classifications developed by State agencies were replaced massively in the documentation for this period to favor the visibility of the dichotomy wild/rebels and voluntarily submitted/presented[5]. Between 1882 and 1885, the military advances that would eventually take place in northern Patagonia achieved a national border crossing system that separated the territorial jurisdiction of both States. By then, a second classification system of indigenous peoples consolidated in terms of their newly imposed national identification. In the official documentation, both in Chile and Argentina, a progressive and fast distinction developed between "Chilean indians" and "Argentinean indians". The paradox here is that while the first classification system operated toward the transformation and extinction of the "wild and rebellious" (as the "subject and presented" were allegedly keen to be civilized and, therefore, less marked as "indigenous"), in the second case although they were "Argentinean" or "Chilean" they still remained marked as "indians".

However both classification systems are intimately related. In Argentina, the first system proposed and maintained over time the stereotype of indigenous peoples condensed in the figure of the "malón" (raid). The "malonero" indian (wild and unruly, whose main form of livelihood would be looting or raids organized to steal property and kidnap people from creole holdings) was built as a threat to property, people and the very integrity of the nation. The "malonero" indian was considered to be not only in a preliminary human stage, but also as a foreigner. The "malonero" indian manufactured by the hegemonic political discourse was supposed to be removed or destroyed. Therefore the continuity of its sociopolitical units becomes inconceivable (in fact they are no longer contemplated within the rations system or the usual treaties shared with the national government through the Ministry of War) after the State conquest. Thus the distinction settled since 1882 between "Argentinean indians" and "Chilean indians" imposes through culture. This builds an internal otherness -the subordinate status of the indigenous population- while maintaining the latency makes its construction as an exception. First of all, those who are classified as such continue to be considered as "indigenous" and therefore potentially "maloneros". The bare idea of the possibility of the manifestation of this condition remains a cultural construct, a myth or a rumor.

5 Ibid., 61–74.

Indeed this normalizing exception[6] manifests itself in the destiny of the indigenous population after the military advance. The campaigns simultaneously operated concentrations, distributions and mass deportations of people onto other regions of the country. In Argentina, most indians were sent to the provinces and separated according to specific demands of labor. The women were distributed as domestic service, children were given as servants, adult males were incorporated into the armed forces and as labor force. The subjugated population in Northern Patagonia was then provided *en masse* as a kind of economic compensation mainly to sugar or wine producers, among any other sectors of society that demanded them. This was articulated and made possible by State institutions and civil society that established a complex system of transfers and deals throughout the country.[7] This process not only meant the dismantling of indigenous sociopolitical units but also of the nuclear family.

The hegemonic discourse proposes to foster both the extinction of the sociopolitical order of life as well as the indigenous families.

> we are engaged in a contest of races in which the indigenous person bears upon himself the tremendous anathema of his disappearance, written on behalf of civilizations. Thus morally destroying this race, springs and annihilate their political organization, their order of tribes disappears if necessary, and divide the family. ("Editorial", *La Prensa,* Buenos Aires 1/3/1878, 1.)

In 1884 the president Julio A. Roca addressed the national parliament asserting and synthesizing the work of the government in the annihilation of the indians. This idea was also reflected in the scientific field, which built a "natural history":

> Argentina is undoubtedly a vast necropolis of lost races, coming from very remote scenarios, driven by the fatal struggle for life in which the strongest prevailed; some arrived victorious and others as losers and they annihilated in our Southern end. (Moreno Francisco, "Editorial". *Revista del Museo de la Plata,* Talleres del Museo de La Plata, I (1890–91), 46)

From this standpoint, the survival of indigenous groups was understood only in terms of "remnants of tribes". By then, these people were "scattered", as they often appear in the official documentation. The only mention of this population in the Argentinean legislation used this concept of remnants. Also the press

6 Walter Delrio et al., "Discussing the Indigenous Genocide", 138–158; Pilar Pérez, *Estado, indígenas y violencia.*

7 About these processes see Enrique Mases, *Estado y cuestión indígena*; Mariano Nagy and Alexis Papazián, "De la Isla como Campo" and Diana Lenton and Jorge Sosa, "La expatriación de los pampas".

referred to them as, for example, "the remnants of the Catriel's tribe, wandering through the Negro river."[8]

Indeed, the expression commonly used until then to define an indigenous tribe was the combination "cacique y su gente" (chief and his people). In this combination it was attributed to a certain "cacique", the power of political influence and over a set of population and territory. Even the Argentine Ministry of War identified the various chiefs according to a State accepted hierarchy among primary and secondary "caciques". This is also expressed in a series of treaties, previous to the military campaigns, that established alliances (and regular delivery of payments and rations) between these sociopolitical units and the nation State. While some authors assume that this hierarchy could be interpreted as a manifestation of a process of concentration of power and segmentation of the indigenous society internal politics.[9] For some others, this would result in the construction of prestige, the creation of information nodes[10] or the same indigenous strategy of negotiating with the State language policy (the police in terms of Ranciere) of the border.[11]

In the post-conquest context and with the hegemonic imposition of the stereotype of "malonero" Indians, these sociopolitical units ceased to be acknowledged as such and from then on, the term used is that of "remnants". In this period, only some chiefs received land for themselves and their family or group of closely related families. This was the experience of chiefs such as Sayhueque, Namuncura and Ñancuche Nahuelquir. They were the once identified either as primary "caciques" or those who, in the context of the campaigns or after their subjugation had served the troops as "baqueanos" (local expert). A handful of them received land for themselves and their family by executive decrees or a special act of Congress and, finally, some others by virtue of being recognized by the State as military veterans.

However, most of the indigenous people subjected in northern Patagonia were concentrated, relocated and distributed throughout the country.

> Then they say that people grew well, the boys back then of 15, 16 ... boys, young... and they carried them, they were older than me, once in Buenos Aires, they let them go on foot, they came on foot, many people walked ... some came back, and many stayed. (Interview with Foyel Quintunahuel, Loma Redonda, Chubut-Argentina 2006)

---

8 *Caras y Caretas Magazine*, Buenos Aires 24/6/1899.

9 Alberto Rex GONZALEZ, "Las exequias de Painé Guor"; Raúl MANDRINI, "La base económica de los cacicatos" and Julio VEZUB, *Valentin Saygueque*.

10 Martha BECHIS, *Piezas de etnohistoria*; Ingrid DE JONG and Silvia RATTO, "La construcción de redes políticas indígenas".

11 Walter DELRIO, "Entrar y salir de la Etnohistoria".

As Nagy and Papazian[12] point out for the case of the concentration camp on the Martín García island, the detainees were imprisoned because they were "Indians". Daily, the authors note, the island authorities received requests for the delivery of Indians for different purposes.

Those who remain in the region after the concentrations or who could flee their destinations or returned to the region would no longer be recognized as tribes by the State. In the following decades, in fact, many official documents State that in Patagonia there were no longer Indians, and that they would have been extinguished by the military campaigns, pests, or have migrated to Chile or could have been integrated with the creole population. Therefore most of the indigenous population in northern Patagonia after years of subjugation, concentration and deportation was constituted by those who were left free by the end of the 1880s, those who had escaped places of concentration and destiny and those who were inducted into military service.

After the first major distribution of land and the creation of private property in the territories conquered this population remain on public lands without recognition of their forms of organization and occupation. Furthermore, the Ministry of Interior ordered the recently named governors of the National Territories that the "Indians living in tribes" should be excluded from the population counts.[13]

In the new society to be built in the occupied territories the right of some, the settlers, is imposed on others. As Wolfe[14] argues, for the Australian case, the deep structure of the settler society is based on elimination in which the indigenous peoples are perceived as their others. However the problem of the "malonero" Indian stereotype, mainly understood as a problem of security, will also be embodied in other social actors who represent a threat to the "goods and people," national integrity and moral health: the bandits, the poor, gipsies, Turks, rustlers, etc. The exception is culturally constructed and a key to the discipline and standardization of the new society.

Communalization in post-genocide contexts

Nonetheless, from the social memory perspective, this precise moment of dismantling is also understood as the origin of new social bonds. At present, there are Mapuche-Tehuelche narratives that remember the long pilgrimages of the

12 Mariano Nagy and Alexis Papazián, "De la Isla como Campo".
13 *Memories of the Ministry of Interior* (MMI), 1900: 21.
14 Patrick Wolfe, "Settler colonialism", 387–409.

elders whether escaping from the death camps or places of detention and concentration, as well as the urban centers where they were confined, in order to finally reach the place of the community of the narrator.

> grandmother used to tell that when (...) she was captive, my grandmother was captive, in Argentina, and then when she was freed, after been captivated when everything settled... there she came out, run away, she left, and came here, and made a family. My grandmother used to mourn. And if I remember, I suddenly remember, because she had her grandmother. The troops took her (...) The people were captivated before, like animals (...) she was captivated during the war, she didn't know her mother when she was grabbed. (Interview with Laureana Nahueltripay, Cushamen, Chubut-Argentina 1997)

In these stories, long pilgrimages take place in an often hostile territory, where hunger is present.[15] After the loss of loved ones, of social ties,[16] in some cases, the ancestors are said to have been accompanied, guided and helped by meaningful animals like a "puma" (cougar) or "ñanco" (eagle like bird). Generally these narratives are linked to new generations, with new bonds that are built on this pilgrimage and new encounters. Afterward, this is remembered through "tayel" (sacred songs) during ceremonies. In these wanderings, people encounter others in similar situations who roam the countryside. The place for arrival is usually the current community.

> They legitimized our father. But it was a person who was neither a relative nor ... he was lost during the war, they do not know who our grandfather was, or us. They were three years old. They say he wept when he was... wrapped with midwives, people before were poor, only cloth around, dressed. They were wrapped with cloth to get dressed, he used to tell while he cried: 'I was that orphan, but I manage to grow up'. (Interview with Carmen Calfupán, Vuelta del Río, Chubut-Argentina, 2003)

---

15 Catalina Antilef, villager of Futahuao, Chubut, in 2004 recalled: "they were taken away, where to... because my grandmother said she escaped, what they call this place? ... Choele Choel. In Choele Choel, she says she escaped. That's where they carried him. He says he saw people who were sick, women who had child, their heads were cut. A cookie says they gave him per week. You could just dye, my grandmother used to cry, my poor grandmother and I ... I grew up with my grandmother ... after they had another elderly, those had returned from the war, these say that when he finished there escaped, some returned, returned alive. One who was called ... Calfucura, late Manuel Antiqueo, those returned when the war ended, they survived, they say they were ... not killed, poor things...".

16 "And there where they dropped it out, or when the grip was that people spilled because the grandmother had sisters here by Languiñeo, so they came looking for family, and parents ... if she was young ... and parents who knows for where they have been able to have died ... I never found the great-grandparents of us, where they went ..." (C. P. Boquete Nahuelpán, Chubut, 2006).

In many cases these new bonds refer to the adoption of children, including those in new family ties. This often was related to "passing over" the name. For example through the mechanism of "lacutun"[17] by which many people were named after -then and through registration in the civil registry- a prestigious person acknowledged as "cacique" by State authorities. Some witnesses also express this process from the perspective of these "caciques", those visible to the same State:

> The late Ñancuche, he had suffered a lot, they ran from side to side, along came the white shooting blood bullets, so... and then he surrendered... the late Ñancuche surrendered to the army, went against indian brothers... to survive (...) Ñancuche and Rafael were talking to the president, and they got the 50 leagues, but for their families. So ... they began to grab family, siblings, they also, not only had to have two, two women every man. So they had to increase. And some families .. Don Ñancuche and ... huh? and who knows, working or help in something ... then gave them place to settle. (Interview with Demetrio Miranda, Cushamen, Chubut- Argentina, 1998)

The meaningful details displayed in this story show the indigenous exception and how indigenous people become dispensable. Also, the submission appears as the only alternative to survive. This also meant in one way or another to go "against indian brothers" in order to survive. Due to these kind of services, and this case is one of the few exceptions, the chief is recognized and rewarded with land for his family. The last step is the extension of that family. As it was mentioned above, through ritual kinship the families are enlarged in order to fill the entire surface of land acquired.[18]

In this case, "going against the indian brothers" does not reduce the chief's prestige and ability to articulate and have a central role in the new processes of indigenous communalization in the post-conquest. His prestige grow even more over time as he was the one who obtained land, sponsored individuals and families and was able to distribute them into the new group. Indeed, many Mapuche-Tehuelche communities in northern Patagonia constituted themselves through these stories of persecution, subjugation and forced dislocation as well as collective efforts deployed in these contexts. New social ties were expressed

---

17 Mapudungun term that has been translated by the narrators as becoming grandfather, pass the name, done in such a family.

18 Indeed the distribution of land and creation of a colony in Cushamen was the result of Julio Roca's presidential decree in 1899. It is an exceptional case where the law 1503 of 1884, also known as Argentine Home Law, originally it applied designed to give land to poor gauchos. The 200 lots of 625ha given the opportunity to the chief Miguel Ñancuche Nahuelquir became the visible head of this communalization process involving not only the growth of social ties in the language of kinship but also access to many family units the earth.

through the language of kinship -for example through “lacutun”- membership and the use of names of spread lineage.

However, in the majority of cases the reconstruction of social ties and new communities involved precarious access to land[19]. Not only these groups were considered as remnants and therefore they were an invisible form of the continuity of indigenous ways of socio-political organization, but their access to land and their rights as residents of public lands was also limited. Simultaneously, the indigenous cultures and ways of life, above all associated with nomadism, became a condemnable stereotype.

In short, the identification of a person or group with an indigenous People’s cultural practices and comunalizations would serve as elements that could potentially enable the exception. However, for some it was worth the risk as some others have received the land due to their ascription as indigenous people.

Invisibilizations and strategic views of the community

Throughout the first half of the twentieth century, the deep structure of the settler society developed in Patagonia. For a group or a person being labeled as indigenous (in Patagonia the labels “Chilean indians” or “araucano” were most usually used) generally operated in favor of expropriation and the denial of rights.

This can clearly be read in the reports written by public land inspectors in which the status of “indian”, “araucano” or “Chilean indian” was reason enough to reject the request for the land. There were also other attributes related to a nomadic life, manifested for example in the fact of living in tents or going hunting, used to dismiss indigenous settlers. They generally included information, such as the consumption of alcohol, to avoid favoring non-indigenous settlers.

Thus we read in a 1940 survey on public land the description about a villager: “his small capital does not allow him to contract with the State, on the other hand, it shows the status of their indolent character improvements as are most Aborigines areas”.[20]

Paradoxically, in some cases after having mentioned that these people lived in tents the description of the dwelling was “walls of mud, bricks and tin roofs”. This type of building, due to the availability of materials, was extended to most

19 Pilar PÉREZ, *Estado, indígenas y violencia.*

20 Instituto Autárquico de Colonización y Fomento Rural de la provincia de Chubut (IAC), Expediente 131718–1938 (2800), Inic. Galván, Pedro Florencio F. 23 y Ss 19/3/40.

inhabitants of that time whether indigenous or not and it is common still today in Patagonia.

In terms of the survival of indigenous sociopolitical units the State had a paradoxical interpretation as well. Whereas until the 1930s the State largely denied that they existed in the National Territories of the south (Patagonia), it not only acknowledged their existence in northern National Territories (Chaco and Formosa) but also sought to displace the influence of mediation functions of the "caciques" and chieftains and even reconvert to the seasonal needs of the labor demand of the industries in those territories, especially for the sugar and cotton harvest. In the north, the State selected certain "caciques" and it also actively modify their role in favor of new purposes[21]. In Patagonia, however, the existence of families who inhabited marginal public lands constituted a sufficient supply of labor designed to work as farm laborers of the landowners stays.[22]

Beyond the State's appreciation, the Mapuche-Tehuelche people organized collective claims, distinguishing themselves, despite the dangers, as local indigenous people and as communities. For example, in September 1911 Juan Napal introduces himself as the representative of 19 indigenous families "in order to claim the land they have been occupying by the current legislation "Ley del hogar"*(Home Law)*. The lands were booked to widen the colony founded in 1899 for the families of Ñancuche Nahuelquir. Assuming the representation of this group (as well as the danger) Napal asserts: "the indigenous people we represent already have the capital required by the State to receive land as an extension to that colony." They tried to pay its first surveying of the land, and then, in the absence of responses, they even offered to buy the land. None of these suggestions were accepted. Finally, they were expropriated or their lands were significantly reduced in order to be given to the non-indigenous settlers, considered "desirable people to inhabit", during the following years.[23]

Facing this case, J. J. de Vedia Army official stressed that it became clear that "the goodwill of the Indians to work as rural labor" and he also suggested that if the government was really interested in civilizing the "Indians" it should also be willing to extend to them "the favors that have been earned until then by a privileged few caciques". De Vedia proposes the case as the beginning of a new official policy "regarding indigenous tribes", that consisted on the delivery of land to "tribes or Indians" under the guardianship or patronage of a State insti-

21 Walter DELRIO, *Memorias de expropiación*, 193–215.

22 Juan Carlos RADOVICH and Alejandro BALAZOTE, "Transiciones y Fronteras Agropecuarias", 63–79.

23 Academia Nacional de la Historia. Archivo Histórico, Fondo Roque Saenz Peña. FRSP, C 65, Doc. 214.

tution.[24] But this claim from Juan Napal was unsuccessful. In Patagonia, indigenous people remain largely invisible as a group and what is worse, liable to be expropriated.

Still, communities implemented negotiation strategies that involve requesting the recognition as a group. In some cases, these claims were silenced by bureaucratic processes such as that presented in 1941 by 54 families from the Comallo stream area in the National Territory of Rio Negro.[25]

At the same time, there were also tensions and conflicts within the indigenous communities themselves regarding what should be recognized as a "tribe". In the case of the so-called Nahuelpán reservation in the National Territory of Chubut, created by a special law for Francisco Nahuelpan as "the chief and his tribe in 1908 in gratitude for the services met". A part of the community, some sons of the "cacique", presents a demand in 1936 to the Land Office in Buenos Aires with the intention of "profiting without acknowledging or taking into account the rights of other members of the family and other members of the tribe." In this direction they indicated that the original distribution of land had been for Nahuelpán "cacique", his family and his tribe and the latter included many more people than the immediate family of that "cacique". They argued that the "fields were agreed for the tribe and that they exploited them communally".[26]

The internal dispute in this case dealt with what should be regarded as a "tribe of a certain cacique". This meant defining who was legitimate: the male's offspring? Those sponsored by that group? Were the females descendants included? Despite these internal conflicts they were overtaken by the presidential decree of Agustín P. Justo of May 5, 1937. It argues that the purpose of the distribution of land in 1908 was meant to "channeling indigenous families" into the "practice of work and progress," "not been achieved due to the lack of working habits of the occupants of these land -as it has been proven by different inspections since 1931-, the natives live precariously and in complete abandonment, accusing absence of methodical work, order and morality." These inspections have suggested that "the maintenance in place of those undesirable elements, is a serious problem for the inhabitants of this rich and prosperous mountainous region".[27].

The immediate eviction of all indigenous families and their relocation to other public lands (lower quality) is ordered with this decree. The families were to be "divided into two groups, in order to avoid a conglomerate of people of the same ideas and customs. Moreover, being thus divided, it would be easier for

24 IAC Expediente 87.567- 1933 (239), F.44-5 26/9/13.

25 Lorena Cañuqueo, Laura Kropff, and Pilar Pérez, "El 'paraje' y la 'comunidad'".

26 IAC, Expediente 5754–1947 (781), tercer cuerpo F. 348 (19/5/36).

27 Ibid. F. 361.

the police authority to monitor them and exercise upon them a more complete control."

Among the arguments used back then, they posed that they were "Araucanian" (from Chile) and that they still practiced hunting and theft. Thus, it is considered that not only the indigenous people were recognized by "not having introduced any kind of improvements or estates worth mentioning, they were only engaged in theft, originating on this occasion extended police raids". They claimed that after they arrested these individuals "after serving short periods of imprisonment, they regain their freedom, they return to pursue their former activities, constituting the threat of the few good people, whom with great efforts have achieved made some capital".[28]

Indeed the actions of the police forces, as it has been addressed by Perez[29], was primarily oriented to discipline, especially, the indigenous population. Being an indigenous person was reason enough to be under the threat of eviction as well as suffering the exception of rights. This is recalled in some narratives:

> Early the next day the police officers were there, they came to look for Dad. They brought him on foot, we were kids, I was only 11 years. Well, they brought him here to the police station of Maitén, he was imprisoned for a month. One month later they brought dad back with the police, the justice officer, the Turkish, all ... and brought down his house, they threw him like a dog out there ... the field of the late Valentin, that's how the Turks did it... with the police, with our very wagon, with our own ax, they cut all the clubs. (Interview with Carmen Jones, Vuelta del Río, Chubut-Argentina 2005)

> During those years there was a total change ... because I myself lived it... I noticed it, Why? because those years there was a change because traders began to fence the fields... and started doing things with people, to send the police ... to check their houses, to bring in the people who were in the houses, made them prisoners. (Interview with Segunda Huenchunao, Vuelta del Río, Chubut-Argentina 2004.)

> there should be one man per house if they had animals, if there were no animals, it was not necessary, if there were animals the males should stay. If not... one whole day making bricks, with these (points to his legs). We worked like horses!! (Interview with Agustín Sanchez, Esquel, Chubut-Argentina 2004)

The memory of these experiences is often associated by people in narratives that link the experiences of expropriation of their grandparents with their parents and themselves. When the more recent stories are told, the old stories that made their grandparents cry and that these generations transmitted to the next generation, emerge in a very specific way to experiment the State.

---

28 Ibid.

29 Pilar Pérez, *Estado, indígenas y violencia.*

Conclusions

> When they threw down my house, it was a big pain, but my soul wasn't thrown down, it was not thrown down… (Interview with Carmen Jones, Vuelta del Río, Chubut-Argentina, 2004.)

As Lenton suggests[30], the idea of community is present in one form or another in the legislation and administrative applications since the same colonial system. Throughout the process of emancipation and State-building in the nineteenth century the concept operated legally, as in the case of the legislation of the province of Jujuy around 1835 that included the prohibition to sale and disposal of Indigenous communities' goods. However, even in this example, this figure was operational at the expropriation of land to indigenous communities in Jujuy as Ana Teruel and Cecilia Fandos analyzed[31], or in the case of the province of Cordoba Jose Maria Bompadre[32]. Lenton marks the differences between the "old provinces" and the new territories, the National Territories, incorporated through the military conquest of 1878. She States that the concept of community remains as how to understand the organization of the broader group that would be the indigenous tribes.

If we select the period of time between the events of submission and incorporation of State in 1878 to the present time, we can appreciate the multiple dimensions of the concept of community. In particular, we study the largest indigenous groups, ways of representing the indigenous socio-cultural organization and the relationship between community and individual.

More specifically we dealt with the period 1883–1943 in which we could distinguish, first, the building of a normalizing exception that involved the implementation of mechanisms operating to detribalization. That is removing the tribe as major socio-political group and in that sense also they administered the dissolution of the indigenous community understood as a form of social and cultural organization. Individuals were liable to civilization and incorporation as a labor force and through this unmarked as indigenous persons. However, the particularities of the advance of capital in different regions of the national territories are uneven. In the national territories of the south and of the north there are clear differences regarding what is expected of those internal others whether they were subjugated or incorporated.

Therefore, in a second moment, in the space of Northern Patagonia new ways of thinking the community are defined. Being related to a broader group which is

---

30 Diana Lenton, "El paradigma de comunidad", 7.

31 Cecilia Fandos and Ana Teruel, "¿Cómo quitarles esas tierras", 209–239.

32 José María Bompadre, "De la preterización", 5.

considered to be extinct makes the ascription to an indigenous people an uncertain arbitrary outcome. At the same time, as a form of representing the organization of social relations, a notable dynamism that leads to new processes of communalization is manifested. In these processes a new post-genocidal context is redefined not only in the relationship between individual and collective, but also the image of that which identifies itself as a particular way of life and identity as a People.

Thus, the community itself is also paradoxical. On the one hand, it remains as something feared by the non-indigenous society. Indeed, the "tribe" will remain part of the stereotype of the Indian as a normalizing exception in the new society of settlers produced after the conquest State. On the other hand, the community will allow people to access land and rights to be able to demand as part of an indigenous People.

Bibliography

Bechis, Martha, *Piezas de etnohistoria y de antropología histórica*. Buenos Aires, Sociedad Argentina de antropología, 2010.

Bompadre, José María, "De la preterización y la extinción a la comunalización contemporánea", *Deodoro. Gaceta de Crítica y Cultura* 4 (2014), 5.

Brow, James, "Notes on Community, Hegemony, and the Uses of the Past", *Anthropological Quarterly* 63 (1990), 1–6.

Cañuqueo, Lorena, Laura Kropff, and Pilar Pérez, "El 'paraje' y la 'comunidad' en la construcción de pertenencias colectivas mapuche en la provincia de Río Negro". In *VIII° Congreso Argentino de Antropología Social*. Salta, Facultad de Humanidades de la Universidad Nacional de Salta, (2006), 210–224.

De Jong, Ingrid and Silvia Ratto, "La construcción de redes políticas indígenas en el área arauco-pampeana: la Confederación Indígena de Calfucurá (1830–1870)", *Intersecciones en Antropología* 9 (2009), 241–260.

Delrio, Walter, *Memorias de Expropiación. Sometimiento e incorporación indígena en la Patagonia (1872–1943)*. Bernal, Editorial de la Universidad de Quilmes, 2005.

Delrio, Walter, "Entrar y salir de la Etnohistoria", *Memoria Americana. Cuadernos de Etnohistoria* 20 (2012), 147–171.

Delrio, Walter, Diana Lenton, Marcelo Musante, Mariano Nagy, Alexis Papazian, and Pilar Pérez, "Discussing the Indigenous Genocide in Argentina: Past, Present and Consequences of Argentinean State Policies toward Native Peoples", *Genocide Studies and Prevention* 5 (2010), 138–159.

Fandos, Cecilia and Ana Teruel, "¿Cómo quitarles esas tierras en un día después de 200 años de posesión? Enfiteusis, legislación y práctica en la Quebrada de Humahuaca (Argentina)", *Boletín del Instituto Francés de Estudios Andinos* 41 (2012), 209–239.

Gonzalez, Alberto Rex, "Las exequias de Painé Guor. El suttee entre los araucanos de la llanura", *Relaciones de la Sociedad Argentina de Antropología* 13 (1979), 137–161.

Lenton, Diana, “El paradigma de comunidad”, *Deodoro. Gaceta de Crítica y Cultura* 4 (2014), 7.
Lenton, Diana and Jorge Sosa, “La expatriación de los *pampas* y su incorporación forzada en la sociedad tucumana de finales del siglo XIX”, *Ieras Jornadas de Estudios Indígenas y Coloniales*, Jujuy, C.E.I.C., (2009), 89–112.
Mandrini, Raúl, “La base económica de los cacicatos araucanos del actual territorio argentino (siglo XIX)”, *VI Jornadas de Historia Económica*. Vaquerías-Córdoba, UNC, 1984, 10–40.
Mases, Enrique, *Estado y cuestión indígena. El destino final de los indios sometidos en el sur del territorio (1878–1910)*. Buenos Aires, Prometeo libros/ Entrepasados, 2002.
Nagy, Mariano, *Estamos vivos. Historia de la comunidad indígena cacique Pincén, provincia de Buenos Aires*. Buenos Aires, Antropofagia, 2013.
Nagy, Mariano and Alexis Papazián, “De la Isla como Campo. Prácticas de disciplinamiento indígena en la Isla Martín García hacia fines s. XIX”. In *XII Jornadas Interescuelas-Departamentos de Historia*. Bariloche, Universidad Nacional del Comahue, (2009), 134–143.
Papazián, Alexis, *El Territorio también se mueve. Relaciones sociales, historias y memorias en Pulmarí (1880–2006)*. Tesis de doctorado, Buenos Aires, Facultad de Filosofía y Letras, Universidad de Buenos Aires, 2013.
Pérez, Pilar, *Estado, indígenas y violencia. La producción del espacio social en los márgenes del estado argentino. Patagonia central, 1880–1940*. Tesis de Doctorado, Buenos Aires, Facultad de Filosofía y Letras, Universidad de Buenos Aires, 2014.
Pérez, Pilar, “Las policías fronterizas: mecanismos de control y espacialización en los territorios nacionales del sur a principios del siglo XX”. In *XII Jornadas Interescuelas-Departamentos de Historia*. Bariloche, Universidad Nacional del Comahue, (2009), 160–175.
Radovich, Juan Carlos and Alejandro Balazote, “Transiciones y Fronteras Agropecuarias en Norpatagonia”. In *Producción Doméstica y Capital. Estudios desde la Antropología Económica*, edited by Héctor Trinchero, Buenos Aires, Biblos, 1995, 63–79.
Wolfe, Patrick, “Settler colonialism and the elimination of the native”, *Journal of Genocide Research* 8 (2006), 387–409.
Vezub, Julio, *Valentin Saygueque y la gobernacion indigena de las manzanas. Poder y etnicidad en patagonia*. Buenos Aires, Prometeo, 2009.

Menara Guizardi

The Boundaries of Self

Reappraising the Social Simultaneity of Transnational Migrant Communities[1]

The Limits of Anthropological Imagination: Community, Otherness and Personhood in the Chilean Northern Border

> According to a famous remark of Karl Popper, "The historicist does not recognize that it is we who select and order the facts of history" (1996, p.269). Popper and other theorists of science inspired by him do not seem to realize that the problematic element in this assertion is not the constitution of history (who doubt it is made, not given?) but the nature of we. From the point of view of anthropology, that we, the subject of history, cannot be presupposed or left implicit. Nor should we let anthropology simply be used as the provider of convenient Other to the we[2].

Back at the end of 2012, when I first started developing ethnographic fieldwork of the daily lives of Peruvian migrant women in Chile's most northern city, Arica[3], my greatest bewilderment was not related to the economic, cultural and political

---

1 The author thanks the Chilean National Commission of Scientific and Technological Research which funded the study that gave rise to this paper through Project FONDECYT 11121177: "Gender conflict, labour insertion and migratory itineraries of Peruvian women in Chile: a comparative analysis between the regions of Arica-Parinacota, Tarapacá and Valparaiso".

2 Johannes Fabian, *Time and the other*, x.

3 Arica is located on the coast of the Pacific Ocean and is the capital city of the Chilean Region of Arica y Parinacota, which limits with Peruvian lands (to the North), and with Bolivian territories (to the Northeast), see José Tomás Vicuña, Menara Guizardi, Carlos Pérez and Tomás Rojas, "Características Económicas", 38. Therefore, Arica y Parinacota constitutes the Chilean side of the Andean Tri-border-Area (TBA) whose geographical milestone (named "El Tripartito") is placed in the Andean Plateau (over 4000 meters above sea level), see Sergio González, "El Norte Grande", 36. Arica is only 30 kilometers from the border with Peru, and 52 kilometers from Peru's most southern city, Tacna, cf. Juan Podestá, "Regiones fronterizas", 127. Both cities enjoy long term historical ties, as observed by Jaime Rosenblitt, *Centralidad geográfica*, 47–81. Arica served as the port of Tacna from the 16th Century (in colonial times), until the division of both between Chile and Peru in 1929, after 46 years of violent litigation – a conflict derived from the Pacific War (1879–1883) – cf. Felipe Valdebenito and Menara Guizardi, "Las fronteras", 288. On its southern most territories, Arica y Parinacota limits with the Chilean Region of Tarapacá, which is followed further south by the Region of Antofagasta. Tarapacá and Antofagasta have

complexity of social interactions between Peruvian, Chilean and Bolivian border territories[4]. Nor was it regarding the relation between contemporary migration, cross border practices, historic patterns of mobility (some of them dating back to pre-colonial periods), and the connection of localities situated on different sides of the national frontiers[5]. The relatively recent establishment of national borders between the three countries[6] reshaped the past patterns of mobility, economic exchanges and communitarian settlement in the Atacama *altiplano* (Andean Plateau) and coastal cities, creating new social boundaries that are experienced by people (dialectically, I would assert) as disruptive and, indeed, constructive of transterritorial (and transborder) social relations. Nevertheless, old routes, commercial circuits, shepherding activities, kinship networks, and the religious or communitarian celebrations in these territories are unflagging reminders of the persistence of cross-border mobility, and its capability in defying military border controls and Nation States' utopias of sovereignty.

This dialectical and conflictive relationship between the Nation State's "Euclidian obsessions" (their anxiety to confine lives, practices and identities to strictly defined bounds), and social mobility (in people's insistence in crossing and reorienting those boundaries on a daily basis) was not a novelty neither in theoretical nor in empirical terms. Many social scientists have described that tension in previous works carried out in different national border territories[7].

---

the most important mining enclaves of Chile, cf. Celina CARRASCO and Patricia VEGA, *Una aproximación*, 16.

4 Cf. Sergio GONZÁLEZ, "El Norte Grande"; Menara GUIZARDI and Alejandro GARCÉS, "Circuitos migrantes"; Marcela TAPIA, "Frontera, movilidad y circulación".

5 Cf. Anne-Laure AMILHAT SZARY, "Are Borders More Easily Crossed Today?"; Broke LARSON, "Andean Communities".

6 The borders between Chile and Peru were defined 46 years after the War of Pacific (1789–1883) through the "Treaty of Lima" (1929), cf. Sergio GONZÁLEZ, "La Presencia Boliviana", 72; Marcela TAPIA, "Frontera y Migración", 181; Felipe VALDEBENITO and Menara GUIZARDI, "Las fronteras", 287. The limits between Chile and Bolivia were established in 1904 by the "Treaty of Peace and Friendship" [Tratado de Paz y Amistad], cf. Sergio GONZÁLEZ, "El Norte Grande", 34 sq.; ID., "La Presencia Boliviana", 73. Both agreements have been contested recently in the International Court of Justice of The Hague: the "Treaty of Lima" was contested by Peru in 2013, and the "The Treaty of Peace and Friendship" by Bolivia in 2014. Regarding the first legal claim, in 2014 the International Court addressed the Peruvian requirement of a change in the definition of the maritime border between Peru and Chile on the coast of Arica. The second claim was still in process when this article was written, but the Court's preliminary decision favored Bolivia's requests for sovereign coastal territory, which the country lost to Chile after the War of the Pacific.

7 For a critical and dialectical perspective on border-studies, see Alejandro GRIMSON, "El puente"; Alejandro GRIMSON, "Fronteras, migraciones y Mercosur"; ID., "Cortar puentes, cortar pollos; Michael KEARNEY, "Borders and Boundaries"; ID., "The classifying"; Markus PERKMANN and Ngai-Ling SUM, *Globalization*; Ngai-Ling SUM, "Rethinking Globalisation".

Accordingly, the study's main findings of interest were related to the contextual and historical particularities through which this "border dialectic" materialized in the Atacama localities where my ethnography took place, and not to the discovery of this dialectical relationship itself.

In fact, my greatest surprise while conducting my research, was the discovery that Chilean social anthropologists, who had been working for decades in the Atacama territories, had paid little attention to the "trans-frontier" life in the border region conformed by the cities of Arica (Chile) and Tacna (Peru), and also between them and the villages of the Andean Plateau found in Peruvian and Bolivian territories. Paradoxically, despite the lack of interest in incorporating observation of the national borders in the ethnographical accounts of the local life in this area, these previous works represent a considerable contribution in the establishment of a critical anthropological perspective on the influence of Nation States' mythologies in the imagination, practice and reproduction of social boundaries. This contribution can be explored in at least three key aspects.

Firstly, these works focused enormously on the social changes of indigenous life within Chilean borders, developing intensive and extensive ethnographies among Aymara groups between 1980 and 2010. In doing so, they created an impressive academic production that articulates a context-coherent anthropological narrative, analyzing three decades of social changes from a regional perspective. In this sense, these previous anthropological works go further than the Chilean national centralism, providing interpretations that challenge the arguments produced by the national intelligentsia located in the Chilean capital, Santiago[8].

8 Indigenous groups in the Chilean Atacama territories suffered an intense cultural and identity violence determined by the improvement of the policy of "Chilenization" (Chilenización); Cf. Alberto Díaz, "Aymaras, peruanos y chilenos"; Sergio González, "El poder del símbolo". This state policy was established after Chilean occupation of the Atacama territories through the military campaign of the War of the Pacific, and was operative until the first decades of the 20th Century; Cf. Felipe Valdebenito and Menara Guizardi, "Las fronteras". Its aim was to nationalize (to "chilenizar" in Spanish terminology) the population of the areas that used to belong to the Peruvian and Bolivian republics before the conflict; Cf. Sergio González, *El Dios cautivo*. For the Chilean government, this nationalization was equivalent to a process of de-indigenization, stating that the difference between Chileans, Bolivians and Peruvians was allegedly a racial, religious and ethnic one, being the latter assumed as indigenous, barbarians, heathen and uncivilized; Cf. Ericka Beckman, "The creolization"; Carmen McEvoy, *Guerreros y civilizadores*. The "Chilenization" was carried out through intensive cultural assimilation of indigenous communities of the Atacama Desert; Cf. Alberto Díaz, "Aymaras, peruanos y chilenos". The anthropological works carried out with these groups at the end of the 20th Century have, therefore, an important political dimension. They weaken the persistence of the conflicts that this violent nationalization created, stressing that the indigenous communities –far from being passive victims of State

Secondly, to produce this "contextually coherent" interpretation, Chilean anthropologists followed the Aymara groups' commercial and transhumance routes[9], their travel circuits between the villages and the port cities, their "urbanization process" (a violent outcome of the policies that encouraged the rural exodus in North of Chile, between 1960 and 1990)[10], and their political organization in urban contexts. They studied with great precision the re-ethnification and cultural changes among these groups[11], especially after the adoption of the law of ethnic recognition in Chile (in 1993)[12], and the conflicts regarding territories and natural resources through which they confront the expansion of mining companies over their lands[13]. Finally, they also examined in detail the changes in gender and kinship patterns these groups experienced[14]. In short, these works state non-essentialist perspectives regarding the conformation of cultural groups. They assume Aymara's social groups are translocal communities (instead of taking for granted that they are statically bound to a territory), that actively construct their ethnicity, and whose social life entails conflicts regarding gender and generational issues.

Thirdly, this anthropological work establishes an incredible critical observation on how macro-structural (national and regional) aspects shaped Aymara groups' social life and their relationship with the Chilean State, the mining industries, and the changes in national politics[15]. By studying the relationship between macro processes and everyday social practices, these works established a close dialogue with historical perspectives, producing a diachronic tension on anthropological praxis that must be recognized as avant-garde in their disciplinary context (from the beginning of the 80's to the early 90's).

After reviewing these previous studies, my first concern was to understand why such an impressively critical anthropology had avoided discussing the national borders, and to resize the focus on ethnic communities by including the

control policies– articulate political, cultural, identity and economic resistance. This resistance contributes to make the "Chilenization" a permanently unfinished political project, assuring the persistence of Atacama's territories social and cultural diversity.

9 Hans Gundermann, "Pastoralismo andino", 293.

10 Hans Gundermann and Héctor González, "Pautas de integración regional", 86; Hans Gundermann and Jorge Vergara, "Comunidad", 122.

11 Hans Gundermann, Héctor González and Jorge Vergara, "Vigencia y desplazamiento".

12 Hans Gundermann, "Sociedades indígenas", 64–68.

13 Id., "Procesos regionales".

14 Ana María Carrasco Gutiérrez, "Constitución de género"; Ana María Carrasco and Vivian Gavilán Vega, "Representaciones del cuerpo"; Vivian Gavilán Vega, "Buscando vida".

15 Hans Gundermann, "Pastoralismo andino"; Id., "Procesos regionales"; Hans Gundermann and Héctor González, "Pautas de integración regional"; Hans Gundermann and Jorge Vergara, "Comunidad".

debate on the relationship between "Chilean" indigenous groups, and the border countries' ethnic policies and native communities. Why had Chilean anthropologists performed their studies of ethnic communities only inside national territories? Why was the intense transborder life between Peruvian, Bolivian and Chilean territories not an object of interest?

I was also surprised by the fact that local historians had devoted much more interest to the impact that the late establishment of national borders in the Andean TBA had on the social life of indigenous and non-indigenous people. Contrasting with the lack of anthropological debates on this issue, there was a prolific historic production devoted to the relation between national projects, militaries campaigns and border policies in shaping the articulation of nationality, ethnicity and political conflicts in the Chilean territories adjacent to the borders with Bolivia and Peru[16]. The process of "Chilenization" of these territories captured the attention of many historians, whose documental and ethno historic research provides an important resource for understanding the Bolivian and Peruvian presence in Chilean Northern lands. Archeologists were also more attentive to the disruptions that national borders caused in the historic patterns of life in the territories located between the three countries. Focusing on the long term human timescales –prospecting the sites of first human groups that lived in these areas (which date from 10.000 to 13.000 years ago), and the latter establishment of Tiwanaku (500–1000 AD) and Inca (1450–1532 AD) societies in these territories–, archeologists were deeply aware that the national frontiers could not be taken for granted, nor suppressed as an element of analysis of the social groups' mobility in Atacama[17].

The question as to why anthropologists did not pay closer attention to the importance of the establishment of national borders when compared to historians and archeologists has one immediate answer, related to the difference of perspectives produced by the focus on the *longue durée* processes of the latter. Although the anthropological studies of Northern Chile articulated the local practices with macro processes –which enforces a sort of historicized perspective of everyday life– their analysis was usually related to the period between 1980 and

16 Cf. Alberto Díaz, "Aymaras, peruanos y chilenos"; Alberti Díaz, Rodrigo Ruz, Luis Galdames, and Alejandro Tapia, "El Arica Peruano de ayer"; Sergio González, "El poder del símbolo"; Sergio González, *Chilenizando a Tunupa*; Id., *El Dios cautivo*; Id., *Arica y la triple frontera;* Id., "El Norte Grande"; Id., "La Presencia Boliviana".

17 Cf., eg., Tom Dillehay and Lautaro Núñez, "Camelids, caravans, and complex societies; Lautaro Núñez and Axel Nielsen, "Caminante, sí hay camino"; Gonzalo Pimentel, Charles Rees, Patricio de Souza and Lorena Arancibia, "Viajeros costeros y caravaneros".

2000 (almost a hundred years after the national borders became a "naturalized" social imperative in this territory).

But it is possible to address other elements when answering that question. Firstly, the temporal perspective adopted by anthropologists produced an unwanted side effect: they avoided relativizing properly the hegemonic ways in which local and national societies build their conceptions of "we" and the "Others". This discussion leads us back to Fabian's[18] quotation that opened this section: defining how the categories of "we" and the "others" that are produced in a given historic moment and in a particular locality should be the point of departure for a critical anthropological perspective, preventing ethnographers from reproducing Nation-State mythologies regarding the alleged homogeneity of a national imagined community, and of assuming inadvertently their own imagination regarding their subjects of study. It is worth noting that the latter is deeply influenced by the archetypical research objects institutionalized by the discipline[19]. Classical social anthropology hegemonized the understanding of the interrelation between the notions of space, community and culture as isomorphic[20], naturalizing the existence of borders that would allegedly frame each social group in a specific "cultural area"[21]. This conceptualization overflowed the national borders' political categories into the anthropological theorization of culture[22], which became hegemonic from the middle of the 19th Century[23]. Anthropologists worldwide projected their objects of study as "the others", defining this category as a social group diverse from the one of the ethnographer: due both to an alleged radically different cultural background, and to their location in someplace else, faraway from anthropologists' own societies[24]. Inspired by Durkheim's theorization of mechanic and organic solidarities, anthropologists understood their own social groups as "societies" in opposition to the "others" social groups, usually understood as "communities".

Following Hannerz's[25] argument, this political (and ethnocentric) bias made anthropology a science obsessed in finding the most "other among others"[26], and

---

18 Johannes FABIAN, *Time and the other.*
19 James CLIFFORD, "Spatial Practices"; Akhil GUPTA and James FERGUSON, "Discipline and practice"; Joanne PASSARO, "You Can't Take the Subway to the Field!".
20 Akhil GUPTA and James FERGUSON, "Beyond 'Culture'".
21 Ulf HANNERZ, "Theory in anthropology".
22 Akhil GUPTA and James FERGUSON, "Beyond 'Culture'".
23 James CLIFFORD, *Spatial Practices.*
24 Kath Weston, "The Virtual Anthropologist".
25 Ulf HANNERZ, "Theory in anthropology", 363.
26 See also Akhil GUPTA and James FERGUSON, "Beyond 'Culture'", 6 and Akhil GUPTA and James FERGUSON, "Discipline and practice", 8.

to narrate this others' social life in a style in which "small is beautiful"[27]. Understanding the social groups as a small and compact unit, classical anthropology had eluded (and in many cases had avoided) to inquire accurately about the relationship between the persons and that social unit. Until the second decade of the 20th Century, the certainty about the primacy of society over the processing capacity of the subjects was a surprisingly hegemonic anthropological assertion. Theoretically, this assertion was provided by the excessive focus on the social cohesion and structure (understood as an ordered system) and on the synchronicity of the "Others" social life[28]. The naturalization of that very idea of the supremacy of communitarian life over the persons' capability to shape social reality has two important consequences. It establishes a dichotomist appreciation of the relation between persons and social groups (between agency and structure, as sociologists address this debate)[29]; and it promotes an anthropologically selective blindness, discouraging ethnographers to deal in detail with the conflictive relation between social customs and hierarchies, and the persons' situational strategies to both reproduce and break this state of affairs[30].

The anthropological studies of Chilean Northern border regions, to the extent they emphasized the Chilean indigenous groups of Atacama as their main object of study, reproduced the epistemological conformation of anthropology as a science dedicated to the "Others", reproducing, simultaneously, the Chilean national imaginary about the indigenous as "non-Chileans" (and therefore, as internal "Others"). The latter conforms what Levitt & Glick-Schiller designate as a methodological nationalism: "the tendency to accept the Nation-State and its boundaries as a given in social analysis"[31]. Dialectically, the results of their work

---

27 In Hannerz's words: "When the concern with otherness comes to dominate anthropology, it turns away from the large-scale, complex Western societies in which it is, after all, intellectually rooted. And it turns to what is not only geographically and culturally most distant but also to the organizationally most different. Consequently, small is beautiful. The anthropology of the Other thrives in the local community where the division of labor is strictly limited, where there is little diversity of experience, and where social contacts are face-to-face, with meanings carried by body movements and spoken words, or by song, dance, and ritual. The particular theoretical problems involving large-scale, high degrees of organizational complexity and varied technologies of intellect and communication are, conversely, low-priority concerns for this anthropology". Ulf HANNERZ, "Theory in anthropology", 364.

28 Johannes FABIAN, *Time and the other*, 25.

29 Jean COMAROFF, *Body of power*.

30 Terry EVENS, "Some Ontological Implications"; Ronald FRANKENBERG, "A Bridge over Troubled Waters".

31 According to the authors, the methodological nationalism entails usually three variables: "ignoring or disregarding the fundamental importance of nationalism for modern societies"; the "naturalization or taking for granted that the boundaries of the nation-state delimit and define

are an outstanding contribution to uproot those very national ideologies, since they pose critical perspectives regarding the national mythologies on indigenous people. But still, their studies also reproduced an excessive focus on social groups –communities, families and societies– as their unit of analysis. The reflection on the role of subjects in the active transformation of these social units tended to be underestimated, or to be centered in the role of masculine leadership.

In 2012, as a foreign ethnographer recently arrived in the North of Chile, I was not sufficiently socialized to the Chilean national ideologies of otherness. Therefore, the phenomenon I first perceived as of anthropological interest was related to the presence of migrants from Peru and Bolivia in Arica and its surroundings, and to the intensity of the daily flows of persons, goods, illegal substances, services, social practices and knowledge that crossed the city, connecting it with Peruvian and Bolivian territories[32]. In complete contrast to that assumed by precedent anthropologists, I could not see anything other than the flow over and connecting national territories. My "anthropological anxiety"[33] was to understand how those transnational experiences rebuild communitarian life and ties. In doing so, my perspective suffered from what could be designated, by contrast, as a methodological transnationalism: the tendency to over-emphasize the border flows, stressing a dematerializing perspective of communities that usually leads researchers to evade (or to avoid) recognizing that Nation-States and national imaginaries are still important determinants of social interactions, even in this neoliberal world[34]. At the same time, my ethnographical imagination was still molded by that classical anthropology anxiety of finding the "most other". As if that was not enough, I imagined this "other" as differentiated by a "national otherness", which is also a very clear methodological nationalism bias. In stressing the social field between Peru and Chile, built by Peruvian migrants' displacements through the national borders, I focused on the social field as a unit of analysis, which prevented me, in the beginning, to go further into the dichotomy between personhood and social group.

Beyond an attitude of self-atonement regarding my own and my colleagues' anthropological insufficiencies, the early discovery of these perspectives' distortions was assumed as a turning point to the production of critical reflections on

---

the unit of analysis", and the "territorial limitation which confines the study of social processes to the political and geographic boundaries of a particular Nation-State", Cf. Peggy Levitt and Nina Glick-Schiller, "Conceptualizing simultaneity", 1007.

32 Menara Guizardi, Orlando Heredia, Arlene Muñoz, Grecia Dávila and Felipe Valdebenito, "Experiencia migrante"; Felipe Valdebenito and Menara Guizardi, "Las fronteras".

33 Georg Marcus, "Ethnography in/of the world system", 99.

34 Alejandro Grimson, "La nación", 39.

anthropological praxis, helping in reorienting the fieldwork strategies[35]. The key point here was not the differences between these perspectives, but the opposite: the things they have in common, the methodological insufficiencies that both generations of anthropologists reproduced in a similar way. Besides the methodological nationalism, my perspective shared with the previous work a careless glance over the relationship between the process of production of personhood and the social changes in communitarian life. When I started my fieldwork, this theoretical statement acquired an impressive coherence, since Peruvian women constantly stressed in our interaction their particular consciousness regarding their capability to partially change the "usual order of things" in the border cities of Arica and Tacna.

The present chapter aims to set up an ethnographical theorization of the relation between the process of production of self, and the ways through which the communitarian life is reinvented in a specific border region. Since my analysis is framed by the debates carried out by the transnational perspective of international migration –which resize classical anthropological assumptions about the isomorphism space-culture–, I will start by briefly defining some theoretical arguments of the transnational studies, especially on the concept of simultaneity. In this instance, the concept will be defined as a dialectical relationship in three dimensions: 1) between two or more local spaces crossed by a national border; 2) between agency and structure, and 3) between two or more dimensions of the migrant's experiences of the "self". To illustrate this last issue, I will narrate examples of situations and dialogues of the fieldwork which evidence the particular "self" constructions of Peruvian migrant woman. Through these examples, I will develop a preliminary reappraisal on the relation between "selves" and "transborder communities".

35 The comparison between the sort of analytic distortions two different generations of anthropologists working on the same territory would be more likely to develop can help in shaping a historical and critical understanding of the transformation of anthropological epistemologies, and social contexts.

Transnational Communities: Reframing Social Simultaneity

Since the 90's, ethnographers have voiced strong criticisms regarding the conceptual triad space-community-culture in social anthropology[36], arguing that the "time-space compression"[37] that characterizes the globalized world required an urgent theoretical review of the limits between identities, subjective action and communitarian spatial structuring[38]. In this context, the concept of border became key and problematic, both in the theorizations regarding the limits of Nation-States and their "cross-border regions"[39], and in the debates about the experiences of social boundaries in these areas. Many researchers started assuming that in these areas, the very concept of "community" could not be taken for granted, since the dialectic relation of limitations and circulations on these spaces demands an analytic perspective attentive to the political tensions between "subject, history and culture"[40]. Therefore, the borders would be plural spaces where Nation States act structurally, while subjects act re-interpreting and negotiating the classificatory hierarchy of the State[41]. In order to articulate these debates to the research carried out in Arica, my theoretical point of departure

---

36 Arjun Appadurai, *Modernity at Large;* Akhil Gupta and James Ferguson, "Beyond 'Culture'".

37 David Harvey, *The Condition of Post-Modernity.*

38 This debate was not a novelty: Barth established a very accurate position on it almost 40 years before. The novelty was, in fact, the surprising generalization of these critics. See: Fredrik Barth, Los grupos étnicos; Alejandro Grimson, "Fronteras, Estados e identificaciones".

39 Markus Perkmann and Ngai-Ling Sum, *Globalization*; Ngai-Ling Sum, "Rethinking Globalisation".

40 Alejandro Grimson, "Disputas sobre las Fronteras", 15. The border areas are crossed by at least three political aspects constituent of their spatiality: 1) literal borders, in the form of political-territorial demarcations; 2) identities, crossed by ethnic, class and nationality variables, and 3) political systems, official and non-official organisations responsible for charting and enforcing political-identity limits, Cf. Michael Kearney, "The classifying".

41 Jorge Brenna, "La mitología fronteriza", 12. Anthropologists of South American border regions have followed these reflections. Grimson, for example, points out that the porosity of borders "does not necessarily imply a modification of identity classifications and national auto-affiliations. Rather the presence of the border allows the organisation of a social system of exchanges between groups that consider themselves different" Cf. Alejandro Grimson, "¿Fronteras políticas", 38. This idea is close to Barth's statement that the fact that people cross social boundaries does not mean that said boundaries disappear; Cf. Fredrik Barth, *Los grupos étnicos*. The judicial, political, economic and identity asymmetries between neighbouring countries promote the emergence of social practices that seek to benefit from these differences, from the liminality between legal and illegal, and between belonging and being uprooted; Cf. Alejandro Grimson, "El puente". These practices use cross-border circularity to achieve those benefits and interests.

is a truistic affirmation: that the city's border condition alters the ways through which agency and social structure builds the "local space"[42]. It also configures a particular communitarian spatiality, since daily life is based on intensive circulation through localities found on different sides of the Chilean-Peruvian national frontier. How should we define "community" and "belonging" in areas where everyday life entails permanent circulation between national spaces?

At present the predominating idea in anthropology is that the cross border condition of migrants materializes as transnational practices that consist of creating social fields that link the countries of origin and destination[43]. "Transmigrants" maintain family, economic, social, organisational, religious and political relationships that cross borders: they take decisions, show interest and negotiate identities with the social networks that connect them with at least two countries[44]. So, transnational migration would lead to a globalisation "from below"[45] that tensions the States limits. Various authors[46] understand this transnational social field as the link between migrants' social and cultural capital, following Bourdieu's debates[47]. Consequently, the communities' axes would be their social networks and not their inscription upon a "cultural area". As Besserer[48] defined it, the transnational spatiality of the migrant communities relies on the frequency of practices that bring the communities together, and not on the literal spatial distance between the members of their networks.

---

42 It implies, at the same time, mutual conformation processes with "global" phenomena; Cf. Michael KEARNEY, "The Local and the Global"; Markus PERKMANN and Ngai-Ling SUM, *Globalization*. This double relationship is inherently dialectic, Cf, Michael KEARNEY, "Borders and Boundaries", and problematic, Cf. John AGNEW, "Borders on the mind", articulating some changes in the borders "in temporal horizons (such as compressed-time and memory-time of nations) and in spatial scales (such as global, regional, national and local scales), Cf. Ngai-Ling SUM, "Rethinking Globalisation", 208.

43 Peggy LEVITT and Nadya JAWORSKY, "Transnational migration studies".

44 Nina GLICK-SCHILLER, Linda BASCH and Cristina BLANC-SZANTON, "Transnationalism".

45 Alejandro PORTES, "Theoretical Convergencies".

46 Donald MASSEY, Joaquín ARANGO, Graeme HUGO, Ali KOUAOUCI, Adela PELLEGRINO, and Edward TAYLOR, "Theories of international migration"; Donald MASSEY, Luin GOLDRING, and Jorge DURAND, "Continuities in Transnational Migration".

47 Social migrant capital is defined as "the aggregate of the actual or potential resources which are linked to possession of a durable network of more or less institutionalized relationship of mutual acquaintance or recognition"; Cf. Alejandro PORTES, "Social Capital", 45. This durable network is not naturally given, but is weaved from strategies aimed at the institutionalization of group relations. Cultural capital would correspond to the knowledge incorporated by the migrants and spread through the networks.

48 Federico BESSERER, *Topografías transnacionales*, 8.

The experience of this sort of communitarian life produces a dialectic relation between "there" and "here" that is defined as an experience of social simultaneity[49]. The spatiality that migrants create in the host society will be updated by practices, relations, identities, affections, desires and social hierarchies that come from experiences in other localities. When practices that are born in specific localities travel as a result of migratory displacement, they cause spaces that may not themselves have a geographical continuity to become superimposed, and so create the transnational social field. The dialectic relation that configures this social field relies on the fact that the social spaces begin to (re)produce through the local ascription of non-local practices. Therefore, at the same time, migrants erode and build the spatial character of the locality. The spatial dialectic of simultaneity allows an explanation of how a community can construct itself not in specific local places, but through the connection between places. This does not mean that the community becomes aspatial, rather that its spatiality will demand an increased mobility from its subjects: in spatial, symbolic and identity terms. So, the concept of simultaneity tensions the notion of spatial communitarian stability (the space-culture isomorphism) and identity stability (the community-identity isomorphism). If the community is built on the crossover of borders and between spaces, then it is logical to think that the people living this lifestyle will form principles of ownership in fitting with this mobility. Following on from this latter point, it is necessary to extend this notion of simultaneity defining it through other dialectic relations.

In the first place, it would be necessary to observe this simultaneity as a specific dialectic relation between agency and structure. The articulation of diverse capital between places that are separated by national borders requires a capacity to mold, overcome, reinvent and articulate given knowledge according to the restrictions, constrictions and structural limits that materialize in local host space. In contradiction, this implies resisting and desisting[50]. So simultaneity also implies an active dialogue –everyday and practical– between the subjects and collectives and the structural restrictions in the contexts of origin and reception. This structural dimension refers to, for example, the construction of macro-economic dynamics of exploiting migrant labor (of a national and global character) and the conformation of political-judicial exclusion that leads to economic

49 Peggy Levitt and Nina Glick-Schiller, "Conceptualizing simultaneity".
50 I subscribe here to Comaroff, for who the dialectic between human action and structural constraint implies taking into account that human beings act determining their own history but through mechanisms that are eminently contradictory "in their everyday production of goods and meanings, acquiesce yet protest, reproduce yet seek to transform their predicament". Cf. Jean Comaroff, *Body of power*, 1.

exclusion[51]. The materialization of these restrictions in the local space is always of a historical character: that recuperates medium and long term processes that made up the local identities and the way in which the locality absorbed and integrated into the construction of the nation. This particular historical relationship of composition of local identities reproduces at the same time as it breaks down self-regulating myths of the cultural and national identity project[52].

Following Grimson[53], we could affirm that border areas such as Arica, make up "cultural configurations"[54]. By thinking in cultural configurations we can concentrate on the context of identity construction as part of a political field. This allows us to investigate the crystallization of culture as an element that, in Arica, particularizes the contents of the Chilean and Peruvian national attributes, as well as the understandings about the relationship between ethnic and gender identities. In addition, this takes us to the debate of Barth[55] about the relationship between interpersonal interactions and the processes of identity change in contexts where a group (in our case Peruvian migrants) are pressed to adhere to the standards of the other group (Arica's society). In these circumstances it is possible that people take on behaviors, symbolism and interpretations from the dominant group without giving up the sense of belonging to their original collective, and without seeing this double affiliation as an identity contradiction[56]. This type of experience specifically describes those social dynamics of cultural configura-

51 Both characteristics are overlapping due to the asymmetrical construction of gender hierarchies and labor markets on a global level, in addition to intersectional impacts because of inequalities based on ethnicity, class, age, and national identity that define the possibilities of social incorporation of women. Cf. Mary Beth MILLS, "Gender and inequality", 42; Kimberle CRENSHAW, "Mapping the margins", 1244.

52 The composition of the national ethnic paradigms is elaborated from the contemporary universalistic enunciation that seeks constitutive homogeneity of the nation. Cf. Rita SEGATO, *La nación y sus otros*. The contextualization of the national in border areas, where the State remembers and is reminded of its limits, destroys this universalism whilst at the same time endowing it with a contextually anchored materiality.

53 Alejandro GRIMSON, *Los límites de la cultura*, 172.

54 Cultural configurations can be defined through four constitutive dimensions. They are "fields of possibility": including institutions, representations and practices that in certain contexts are possible, those which in this same context would be impossible, and those that become hegemonic. Secondly, they suppose that, in a certain context, actions, ways of being and expressing, relationships, experiences and knowledge have some level of interrelation amongst themselves, but this does not lead to a constitutive homogeneity. Thirdly, to articulate itself, a common symbolic fabric is needed, that allows linking (even when heterogeneously) common and shared manners (those which make up its fourth dimension). Cf. Alejandro GRIMSON, *Los límites de la cultura*, 172–177.

55 Fredrik BARTH, *Los grupos étnicos*, 30 sq.

56 Ibid., 31.

tions in which some dialogical and dialectic interaction constructs similarities without implying the disappearance of diversity, hierarchy and differences. The analytical focus will be posed, therefore, on the way in which groups construct the limits between identities and not on the supposed cultural contents that these enclose[57].

That said, the operation of situational adaptations by migrants does not cause complete erosion of their subordination: structural limits do not desist nor do their effects cease. Simultaneously, neither does the persistence of structural limitations imply that the subjects' capacity to act is annulled. In Arica, the fact that the Chilean State converted the city in one of the principal national military enclaves[58]; the fact that mechanisms regulating customs operate discretional selection, making the mobility process of certain Peruvian social groups to Chile difficult; or the fact that the labor market in Arica institutionalizes illegal migration as a way of gaining capital, are all structural aspects that impact on the local migrant experience. However, migrants are also active: crossing unmanned borders, plotting collective negotiations on the cost of illegal labor[59], setting up businesses adorned with the colors of their national flag right next door to military regiments (so that army personnel consume the services and products on offer).

But even taking into account this second dialectic of simultaneity –between agency and structure in Arica's conformation as a cultural configuration– we are still left with one question to answer: What does the everyday articulation of this simultaneity mean to those people who live out these communitarian transborder experiences? Throughout my field work with Peruvian migrants in Arica, I observed that the condition of transnational simultaneity carries specific forms of production of the relationship between subject, space and identity. And so we come to a third dialectic relationship that makes up transnational simultaneity, that which refers to the formation of the "self". However, as means of an introduction to this debate I will first share some ethnographic accounts gathered in Arica.

---

57 Ibid., 17.

58 Dona HOLAHAN, "El uso de minas terrestres en Chile".

59 Menara GUIZARDI, Orlando HEREDIA, Arlene MUÑOZ, Grecia DÁVILA and Felipe VALDEBENITO, "Experiencia migrante".

Configurations of the "self"

Despite the fact that the international borders of the Andean TBA lie in the mountains inland, the commercial and human networks that cross them extend towards the Pacific coast, in particular to Arica (Chile) (with 190,000 inhabitants), and Tacna (Peru) (with more than 300,000 people), between which some 6 million border crossings are made every year[60]. This creates a complex dialogue with the territory's vast history of human flows, and with the current intensification of migratory and commercial pathways between Chile, Peru and Bolivia, in which Arica takes an active part. The migration of Bolivians and Peruvians to Chilean northern localities has intensified and been feminised since the 90's[61], a phenomenon associated with the economic dynamism of two Tax Free Zones: ZOFRI, in Iquique (Chile) and ZOFRA in Tacna (Peru)[62]. In addition, the economic boom of Chilean mining in the Atacama Desert increased the demand for labour in mining and its associated activities. In Arica's valleys, Peruvian migrant networks were involved in labour enclaves linked to agriculture. In the urban setting, they work in homes, restaurant kitchens, packaging enterprises, as sellers in small businesses, in construction companies, and even in the textile industry. In all those occupations, they are exposed to higher levels of labor vulnerability than Chileans. This is related to the ethnic discrimination they face from Chileans, and also to the conditions of poverty in their localities of origin[63].

Currently, Arica is an urban scenario where the construction of a national border is an ongoing process that is evident in the militarization of the city and its periphery, and in the inherent contradictions of imaginaries that separate Chileans from the "others" (foreigners, especially Peruvians and Bolivians). Since the national borders are relatively recent in this area (established in 1929), and since everyday life between Tacna and Arica entails constant transnational mobility for the inhabitants of both cities, the social imaginaries regarding the relationship between local spaces, nationality and communitarian (or family) life acquire particular forms. My lack of awareness of these particularities led me to some curious dilemmas. For example, my efforts to reconstruct the nationalities of members

60 José Tomás Vicuña, Menara Guizardi, Carlos Pérez and Tomás Rojas, "Características Económicas".

61 Cf. Menara Guizardi and Alejandro Garcés, "Circuitos migrantes"; Marcela Tapia, "Frontera, movilidad y circulación".

62 These have increased the massive movement led by Peruvian and Bolivian women who travel to buy merchandise in Iquique and Tacna, and then cross the borders of Chile, Peru, Bolivia and Argentina. Cf. Menara Guizardi and Alejandro Garcés, "Circuitos migrantes".

63 Most of Peruvian migrants in Arica come from small, rural and poor villages of the southern regions of Peru, located mainly in the regions of Puno and Tacna.

of family networks of Peruvian migrants were frustrated time and again. This ethnographic exercise revealed two things, the methodological nationalism and the notable analytical naivety of my initial assumptions on the way in which the subjects configure limits between identities and border territories. Orlando, a Peruvian who lived in constant transit between Tacna and Arica forewarned, "my grandfather Juan, who I never knew, was from Las Condes, in Santiago. [...] My other grandfather who I also never knew [is] from up there in the highlands, putreño[64]". And when asked if this grandfather born in Putre was Chilean:

> I think so because Putre at that time would be, was Chilean. Let's see ... we're talking about around 1900 at least [...] So my father was brought up, lived all his life in Peru, in Tacna [...]. My grandmother, who is about 102 or 103, is from Putre but born in Putre she was; but she settled in [the] triple border area [in the Andean TBA]. [...] She now lives in Tacna, she has residency. Well my father, and I thought he was from Tacna, but he my father would have been from Putre. [...] But I was born here in Arica. (Orlando, Peruvian, 58. Arica, January 2013).

Orlando did not tacitly distinguish the nationalities of his grandparents and his father. After the War of the Pacific, the demarcation of the borders between Chile and Peru created a migrant population from people who had not moved[65]. Orlando's grandfather and grandmother born in Putre at the beginning of the 20th Century were probably Peruvians who moved between Putre, Tacna and Arica without this entailing an international migration. With the definition of the Chilean-Peruvian border, to be born in Putre or in Tacna then marks the difference of being Chilean or Peruvian. This difference is blurred when it comes to defining the nationality of Orlando's grandparents' generation. But assigning nationality to his father's generation poses another problem for Orlando. His father considered himself Peruvian, since he was born in Putre before this locality was permanently integrated to Chile. So, for Orlando's generation, these confusions have another dimension: he is very clear about which locality means you belong to Chile and which means you belong to Peru for those of his age (those born in Tacna are Peruvian, those born in Putre or Arica are Chilean). Given these identity definitions that Orlando's generation have incorporated, he himself could not understand why his father considered himself Peruvian if he was born in Putre. Orlando describes that for many years he believed his father had actually been born in Tacna. In the space of three generations, a substantial difference is produced in the way social representations assign the nationality of the people

64 "Putreño" is the adjective used for people born in Putre, a district located at an altitude of 4000m in the Andean Plateau of the present day region of Arica y Parinacota, Chile.
65 Marcela Tapia, "Frontera, movilidad y circulación".

in relation with their localities of origin: the institutionalization of the border changed the identity representations within the families.

Accounts like Orlando's are common not only among Peruvians in Arica, but also amongst Chileans who always told me about the nationalities of their ancestors with the same uncertainty. Uncertainty and lack of definition about the nationality of family networks, that caused me such analytical anxiety, was not a problem for them. What is more, Orlando considers himself Peruvian, despite having been born in Arica and also having Chilean nationality. Many pointed out to me how curious my anxiety to elaborate a clear family tree that included nationality for the family members was. And so, we find ourselves with the first dialectic contradiction in the way in which the subjects experience the meanings and limits between community and national identities in Arica. In an area in which nationalization through militarization of the zone has not ceased since the 19th Century, in which the States have invested a significant effort in producing the sense of differentiation of the nationalities, it would seem that maintaining a sense of affiliation to localities, communities and families has meant that people have created selective ways of blurring the identification of their own ancestors who had their identities split by the national border. The recurring classification and labeling of the nationality of the people in Arica's everyday social experiences is simultaneous to a contrary exercise: the breaking up of the memory of past identities. That is why the fixation with classifying the nationality of ancestors seemed, to Chileans and Peruvians, such a strange exercise; that only made sense to a foreign anthropologist who was not capable of considering the particular ways in which the memories of nationhood are produced in that border region.

But this was not the only compositive dialectic of the relationship between the subjects and the cultural configuration of the border that I found. The first space used by migrants that I frequented in Arica –from November 2012– was the International Bus Station. There I spent mornings and afternoons sat in the terminal where buses that connect Arica and Tacna leave from and arrive to. Sat for several hours a day on a medium height wall where people rested, waiting for the bus and in the queue for the bathroom (just behind) I spoke, from the first day, to Peruvian and Bolivian women and men of both nationalities[66]. A Peru-

66 In the spaces of migrant concentration in Arica there is a gender space distribution in operation: Peruvian men and women frequent separate spaces. Cf. Menara Guizardi, Orlando Heredia, Arlene Muñoz, Grecia Dávila and Felipe Valdebenito, "Experiencia migrante". My own gender deprived me of interacting more often with men, facilitating my access to female spaces. The bus station wall was an exceptional space in this sense: there men and women had to be together, given the queue that formed for the use of the toilets and showers. The shared wait generated certain permissiveness in dialogue and contact between the genders.

vian woman was the first to approach me, some fifteen minutes after I sat down. She asked me if I could help her holding her daughter while she prepared her something to eat. She was taking the little one back to Tacna where she would be looked after by her older sister, because she, the mother, was working as a maid in a house in Arica. I accompanied her until her bus left, and while we chatted –I was asking her about her experience as a mother of daughters, and in particular her experience of having them split between two cities–, other Peruvian women who were also sat on the wall joined in the conversation. They told me that they also had similar realities and about how they solved the difficulties that such a family lifestyle presented.

The women questioned me in the same way, if not more, than I did them. The woman whose child I had looked after asked me if I had children. On getting a negative answer she enquired as to why not. She interrogated me about my love life, my partner's nationality, about whether I had managed to legalize my documents in Chile, and how Chileans treated me in the city and at my place of work. This conversation turned out to be the first in-depth interview of my ethnography and the person interviewed was not the Peruvian woman who –to use anthropological terms– was the "priority subject of my research". The first in-depth interview of the study was carried out by the Peruvian woman on me, and she asked me everything that I had intended to find out about her. On hearing my responses, the women seated next to us began giving their opinions, and asking me additional questions. Thanks to their comments and advice about the documentation issue, I realized that they were going to help me a lot more than I had previously thought. I asked them for authorization to take notes about the information they were giving me about legalizing my immigration papers: about the places to go to, and about what procedures to do. And so the first recordings to be found in my field work diary are notes about how to solve my immigration problems in Chile, not exactly what qualifies as an in-depth description of the study's subjects. This conversation was followed by many others with Peruvian women. Time and again they questioned me about my life, and our conversations always started with questions about my immigration status. Moving on from my answers they told their own stories about immigration and the transborder lifestyle, but I rarely brought up the topic first. In addition, much of what they asked me had never occurred to me as being a relevant topic. They were giving me guidelines as to how they structured their concerns, actions and fears. And their guidelines made me pay attention to my own concerns, actions and fears about my immigration situation in Arica: their awareness and rationalization about this issue was undoubtedly more accurate that mine.

I remember having finished my second day of field work with the sensation of having narrated my life story more times than I could possibly count. In contrast to my experience with Chileans in Arica[67], my phenotype inspired trust in the Peruvian women. And the reasons for this were even more interesting: "well, it's obvious that you're not Chilean". Chileans and Peruvians had labeled me by my phenotype, associating it with that of a foreigner in Chile. But whilst for the former this made me a person to be mistrusted; for the Peruvian women this turned me into someone they could trust.

Several days sitting on that wall also gave me an understanding of the circular dynamics of the transborder lifestyle: of the women who came and went, always on certain days of the week, of the women who moved contraband second hand clothes from Chile to Peru in their up to six daily border crossings. Over the weeks, on finding myself meeting the women who circulated in that space on set days, they also began to take an interest in my presence, and soon my work in the ethnographic field was no longer a novelty. But what interested them most was my story: every week they asked me about my paperwork, about the bureaucratic problems that I had with my work contract and the difficulties in my personal experience. Little by little, these women began inviting me to their homes, to see the rooms they shared with other Peruvian women in Juan Noé, a poorer, precarious neighborhood (next to the bus station). In their houses they told me about their lives, they introduced me to other women and adopted me in their networks. My sense of refuge when with the Peruvian women contrasted with the tense feeling my interactions with Chileans in the city left me with. These circumstances, going beyond the mere ethnographic cliché –of an anthropologist who naively identifies with her study subjects– have had a singular impact on the way I have come to theorize on transnational simultaneity in the framework of Peruvian women migrants in Arica.

The Peruvian women, with whom I began to socialize, believed that I was not clear enough about how to manage the difficulties of border life. Their experience of circulating from one side to another of the border meant that they developed the capacity of structuring a "self" that constituted central resources for the reproduction of their migrant networks. The first of these abilities was the capacity of taking or leaving the position of the "other". They developed, little by little, the ability to transit from the relation between "others and oneself" projected by

67 In Arica I felt a constant tension from Chileans in relation to my different phenotype. They would ask me on a daily basis about my nationality, almost always assuming that I was European or American given the color of my skin, eyes and hair. What initially seemed to me to be curiosity towards difference then began to feel like an aggressive label: like a narrative of my inappropriateness to the local context. Along with the remarks about my different phenotype, I was constantly asked how much longer I intended staying and why didn't I go back to my own country. These questions were repeated at least three times a day in the widest range of spaces.

Chileans towards them, and the relation between "others and oneself" that they themselves projected towards the Chileans. This is not exactly new: it is something that is verified in the migrant experience in different social and national contexts. But what is noteworthy is that the Peruvian women migrants make this transition several times a day and sometimes in the context of the same situation. And they also did it in their communication processes with their spaces of origin which, in the case of Arica, are not only virtual, but also derive from physical displacement: the women travel to Peru every week and, in some cases several times a week. So I began to pay special attention to the social situations in which they took on the role of otherness and to the tensions this identification as a way of fulfilling, strategically, the identity demands that were positioned by the "other".

For example, in the morning, the Peruvian women who work for a daily wage wait for Chilean contractors sitting in the corners of the bus station. Men, who arrive on foot having parked their cars on nearby streets, come and offer the women work. They normally treat the women aggressively. They ask them if they shower or brush their teeth and they reject those that they consider "dirty". With indigenous women their manner is often even harsher. On many occasions I heard exchanges such as: "speak properly Indian, can't you speak Spanish?". When one of these men arrives, the women run to surround him to negotiate work as a maid, cooking, or in packing or textile companies[68].

Their way of interacting is always very submissive: they do not make eye contact; they talk very softly to them and, if the wage offered is considered unfair, they will not state that openly. They avoid that conversation leaving the offer open to be accepted by women who consider it acceptable. When I questioned them about this practice, they told me that the contractors were idiots, that they expected the women to be docile, and whilst they continued to give them jobs, they were willing to adapt to this (inexact) image that they had of them. By taking this silent step back, the women personified their right to reject the work conditions –an act of choice that involves a certain amount of power and agency–, but without breaking the submissive relationship with the contractor that is expected of them. Halfway between resistance and submission, the women entered and exited, this one scene, of the relationship of otherness and constituted difference with the other Chilean.

On questioning them about this capacity to adapt or not adapt, thousands of other scenes began surfacing in their stories. Many of them referred to the work environment, reiterating the necessity of this flexibility when relating with the locals. The women talked about these strategies amongst themselves, especially

68 Agricultural work is the last option: the women described it as lower paid in comparison to city work, and much more demanding physically and in unhealthy conditions.

in these female spaces while waiting for work. In these conversations, I confirmed how they interpreted their contact with the natives, about their different perspectives and about their recognition of the exploitative mechanisms to which they were being subjected. Genny told me, for example, that "Chilean people were stabbing them in the back", they made them work long hours. To their face they told them that "great, everything is fine". But behind their backs they said "Peruvian women don't work well, they're lazy, they're no use, they don't do a good job". And Genny continues, "Peruvian women are scared by the way Chilean women do things in the kitchen":

> I worked in Chilean restaurants here, chopping potatoes, chopping vegetables, washing dishes. Chilean cooks don't wash the vegetables: they tell us to chop, and after we chop they cook them like that, without washing them. I worked here in restaurant X [name removed], and there the cook is a Peruvian woman, so the food there is clean, nice and clean, because she made us wash the vegetables two, three times before cutting them, that you have to wash the dishes very well, well. (Genny, Field notes. December 2012).

In Genny's story we see the conformation of a determining discourse about the "being" of the Peruvians and the "being" of the Chileans that is radically different from the local discourse (equally essentialist) about the differences in these national identities. Beyond the essentialist limitations of one or other of these visions, the Peruvian women had developed ways of self-fulfilling the determining ideal of the Chilean view of them, without abandoning the construction of their own identity archetype. The interesting thing was exactly this: the social and psychological capacity to experience asymmetric ways of configuring the "self", moving between them without suffering from personality incoherence that this movement might cause.

This type of ability required certain physical changes: in hairstyles, ways of dressing, talking, looking, walking and eating. They instilled a certain capacity to perform the "self" that varied from the slightest change in the way of speaking, to more complex and carefully planned changes in their physical appearance. One example of this is the adaptation that they go through at the moment of crossing the border. The region of Arica y Parinacota had (until July 2013) its own legislation for the movement of people with Peruvian nationality. There was a treaty known as "the seven-day agreement" in place, which allowed Peruvian migrants to enter the region without a tourist visa or a residency visa[69]. By crossing Cha-

---

69 This is the "Convenio de Tránsito de Personas en la Zona Fronteriza Chileno-Peruana de Arica-Tacna" (Agreement for the Free Movement of People in the Chilean-Peruvian Border Area of Arica-Tacna), a special piece of legislation that allowed entry to Chile but only within the border of the region of Arica y Parinacota. It was signed in Lima on 13 of December, 1930, enacted in

calluta –the Chilean border control– Peruvians received a document called "Salvoconducto" (a kind of laissez-passer), with which they could supposedly move around and be a tourist in the region of Arica y Parinacota, but it did not allow them to work. Nor could they cross the southern border of the region with this document. At the Chilean customs of Cuya (between the regions of Arica y Parinacota and Tarapacá), Peruvians had to show a tourist visa, or even a residency visa (temporary or permanent). If not, there was no access to the regions further south, where the most important mining industries of Chile are found.

This caused a marked difference in the profiles of those Peruvian migrants who were headed to Arica from those who were headed to mining cities such as Iquique and Antofagasta. To get a tourist visa, Peruvians need to have a passport from their country and travel money from 500 to 1500 US dollars. As such, this process is usually out of reach for rural workers or informal urban workers from the south of the country, given their low-income profile[70]. The procedure requires, at the same time, knowledge on how to talk to State employees, how to do the paperwork and especially on how to present yourself when crossing the border as a tourist. So to obtain a Peruvian passport and get a tourist visa for Chile requires legal, documentary, discursive and even performative knowledge which can be gleaned from social networks of migrants who have already gone through the process.

The Peruvian women told me it was very difficult to enter Arica as a tourist, even when you had a passport and travel money. Of the 42 Peruvian nationals interviewed in the city only one man was able to enter as a tourist on his first attempt. For the rest it took between two and five years to get the tourist visa: some of whom crossed the border, passport in hand approximately once a week for at least a year. The women told me of some ladies who work selling services to Peruvians who want to enter as tourists, teaching them exactly how to dress, what to say, and what to take when going to confront the border official in Chile:

> I spent two years here [in Arica] in clothes [contraband second hand clothes] and I got my passport and I went to Santiago with my sister. And they stamped my passport [...]. In those years, yes, in those years I found a woman who lent me the travel money, I don't know if you've heard about that [...]. In those years the excuses of the PDI [Criminal Investigative Police of Chile] were –'what is your purpose?'. –'I'm in retail', I told them. –'Ah!', and he asked: –'And, where are you going?'. –'To Iquique', I said –'to do what?'...Well, we had

Chile on 20 of February, 1931 and brought into force in that country in 1983. From July 2013 the Chilean government suspended its application. From that date all Peruvians who enter Arica y Parinacota must request a tourist visa.

70 These workers ended up developing a transborder lifestyle, working illegally in Arica and returning to Tacna at the end of the seven days.

> already planned everything: –'I'm going to buy trainers', I said. –'What trainers?'. 'Reebok, Nike'... you had to learn a script [...]. –'Ok, and how much money have you got?'. –'I've got about a thousand dollars'. –'Show me', they said. And we... –'No, no stop there', he said. Because it's typical for Peruvians, for traders to put [money] there... Inside their bra. And so they began: –'No, no stop there', they said. And besides the woman who lent us the money, because they also started to lend, they rent you practically that amount just in case you are asked 'show me' by the PDI. So [the woman to us] said 'you are going to go as traders, you are going to go with only one small bag', 'you are not going to go with your suitcases because it's all psychological. Let's see, like a trader with a little case, only the basics, so even if they open [if they check the baggage] they're only going to see underwear, one change of clothing, because you're only going for two, three days no more' [...]. Everything was worked out, somebody else brought our cases across [...]. With the years that I worked here [in Arica] well, yes I got to Santiago. I also had the full address of where I had to go to in Santiago. (*Meche*, Peruvian, 31 years old. January 2013).

We can pick out several key elements from *Meche* in order to understand the documentation, social and economic dynamics of crossing the Peru-Chile border. Firstly, as an informal worker in Peru, *Meche* had to go to Arica, save money, after two years between Arica and Tacna with the *salvoconducto*, get her passport and make an attempt to travel to Santiago. Secondly, her experience in Arica was what enabled her, not only financially, but through the social networks and shared knowledge, to cross the border (by contracting a woman who "rented" her the dollars and trained her on how to show up at the Chilean border control). Thirdly, it highlights the inconsistencies of the Chilean immigration legislation that does not have a specific visa for foreigners who want to enter the country for specific commercial activities. Due to the lack of adequate immigration legislation, traders have to enter using a tourist visa, a document that in principle should not be used for economic purposes[71]. Migrants understand this "ambiguous" point in the Chilean legislation, and play with the possibilities that this situation allows them at the moment of crossing into the country. Fourthly, this story allows us to observe the operation of sui generis patterns in the giving of visas at the border control. The tourist visa is given at the border by Chilean PDI officials, but the criteria for granting them are not explicit and legislation gives discretional powers to the authority. The maximum length of stay is 90 days, but less can be given if there any doubts about the foreigner's purpose. So, as well as having to fulfill all the legal requirements for the visa, the Peruvian citizen also has to convince, with their performance, the official on duty. We observed in our ethnography that the authorities' decisions appeared to be taken with pheno-

---

71 On the anachronisms of the Chilean Ley de Extranjería (Immigration Law) (Decreto Ley 1094/1975), see María Florencia JENSEN, "Inmigrantes en Chile", 106.

type criteria (this is borne out by those interviewed too), and "non-streamlined" methods of selection are applied.

In this way, convincing the authority at the border is a crucial step, that implies learning language forms (as well as putting together a believable story, without contradictions about your reason for travel); dressing in a way that places you with urban workers (indigenous and peasants seem to be rejected more frequently); and carrying yourself in a way that shows no fear, uncertainty or anxiety. Crossing the border as a tourist requires the women to rehearse performances, to be played out in front of the other Chilean, and that serve the selective expectations of this other. Here, as an example of what I related about the women offering their work in the bus station, the migrants have to develop a flexible ability to move in and out of character that allows them to move around or stay in one space or another. In the case of crossing borders, however, this characterization is more intense, sharper, deriving from theatrical forms of being in relation with "the other" (in this case the customs official). All this implies that they manage to live together with the dialectic personification of their "self" in at least two simultaneous registers: the "self" that satisfies local relationships with Chileans; and the "self" that satisfies their own relationships with their families at home and their social networks with the Peruvian migrants in Arica.

Final Remarks: The Boundaries of the "Self"

The stories and scenes described in the previous section are the background from which I will explore the relationship between the social construction of the border context, the transnational and circulative spatiality of the migrant communities, and the establishment of the principles of subjectivity that allow that these realities equip themselves with social coherency. I refer to the affirmations of Appadurai that state that "the relationship between the production of local subjects and the neighborhoods in which such subjects can be produced, named, and empowered to act socially is a historical and dialectical relationship"[72]. If we think of the border area as a cultural configuration; and if the transnationalized communities emerge in these spaces turning them into transborder fields, then no historized explanation of this context could ignore, as part of the dialectics of production of the locality, the question about the subjective dimension of the phenomenon:

72 Arjun APPADURAI, *Modernity at Large*, 181.

> Put summarily, as local subjects carry on the continuing task of reproducing their neighborhood, the contingencies of history, environment, and imagination contain the potential for new contexts (material, social, and imaginative) to be produced. In this way, through the vagaries of social action by local subjects, neighborhood as context produces the context of neighborhoods. Over time, this dialectic changes the conditions of the production of locality as such. Put another way, this is how the subjects of history become historical subjects, so that no human community, however apparently stable, static, bounded, or isolated, can usefully be regarded as cool or outside history.[73]

These debates lead us to the third proposed dialectic relation of transnational simultaneity, in complement to the previous two that I already discussed in the second part of the present chapter. This third dimension refers to the construction of the migrant "self" as an experience repeatedly marked by sameness and selfhood[74], as I will detail hereafter. The transnational and transborder communitarian life requires migrants to be able to adapt themselves to the host context, somewhere half way between being contained by and overcoming the structural constraints. To do this, they must constitute a sense of appropriate actions (rational or not, conscious or not) in order to integrate in their daily life the two dimensions of their "self". On the one hand, they build a socially referent meaning to their experience as belonging to a place, in other words, their experience as a native belonging to a certain social space (even when this belonging implies a marginal position in the social hierarchy and distribution). On the other hand, they develop a socially referent meaning to their experience as an "other" in the host context. Both refer to the construction of the processes of national otherness and to the materialization and crystallization of the contexts of migrant emission, transit and reception.

In Arica the way in which the otherness of the Peruvian migrant will be built and experienced in the local space has to do with long running processes: The War of the Pacific, the late establishment of the border between Chile and Peru, and the identity ideologies of being Chilean, for example. These processes, given that they make up Arica in terms of cultural configuration, put a structural pressure on the assignation of identities of the Peruvian others, who are narrated as external subjects: contrasting as a bipolar and negative opposite to the national community's ideals. For the migrants, the experience of being "the other" in Arica's society and, simultaneously being assigned to a specific locus (and a source of distinction and differentiation) in the hierarchy of those considered "their equals", operates a process of tension of the self that takes on a certain importance. The Peruvian women's accounts of their adaptation to Arica,

73 Ibid., 185.
74 Paul Ricoeur, *Sí mismo como otro.*

included narratives about illnesses at crucial moments in those who had not managed to resolve coherently their position between their own "otherness" and that which was "added in Chile". The adaptation processes are narrated repeatedly as processes in which the migrants learn how to navigate for one position or condition to another.

In this sense, the third dialectic dimension of simultaneity refers to the construction, in migrant subjects, of social and psychological mechanisms that allow them to live synchronously their constitution as a "self" and an "other". This constitution, beyond any analytical bipolarity, does not operate as a dichotomy in the personality composition; rather it operates from the dialectic experience of otherness as something coherent and constitutive in a particular identity. If migrants do not find meaning in these transitions –emotional, rational, psychological and corporal–, then it will be very difficult for them to establish a situational adaptation process in Arica of the knowledge, practices and ways of being that they have incorporated in the origin. The same principle operating here is that which Barth[75] observed when he talks about communities that, in identity assimilation processes, were able to take on the identity imposed by the otherness relations with the dominant group and their own without seeing this experience as a fragmentation.

This dialectic composition process of the subjective principle of identity, however, is not something specific to the transnational migrant simultaneity process. According to Ricouer[76] we can consider identities as being composed of a double comparison process, generated by the dialectic relationship between the two facets of subjects' existence. On the one hand, we have the idem, which would be the subjects' (or groups') identity, created when they compare themselves to others whom they consider "equals", a process Ricouer calls sameness. On the other hand, we have the ipse, which would be the subjects' (or groups') identity, constituted from what they consider their singularities: the particularities that have no parallel with what is shown by the external other, constituting what the author coined as selfhood[77]. Identity would be a game of combination and intersection of the elements resulting from the comparison of idem, and the results of the particular understanding of oneself (ipse). A game of sameness and selfhood. If Ricouer is right, then all identity is a projection of meaning and simultaneity between the experience of "oneness" and "otherness". This simultaneity that we find in the experience of the "self" of the transborder migrants

---

75 Fredrik BARTH, *Los grupos étnicos.*
76 Paul RICOEUR, *Sí mismo como otro.*
77 Juan CLAVEL, Tomás MORATALLA and Alberto VELILLA, *Lecturas de Paul Ricoeur*, 152; Muniz SODRÉ, *Claros e escuros*, 42.

would not be qualitatively different from that which we all experience, migrants or not, in our identity processes.

The particularity of the identity experience of the Peruvian migrants in Arica is found in three elements. Firstly, this simultaneity operates from the strategic use of the tension between lawful and unlawful in the configuration of the border area. Secondly, it uses the transborder's mobility and circularity as a fundamental mechanism to achieve benefits and interests (both individual and group or collective) derived from the differences between the territories on one and other side of the border. Thirdly, this simultaneity of the articulation of the otherness identity by the migrants displays itself and operates with particular intensity and frequency. The social experience in border areas puts pressure on the migrants to constantly update the way in which they live the processes of sameness and selfhood. What is more, it forces them to quickly move in and out of different ways of structuring this relation. To paraphrase Ricouer[78], transnational simultaneity implies the constitution from the migrant self in which the subject is radically reminded of their condition as "*Oneself as Another*".

But in order to understand this process, it is necessary to go beyond Ricouer himself, and his consideration that the way of structuring the relation between the "self" and the "other" in the identity process requires and depends on a diachronic process of narrative enunciation[79]. Far from denying the importance of discourse in the composition of identity, I go back to the reflections of Comaroff[80]. As the author points out, the anthropological interpretation of the dialectical relationship between human action and structure, and between them and the process of building the "self" have been impacted by the concept of ideology in Marxist intellectual tradition, which articulates relationship between social practice, context, awareness and intentionality[81]. However, a more rationalist reading has taken precedence, which interprets the capacity to intervene in the social reality as mediated by rational forms of consciousness able to be constructed as (or break away from) the hegemonic ideologies.

---

78 Paul RICOEUR, *Sí mismo como otro.*

79 Ricouer considers that the relationship between sameness and selfhood is unstable and requires a type of discursive institutionalization. In this discursive process, narrativity would have a fundamental function, since exemplary images revived from narration would be threads leading from those from which a synthesis of identity principle operates. Cf. Muniz SODRÉ, *Claros e escuros*, 45.

80 Jean COMAROFF, *Body of power.*

81 Ideology appears in Marx's later works as double dimensional: on the one hand, as the rational management of the consciousness of a social class and, on the other hand as derived from a lived experience, a "practical consciousness". Cf. Ibid., 4.

As Comaroff[82] points out, Foucault and Bourdieu realized the limitations of this concept exploring in the methods how the structural ways were able to produce and reproduce themselves from the subjects. This led them to see that consciousness is produced in everyday life, in social practices, forging subjects that are built by social and cultural forms external to the individual[83]. In both one and the other, the forms of internalization acquire a central bodily character. In Foucault[84], given his consideration of the body as recipient and vehicle of forms of power. In Bourdieu[85], for the understanding that social, cultural, symbolic, political and economic capitals, can only (re)produce itself if subjects internalize it in their habitus mode[86]. Each in their way, Foucault and Bourdieu have overshadowed in their arguments the role of the rationalization of the consciousness in shaping the social experience, in mediating between a body that adheres and transforms the context and that at the same time produces and reproduces it[87]. From my point of view, indebted to Comaroff's analysis, the body and consciousness are dialectically intertwined dimensions in the construction of the "self", and both have specific weight in the social and personal construction process of the meaning of social practices.

The examples given in the previous section point to exactly that. In the accounts given, the women offered me a lucid analysis about how they had been protagonists in processes of conformation of their "self" in front of the other. They understood and enunciated rationally the need for these situational adaptations. They learned from them, and shared them amongst women and, in some cases they even transformed this knowledge into business: offering to teach and train other Peruvian women in this art, in order to cross the border and get the desired visa. In these processes, the game of identifying "the others" from the vision of "oneself" does not disappear. But understanding the difference between this vision and that held by the natives becomes a central point for social survival.

---

82 Ibid., 4 sq.

83 Foucault understands this process as part of the composition of the self-individual in modernity, the result of centuries of violent constructions of disciplinary processes that have in effect internalized the social contentions and limits and inscribed them in the individual consciousness. Cf. Michel FOUCAULT, *Vigiar e punir*. Bourdieu has explored the articulation of the internationalization with the possibility of any agency that, mutatis mutandis, will also change the social structure. Cf. Pierre BOURDIEU, *Las estrategias*.

84 Michel FOUCAULT, *Vigiar e punir*.

85 Pierre BOURDIEU, *Las estrategias*.

86 As "lasting dispositions, lasting manners of maintaining and moving around, of talking, walking, thinking and feeling that are presented with all of nature's appearance. Cf. Alicia GUTIÉRREZ, "Poder, habitus y representaciones", 293.

87 Jean COMAROFF, *Body of power*, 5.

More than just understanding the native vision, the migrants learn and develop setting strategies that allow them to feel like a coherent "self" in the adaptation processes.

This is not a minor detail for my argument about the simultaneity of the construction of the self in the experience of the transnational communitarian life. When we understand transnational simultaneity as a phenomenon that inscribes in and from the bodies of the migrants, we see that the conceptual dichotomies are an analytical mirage: they never settle as equal opposites in the practice of people. And this comes exactly from the dialectical dimension of the bodily experience, halfway between objectivity and subjectivity[88]. So we face the need to move towards an understanding, ever more corporal in those phenomena that, in migration studies, we call transnational.

That said, the reference to the boundaries of the "self" in the title of this article takes into account that these boundaries enclose a dialectical relation between control and lack of control in the transborder experience of Peruvian women in Arica. Controlling the "self" for the women reproduces, in many aspects, the complete lack of choice in certain points of their migratory experience. Consequently, it is not about total control. Supposing that would be the equivalent of proposing the subject's complete autonomy in relation to macro-structural aspects that culturally configure Arica. The "self" strategies allow the women to take some level of agency, but it does not imply a break away from the conditioning of their marginal position, their vulnerability and the effects of intersecting classifications (of gender, ethnic identity and class) that they suffer. Additionally, these "maneuvers" of the personification of their "self", even when restricted to specific micro social interactions, could be interpreted by the States of the border region as an indicator of their loss of power over the women. And they could be interpreted as a humiliation on the other side of the border in their communities of origin. Thus, the definition of how much control or lack of control there is in these strategies is an analytical exercise that forces the researcher to define (also dialectically) the situatedness and the contextual conditioning of her own viewpoint.

Bibliography

Agnew, John, "Borders on the mind: re-framing border thinking", *Ethics & Global Politics* 4 (2008), 175–191.
Amilhat Szary, Anne-Laure, "Are Borders More Easily Crossed Today? The Paradox of Contemporary Trans-Border Mobility in the Andes", *Geopolitics* 1 (2007), 1–18.

88 Ibid., 6 sq.

Appadurai, Arjun, *Modernity at Large. Cultural Dimensions of Globalization*. Minneapolis-London, University of Minnessota Press, 1996.
Barth, Fredrik, *Los grupos étnicos y sus fronteras: La organisación social de las diferencias culturales*. Ciudad de México, Fondo de Cultura Económica, 1976.
Beckman, Ericka, "The creolization of imperial reason: Chilean state racism in the war of the Pacific", *Journal of Latin American Cultural Studies* 1 (2009), 73–90.
Besserer, Federico, *Topografías transnacionales. Hacia una geografía de lavida transnacional*, Ciudad de México, Plaza y Valdés Editores, 2004.
Brenna, Jorge, "La mitología fronteriza: Turner y la modernidad", *Estudios Fronterizos* 24 (2011), 9–34.
Bourdieu, Pierre, *Las estrategias de la reproducción social*. Buenos Aires, Siglo Veintiuno Editores, 2011.
Carrasco Gutiérrez, Ana María. "Constitución de género y ciclo vital entre los aymarás contemporáneos del Norte de Chile", *Chungará* 1 (1998), 87–103.
Carrasco, Celina and Patricia Vega, *Una aproximación a las condiciones de trabajo en la Gran Minería de Altura*. Santiago, Dirección de Trabajo del Gobierno de Chile, 2011.
Carrasco Gutiérrez, Ana María, and Vivian Gavilán Vega, "Representaciones del cuerpo, sexo y género entre los aymara del norte de Chile", *Chungará* 1 (2009), 83–100.
Clavel, Juan, Tomás Moratalla and Alberto Velilla, *Lecturas de Paul Ricoeur*. Colección de Estudios. Madrid, Universidad de Comillas, 1998.
Clifford, James, "Spatial Practices: Fieldwork, Travel, and the Disciplining of Anthropology". In *Anthropological Locations. Boundaries and Grounds of a Field Science,* edited by Akhil Gupta and James Ferguson. Berkeley/Los Angeles, University of California Press, 1997, 185–222.
Comaroff, Jean, *Body of power, spirit of resistance: The culture and history of a South African people*. Chicago, University of Chicago Press, 1985.
Crenshaw, Kimberle, "Mapping the margins: Intersectionality, identity politics, and violence against women of color", *Stanford Law Review* 6 (1991), 1241–1299.
Díaz A., Alberto, "Aymaras, peruanos y chilenos en los Andes ariqueños: resitencia y conflicto frente a la chilenización del norte de Chile", *Revista de Antropología Iberoamericana* 2 (2006), 296–310.
Díaz A., Alberto, Rodrigo Ruz Z., Luis Galdames R., and Alejandro Tapia T. "El Arica Peruano de ayer. Siglo XIX", *Atenea* 505 (2012), 159–184.
Dillehay, Tom D., and Lautaro Núñez, "Camelids, caravans, and complex societies in the south-central Andes". In *Recent studies in pre-Columbian archaeology*, edited by Nicholas J. Saunders and Olivier Montmollin (B.A.R. International Series, 424). Oxford, B.A.R., 1988, 603–634.
Evens, Terry M.S., "Some Ontological Implications of Situational Analysis". In *The Manchester School. Practice and Ethnographic Praxis in Anthropology*, edited by Terry M.S. Evens and Don Handelman. New York, Berghahn Books, 2006, 49–63.
Fabian, Johannes, *Time and the other: How anthropology makes its object*. New York, Columbia University Press, 2002.
Foucault, Michel, *Vigiar e Punir: Nascimento da Prisão*. Petrópolis, Vozes, 2004.
Frankenberg, Ronald, "A Bridge over Troubled Waters, or What a Difference a day Makes. From the Drama of Production to the Production of Drama". In *The Manchester School. Practice and Ethnographic Praxis in Anthropology*, edited by Terry M.S. Evens and Don Handelman. New York, Berghahn Books, 2006, 202–222.

Gavilán Vega, Vivian, "Buscando vida: Hacia una teoría aymara de la división del trabajo por género", *Chungará* 1 (2002), 101–117.

Glick-Schiller, Nina, Linda Basch and Cristina Blanc-Szanton "Transnationalism: A new analytic framework for understanding migration", *Annals of the New York Academy of Sciences* 1 (1992), 1–24.

González M., Sergio, "El poder del símbolo en la chilenización de Tarapacá. Violencia y Nacionalismo entre 1907–1950", *Revista de Ciencias Sociales UNAP* 5 (1994), 42–56.

González M., Sergio, *Chilenizando a Tunupa. La escuela pública en el Tarapacá andino 1800–1990*. Santiago, DIBAM, 2002.

González M., Sergio, *El Dios cautivo; las Ligas Patrióticas en la chilenización compulsiva de Tarapacá (1910–1922)*. Santiago, LOM, 2004.

González M., Sergio, *Arica y la triple frontera, Integración y Conflicto entre Bolivia, Perú y Chile*. Iquique, Aríbalo Ediciones, 2006.

González M., Sergio, *La llave y el candado. El conflicto entre Perú y Chile por Tacna y Arica (1883–1929)*. Santiago, LOM, 2008.

González M., Sergio, "El Norte Grande de Chile y sus dos Triple-Fronteras: Andina (Perú, Bolivia y Chile) y Circumpuneña (Bolivia, Argentina y Chile)", *Cuadernos interculturales* 13 (2009), 27–42.

González M., Sergio, "La Presencia Boliviana en la Sociedad del Salitre y la nueva Definición de la Frontera: Auge y Caída de una Dinámica Trasfronteriza (Tarapacá 1880–1930)", *Revista Chungará* 1 (2009), 71–81.

Grimson, Alejandro, "El puente que separó dos orillas. Notas para una crítica del esencialismo de la hermandad". In *Fronteras, Naciones e Identidades. La periferia como centro*, edited by Alejandro Grimson. Buenos Aires, CICCUS, 2000, 201–230.

Grimson, Alejandro, "¿Fronteras políticas versus fronteras culturales?". In *Fronteras, Naciones e Identidades*, edited by Alejandro Grimson. Buenos Aires, CICCUS, 2000, 9–40.

Grimson, Alejandro, "Fronteras, migraciones y Mercosur. Crisis de las utopías integracionistas", *Apuntes de Investigación del CECYP* 7 (2001), 15–35.

Grimson, Alejandro, "La nación después del (de) constructivismo", *Nueva Sociedad* 184 (2003), 33–45.

Grimson, Alejandro, "Disputas sobre las Fronteras". In *Teoría de la frontera: los límites de la política cultural*, edited by Scott Michaelsen and David Johnson. Barcelona, Gedisa, 2003, 13–23.

Grimson, Alejandro, "Cortar puentes, cortar pollos: conflictos económicos y agencias políticas en Uruguayana (Brasil) - Libres (Argentina)". In *Nacionalidade e Etnicidade em Fronteiras*, edited by Roberto Cardoso de Oliveira and Stephen Baines. Brasília, UNB, 2005, 66–76.

Grimson, Alejandro, "Fronteras, Estados e identificaciones en el Cono Sur". In *Cultura, política y sociedad Perspectivas latinoamericanas*, edited by Daniel Mato. Buenos Aires, CLACSO, 2005, 127–142.

Grimson, Alejandro, *Los límites de la cultura. Crítica de las teorías de la identidad*. Buenos Aires, Siglo Veintiuno, 2011.

Guizardi, Menara, and Alejandro Garcés, "Circuitos migrantes. Itinerarios y formación de redes migratorias entre Perú, Bolivia, Chile y Argentina en el norte grande chileno", *Papeles de Población* 78 (2013), 65–110.

Guizardi, Menara, Orlando Heredia, Arlene Muñoz, Grecia Dávila and Felipe Valdebenito, "Experiencia migrante y apropiaciones espaciales: una etnografía visual en las

inmediaciones del Terminal Internacional de Arica (Chile)", *Revista de Estudios Sociales* 48 (2014), 166–175.
Gundermann K., Hans, "Pastoralismo andino y transformaciones sociales en el norte de Chile", *Estudios Atacameños* 16 (1998), 293–319.
Gundermann, Hans, "Procesos regionales y poblaciones indígenas en el norte de Chile. Un esquema de análisis con base en la continuidad y los cambios de la comunidad andina", *Estudios Atacameños* 21 (2001), 89–112.
Gundermann K., Hans, "Sociedades indígenas, municipio y etnicidad: La transformación de los espacios políticos locales andinos en Chile", *Estudios atacameños* 25 (2003), 55–77.
Gundermann K., Hans, and Héctor González C., "Pautas de integración regional, migración, movilidad y redes sociales en los pueblos indígenas de Chile", *Universum* 1 (2008), 82–115.
Gundermann K., Hans, Héctor González C., and Jorge I. Vergara, "Vigencia y desplazamiento de la lengua aymara en Chile", *Estudios filológicos* 42 (2007), 123–140.
Gundermann K., Hans, and Jorge I. Vergara, "Comunidad, organización y complejidad social andinas en el norte de Chile", *Estudios Atacameños* 38 (2009), 107–126.
Gupta, Akhil and James Ferguson, "Beyond 'Culture': Space, Identity, and the Politics of Difference", *Cultural anthropology* 7 (1992), 6–23.
Gupta, Akhil and James Ferguson, "Discipline and practice: "The field" as site, method, and location in anthropology". In *Anthropological locations: Boundaries and grounds of a field science*, edited by Akhil Gupta and James Ferguson. Berkeley/Los Angeles: University of California Press, 1997, 1–47.
Gutiérrez, Alicia, "Poder, habitus y representaciones: recorrido por el concepto de violencia simbólica en Pierre Bourdieu", *Revista complutense de educación* 1 (2004), 289–300.
Hannerz, Ulf, "Theory in anthropology: Small is beautiful? The problem of complex cultures", *Comparative Studies in Society and History* 2 (1986), 362–367.
Harvey, David, *The Condition of Post-Modernity: An Inquiry into the Origins of Cultural Change*. Oxford, Blackwell, 1989.
Holahan, Dona, "El uso de minas terrestres en Chile. Hacia una teoría de la frontera militar", *Civitas* 2 (2005), 343–351.
Jensen, María Florencia, "Inmigrantes en Chile: la exclusión vista desde la política migratoria chilena". In *Temáticas migratorias actuales en América Latina: remesas, políticas y emigración*, edited by Eduardo Bologna. Rio de Janeiro, ALAP, 2009, 105–130.
Kearney, Michael, "Borders and Boundaries of State and Self at the End of Empire", *Journal of Historical Sociology* 1 (1991), 52–74.
Kearney, Michael, "The Local and the Global: The Anthropology of Globalization and Transnationalism", *Annual Review of Anthropology* 24 (1995), 547–565.
Kearney, Michael, "The classifying and value-filtering missions of borders", *Anthropological Theory* 2 (2004), 131–156.
Larson, Broke, "Andean Communities, Political Cultures, and Markets: The Changing Contours of a Field". In *Ethnicity, markets, and migration in the Andes: at the crossroads of history and anthropology*, edited by Olivia Harris, Larson Broke and Enrique Tandeter. Durham, Duke University Press, 1995, 5–54.
Levitt, Peggy, and Nadya Jaworsky, "Transnational migration studies: Past developments and future trends", *Annual Review of Sociology* 33 (2007), 129–156.
Levitt, Peggy, and Nina Glick-Schiller, "Conceptualizing simultaneity: A transnational social field perspective on society", *International Migration Review* 3 (2004), 1002–1039.

Marcus, Georg E. "Ethnography in/of the world system: The emergence of multi-sited ethnography", *Annual review of anthropology* 24 (1995), 95–117.
Massey, Donald, Joaquín Arango, Graeme Hugo, Ali Kouaouci, Adela Pellegrino, and Edward Taylor, "Theories of international migration: a review and appraisal", *Population and Development* 3 (1993), 431–466.
Massey, Donald, Luin Goldring, and Jorge Durand, "Continuities in Transnational Migration: An Analysis of Nineteen Mexican Communities", *The American Journal of Sociology* 6 (1994), 1492–1533.
McEvoy, Carmen, *Guerreros y civilizadores. Política, sociedad y cultura en Chile durante la Guerra del Pacífico.* Santiago, Ediciones UDP, 2011.
Mills, Mary Beth, "Gender and inequality in the global labor force", *Annual Review of Anthropology* 32 (2003), 41–62.
Núñez, Lautaro, and Axel Nielsen, "Caminante, sí hay camino: Reflexiones sobre el tráfico sur andino". In *En Ruta. Arqueología, Historia y Etnografía del tráfico sur andino*, edited by Lautaro Núñez, and Axel Nielsen. Córdoba, Encuentro Grupo Editor, 2011, 11–41.
Passaro, Joanne. "You Can't Take the Subway to the Field! 'Village' Epistemologies in the Global Village". In *Anthropological locations: Boundaries and grounds of a field science*, edited by Akhil Gupta and James Ferguson. Berkeley/Los Angeles: University of California Press, 1997, 147–162.
Perkmann, Markus, and Ngai-Ling Sum, *Globalization, regionalization and cross-border regions: scales, discourses and governance.* London, Palgrave Macmillan, 2002.
Pimentel, Gonzalo, Charles Rees, Patricio de Souza and Lorena Arancibia, "Viajeros costeros y caravaneros. Dos estrategias de movilidad en el Período Formativo del desierto de Atacama, Chile". In *En Ruta. Arqueología, Historia y Etnografía del tráfico sur andino*, edited by Lautaro Núñez, and Axel Nielsen. Córdoba, Encuentro Grupo Editor, 2011, 43–81.
Podestá, Juan, "Regiones fronterizas y flujos culturales: La peruanidad en una región chilena", *Universum* 26 (2011), 123–137.
Portes, Alejandro, "Social Capital: Its origin and applications in Modern Sociology". In *Knowledge and Social Capital: Foundations and Applications* edited by Eric Lesser. Woburn, Butterworth-Heinemann, 2000, 43–57.
Portes, Alejandro, "Theoretical Convergencies and Empirical Evidence in the Study of Immigrant Transnationalism", *International Migration Review* 3 (2003), 874–892.
Popper, Karl, *The Open Society and Its Enemies.* Princeton, Princeton University Press, 1996.
Ricoeur, Paul. *Sí mismo como otro.* Buenos Aires, Siglo Veintiuno, 2006.
Rosenblitt, Jaime, *Centralidad geográfica, marginalidad política: La región Tacna-Arica y su comercio, 1778–1841.* Santiago, Centro de Investigaciones Barros Arana, 2013.
Segato, Rita Laura, *La nación y sus otros: Raza, etnicidad y diversidad religiosa en tiempos de Políticas de la Identidad.* Buenos Aires, Prometeo, 2007.
Sodré, Muniz, *Claros e escuros. Identidade, povo e mídia no Brasil.* Petrópolis, Vozes, 1999.
Sum, Ngai-Ling, "Rethinking Globalisation: Re-articulating the Spatial Scale and Temporal Horizons of Trans-border Spaces". In *State/Space: A Reader*, edited by Neil Brenner, Bob Jessop, Martin Jones and Gordon MacLeod. Oxford, Blackwell Publishing, 2003, 208–224.
Tapia L., Marcela, "Frontera y Migración en el norte de a partir del Análisis de los censos Población. Siglos XIX- XXI", *Revista de Geografía Norte Grande* 53 (2012), 177–198.
Tapia L., Marcela, "Frontera, movilidad y circulación reciente de peruanos y bolivianos en el norte de Chile", *Estudios Atacameños* 50 (2015), 195–213.

Valdebenito T., Felipe, and Menara L. Guizardi, "Las fronteras de la modernidad. El espacio Tacnoariqueño y la nacionalización del Norte Grande chileno (1883–1929)", *Estudos Ibero-Americanos* 2 (2014), 277–303.
Valdebenito T., Felipe, and Menara L. Guizardi, "Espacialidades migrantes. Una etnografía de la experiencia de mujeres peruanas en Arica (Chile)", *Revista Gazeta de Antropología* 1 (2015): s/n.
Vicuña, José Tomás, Menara L. Guizardi, Carlos Pérez, and Tomás Rojas, "Características Económicas y socio-demográficas de la Región de Arica y Parinacota". In *Migración internacional en Arica y Parinacota: Panoramas y tendencias de una región fronteriza,* edited by José Tomás Vicuña and Tomás Rojas. Santiago, Editorial de la Universidad Alberto Hurtado, 2015, 37–48.
Weston, Kath, "The Virtual Anthropologist". In *Anthropological locations: Boundaries and grounds of a field science*, edited by Akhil Gupta and James Ferguson. Berkeley/Los Angeles: University of California Press, 1997, 163–184.

Ana María Vara

When a Rebel Finds a Cause, a Discourse, and a Homeland

Rafael Barrett and Latin America

There is something peculiar regarding Rafael Barrett's nationality. Something beyond the mere fact of his being born in Spain –in Torrelavega, Santander, in 1876– from a British father, which made him a British national according to the rule of *jus sanguinis*. The US Library of Congress catalogues his work among that of the Spanish authors of the Generation of 1898 such as Azorín, Pío Baroja, Ramón del Valle Inclán, Miguel de Unamuno. But he is also mentioned as a key author of Paraguayan literature in a prestigious dictionary on the subject;[1] and as part of the Argentine anarchist movement in a reputed dictionary on the Argentine left.[2] Was he Spanish, or British? Or Paraguayan? Or Argentine? Was he a writer, or an activist? Where, in which national tradition should we locate him? How and for what should he be remembered?

Undoubtedly, Barrett represents a paradoxical character. He is first a rebel with no apparent cause, and then becomes a highly committed political subject. And he challenges and redefines what should be considered a community, beyond national borders. He abandoned his home country, Spain, in 1903, as a result of an apparently banal incident –he was denied the possibility of engaging in a duel. He arrived in Buenos Aires as part of the migration masses, in spite of being connected to the Spanish nobility. He took part in the debates on Argentina's migration policies and, again, was involved in an incident that had to do with a duel. He fled one more time, and arrived in Asunción in 1904 in the middle of a revolution. He was welcomed in Asunción's salons, and began to publish successfully in local and regional periodicals. But his commitment to justice –which he had already showed in some publications in Buenos Aires– induced him to join the anarchist movement. Most importantly, he decided to denounce the exploitation of men and nature in the extraction of yerba mate in a leaflet that would become his most famous text.

Lo que son los yerbales paraguayos (*About the Paraguayan yerbales*),[3] published in 1908, eventually caused his ousting from Paraguay. Barrett may be con-

1 Pérez Maricevich, *Diccionario de la literatura paraguaya*, 77–86.

2 Horacio Tarcus, *Diccionario biográfico de la izquierda argentina*, 50 sq.

3 Since Barrett's books have not been translated into English, all of the titles and excerpts presented here have been translated by the author. The same with quotations from bibliography originally published in Spanish.

sidered a rebel and a pariah in three countries; but also a citizen by choice. A citizen of Paraguay, the country he learned to love, where he got married, had his only child, and deeply engaged in politics to the point of risking his life. A citizen of Latin America, since he became aware that the "dolor paraguayo", the Paraguayan sorrow he empathized with, was also a Latin American grief. And finally, a citizen of the world, a true cosmopolitan, since he understood that the injustices endured by the disadvantaged masses in Latin America had to do with an unfair, colonial and neocolonial world.

From Madrid to Buenos Aires

Barrett seems to be born not just once but several times. To public life, certainly in Madrid at the turn of the nineteenth century, as a young dandy, an early contributor to periodicals, and probably a gambler. He was introduced to brilliant salons thanks to his nice looks, his education, his wittiness, and his mother's family ties to the dukes of Alba. But he seemed to have gone bankrupt, and all of a sudden he became a fallen angel in Madrid's milieu. In order to justify his ousting from good society, rumors were spread that he had committed "vice against nature", a euphemism that implied a condemnation both religious and social —not no mention legal and theological. He rejected the accusation, and requested to be examined by doctors of the Protomedicate, who exonerated him and his friend. Certificate in hand, he confronted a prominent figure in public, accusing him of being the source of the calumny and attacking him with a whip. But that was not the end of shame: the Tribunal of Honor that arbitrated on duels denied him permission to engage in one, due to the very calumny that caused his fury. We know all of these thanks to Ramiro de Maeztu's testimony,[4] who later on would become Barrett's admirer, and would write a Preface to one of the many editions of his most famous work, *Lo que son los yerbales paraguayos* (*About the Paraguayan yerbales*).

Much has been written on duel as a political institution at the center of social turmoil in Spain at the turn of the century. Francisco Corral characterizes duel at those times as "one of the ideological battlefields in which the Spanish society's crisis becomes apparent; on a philosophical level, it corresponds to the 'crisis of conscience' of the transition period between the nineteenth and the twentieth

4 Ramiro de MAEZTU, "Rafael Barrett en Madrid", 10.

centuries".[5] A new generation was challenging the old order, and the establishment reacted.

Barrett left Spain and headed to Buenos Aires in 1903. Many authors link this episode of his youth not just with his decision to emigrate but also to a sudden radicalization of his thought. Maeztu claims: "Undoubtedly, the injustice done to him opened his chest to feeling social injustice".[6] Pointing at the depth of his change, Fernández talks of an "existential rupture".[7] Barrett himself admitted: "Since I am a disgraced person, I love disgraced people, the fallen, the *pisados*".[8]

But again, biography meets history, and personal decisions become political. Corral understands Barrett's journey as a "non-destructive escape", and considers the idea of leaving Spain as characteristic of his generation.[9] In contrast, David Viñas points to the future instead of to the past when he comments on Barrett's decision. And he connects his conversion not so much with his leaving Spain but with his arriving in Buenos Aires as part of the immigrant masses that were disembarking in the Americas. He sees him among immigrant laborers travelling from Europe to the Americas, and speculates with some kind of identification process taking place in his soul:

> Barrett is not an anarchist in his own country; he becomes a libertarian as he is part of immigration. That is, in his departure from Spain he becomes embedded in a mass of people exiled because of political and, mostly, economic reasons. Social 'climate' that helps him become aware of the conditions at the immigration ports —depart or arrival ones—, the itinerary followed by grassroots *gayegos*[10] and, at the final stage, the depressed labor conditions endured by those who were following this trail.[11]

Soon after arriving in Buenos Aires in early 1903, Barrett resumed his writing in periodicals. His first article was published on August 1st in the magazine *Ideas*, directed by Manuel Gálvez. He would then become a contributor to *El Tiempo* newspaper, and to *El Correo Español*, a magazine published by the Republican Spanish community where Barrett would harshly criticize the Spanish monarchy.

5 Francisco Corral, *El pensamiento cautivo de Rafael Barrett*, 14.
6 Ramiro de Maetzu, "Rafael Barrett en Madrid", 11.
7 Miguel Ángel Fernández, "Introducción", 13.
8 Quoted in Francisco Corral, *El pensamiento cautivo de Rafael Barrett*, 20; emphasis in original. The "pisados" means literally "those who are walked upon;" metaphorically, the humiliated.
9 Ibid., 12.
10 Since most Spanish immigrants who arrived in Buenos Aires at the turn of the century were born in the region of Galicia, they were carelessly called "gallegos". The spelling "gayegos" in Viñas' quotation, marked in italics, mimics the local pronunciation, evoking the derogatory connotation of this naming.
11 David Viñas, *Anarquistas en América Latina*, 30.

Writing for this magazine, he would engage again in a discussion that would end up in a duel challenge, frustrated one more time for the accusations he suffered in Spain.[12] This incident would induce him to leave for the second time. He left Buenos Aires and headed to Paraguay.

Barrett's brief stay in Buenos Aires could be seen as a moment of transition, in which he looks back to Spain, taking part in some discussions at home from a distant position. But that would be a mistaken view, since during most of 1904 Barrett writes profusely for the Spanish community in Buenos Aires, and takes part in important local political debates. In his writings, he shows an international understanding of the problems, and a characteristic sensibility to inequality between countries and social groups.

The city of Buenos Aires was undergoing a radical transformation: hundreds of thousands of immigrants were arriving from Italy, Spain and Eastern Europe, among other regions, and would end up living in very poor conditions. Between 1869 and 1910, Buenos Aires population quintupled, from 200,000 to one million people. At the same time, a modernization process was taking place: new trains and tramways, electricity, new public buildings were changing the face of the city. Without an agrarian reform, there was no land for distribution to the immigrant laborers, who remained at the city working in factories and the service sector. And there was no welfare state. This context was favorable for the arrival of anti-establishment ideologies like anarchism.[13] The increasing presence of anarchist groups and their strikes and violent actions from 1880 on, induced a repressive reaction from Argentina's government. One of the key norms was the Law of Residence, passed in 1902. It allowed to forbid the entrance of immigrants, as well as to deport them due to political reasons.

Barrett wrote against the Law of Residence, showing his acquaintance with political discussions in Argentina. And he also wrote about social problems in Buenos Aires, about the poverty and lack of protection endured by some immigrant laborers, about the inequality he saw. One of the key short essays of this period is precisely called "Buenos Aires". It has been considered a masterpiece by many critics, who quoted it profusely throughout the twentieth century. Barrett

---

12 A speech by the Republican leader Ricardo Fuente at the San Martín theatre in Buenos Aires, on April 17th, 1904, was followed by several critical responses. In turn, Barrett responded to one of them, signed by a Spanish military officer, Juan de Urquía; Barrett defended Fuente's honor writing at *El Correo Español*. The aggressive exchange ended up in a duel challenge, which was suspended due to De Urquía's allegation regarding Barrett's previous disqualification. He then looked for De Urquía at a hotel, and hit someone else by mistake. Francisco Corral comments: "It is pathetic to find Barrett engaged again in a public fight, and at the mercy of the Trial of Honor". Cf. Francisco CORRAL, *El pensamiento cautivo de Rafael Barrett*, 29.

13 Juan SURIANO, *Anarquistas*, 18.

himself chose it for his first anthology, *Moralidades actuales* (*Current Customs*),[14] the only one prepared by him and published before his death in 1910. "Buenos Aires" talks about a sad awareness: that the same poverty and inequality characteristic of old Europe is present in the promising lands of the Americas. It represents a denunciation of the exploitation of many immigrants, and the intolerable sufferings endured by the most unfortunate. It depicts Avenida de Mayo, one of the most elegant streets in the city, in the early hours of the day, when tired workers leave home toward their work, and young children walk the streets selling newspapers. At a crucial moment, a starving homeless man approaches a garbage bin, and finds a bone with some meat left which immediately begins to bite. The ending paragraph is a cry of horror, and an exaltation of rebellion: "*Also in America*! I felt the infamy of the species in my heart", claims the chronicler. "At that instant, I understood the greatness of the anarchist actions, and admired the magnificent rejoicing when dynamite blows open the despicable human anthill".[15]

According to Fernández' analysis of "Buenos Aires", during his stay in Argentina Barrett began "to see social reality, and to perceive the deep contradictions that were shaking a society founded in human misery.[16] More interestingly, during this time, Barrett reflected on imperialism, and wrote about the receding influence of the British Empire in Latin America, and the increasing presence of the United States in the region. In this sense, he could be included next to José Martí, Rubén Darío and Enrique Rodó among the intellectuals that ushered in "a new era" of discourse on the US in Latin America, as it has been analyzed by authors such as José de Onís. His vision would be eloquently summarized in an epiphonema on Monroe's doctrine: "Monroe —'America for Americans'. Very nice, but a bit too vague. 'North America for North Americans' would have completely reassured me".[17]

14 The Word "moralidades" refers to a Medieval allegorical genre, but also evokes the Latin word "mores", customs.

15 "*¡También América!* Sentí la infamia de la especie en mis entrañas. Sentí la ira implacable subir a mis sienes, morder mis brazos. Sentí que la única manera de ser bueno es ser feroz, que el incendio y la matanza son la verdad, que hay que mudar la sangre de los odres podridos. Comprendí, en aquel instante, la grandeza del gesto anarquista, y admiré el júbilo magnífico con que la dinamita atruena y raja el vil hormiguero humano". Cf. Rafael BARRETT, *Obras completas II*, 29. Emphasis in original.

16 Miguel Ángel FERNÁNDEZ, "Introducción", 11. Emphasis in original.

17 Rafael BARRETT, *Obras completas II*, 313.

In Asunción: literature and denunciations

In spite of the significance of the time Barrett spent in Buenos Aires, it is his years in Paraguay that represent a turning point in his life and work. Why did he choose Paraguay? Francisco Corral suggests that it might have been due to the influence of Carlyle's, an author admired by Barrett who had showed interest in the tragic history of this country.[18] In any case, there was an immediate reason, since he arrived in Asunción in October 1904 as a correspondent of *El Tiempo* newspaper to cover a coup. The so called Liberal Revolution had begun in August, and was supported by Argentina, a country with an almost imperialistic involvement in Paraguayan politics. He sympathized with the revolutionaries. In his only writing on the revolution, titled "La revolución de 1904" ("The 1904 revolution"), he warns on the importance of the process of change it might represent, showing a vision of Paraguay's strategic situation in relation to its natural resources and foreign actors —although he does not explicitly identify them.[19]

Barrett entered Asunción most probably on Christmas Eve, accompanying the triumphant forces, whom he had joined as technical support. As soon as January 1905, he was designated assistant at the General Statistics Office, and a few months later he was promoted to supervisor, but he resigned before the end of the year. He also worked for the railroad company, but also resigned, this time in protest for the way workers were mistreated.

In Asunción he met Francisca López Maíz, a member of the local bourgeoisie who would become his wife and mother of his only child, Alex. In the beginning, everything went fine, but a third incident related to a duel changed this situation.[20] It aroused the animosity of Albino Jara, a military officer who would eventually come to power in 1908, after a new coup.

Barrett published his first article in Asunción in January 1905, in *El Diario* newspaper. Francisco Corral counts as many as forty articles that same year. He would also write for other Paraguayan publications, such as *Los Sucesos*, *La Tarde*, *El Paraguay*, *El Cívico*, *Alón*.[21] In late 1906 he would consider depending on his writings as his only means of living, due to their increasing prestige. At that

18 Francisco Corral, *El pensamiento cautivo de Rafael Barrett*, 31.

19 Rafael Barrett, *Obras completas IV*, 60.

20 This is what happened: two young liberals, Gomes Freire Esteves y Carlos García, engaged in duel. The latter was hurt, and died soon thereafter. Barrett published an article accusing his seconds of incompetence, since the deceased suffered of incapacitating myopia. One of them was Albino Jara. Fernández comments: "In these incidents we can see one (but only **one**) of the reasons of Jara's rage against Barrett..." Cf. Miguel Ángel Fernández, "Introducción", 14. Emphasis in original.

21 Francisco Corral, *El pensamiento cautivo de Rafael Barrett*, 39.

time, he also took part in the founding of La Colmena group, a literary gathering like those fashionable in Madrid at the time; among the attendees there were important Paraguayan intellectuals such as Viriato Díaz-Pérez and Juan O'Leary.[22]

When is it that Barrett stops playing the good visitor, the noble, educated Spanish gentleman, the literary champion, the good Samaritan, and becomes an anarchist intellectual and activist? Francisco Corral situates that moment in 1906, coincidentally with a series of strikes throughout Paraguay.[23] But he also quotes José Concepción Ortiz, who situates this transformation two years later. And we have also seen that Barrett had feelings regarding the hard conditions endured by immigrant workers when he was already in Buenos Aires, and he even praised anarchist bombings in "Buenos Aires". We would come back to this second transformation in Barrett's subjectivity and social role later on.

In any case, Barrett began giving speeches at the request of Unión Obrera (Worker's Union, a Paraguayan anarchist union) in 1907, when he also founded *Germinal*, a periodical aimed at Paraguayan workers, with the Argentine anarchist Guillermo Bertotto. Ángel J. Cappelletti stresses the importance of *Germinal* for anarchism in Paraguay, and points at Barrett's relationship with the anarchist Federación Obrera Regional Paraguaya (FORP, Paraguayan Worker's Federation), founded in 1906 in Asunción, with the support of the strong Federación Obrera Regional Argentina (FORA, Worker's Federation).[24]

Predictably, Barrett's activities aroused concern among good society. The doors of the Instituto Paraguayo and Teatro Nacional were closed for his social talks. He and Bertotto would have to find other places for their gatherings, among those the one on "La infamia de los yerbales" ("The infamy of the yerbales"). At the same time, Barrett would publish a series of articles on the same topic in *El Diario*. Collected and republished in full with the title *Lo que son los yerbales paraguayos* (*About the Paraguayan yerbales*), these texts would become his most important contribution to Latin American essay tradition.

22 Francisco Corral also mentions Juan Casabianca, Manuel Domínguez, Arsenio López Decoud, Modesto Guggiari, Ignacio A. Pane, Juan Silvano Godoy, Fulgencio R. Moreno, José Rodríguez Alcalá, and Ricardo Marrero Marengo. Ibid.

23 Ibid., 40.

24 Cappelletti considers *Germinal* "the most significant expression of the libertarian movement at those times" in Paraguay, along with *El Despertar*. *Germinal* was published between August 2 and October 11, 1908. Cf. Ángel J. Cappelletti, "Anarquismo latinoamericano", LXXVIII-LXXXI.

Barrett accuses

It is not enough to say that *Lo que son los yerbales paraguayos* is Barrett's main contribution to Latin American literature and politics: it may be considered one of the founding texts of a new way of talking about imperialism in the region – and, more generally, on the very history of the region— since it represents a key text in the construction of a discursive matrix that would shed light on the issue throughout the twentieth century and beyond.[25] This discourse represents a key element when trying to understand Barrett's belonging to a community. It is a community of rebels, a counter-hegemonic community, that speaks in favor of the exploited and the humiliated in Latin America.

Los yerbales was published as a series of articles between the 15th and the 27th July, 1908, in *El Diario*, just a few days before a new coup led by Albino Jara, which in turn would be followed by a repressive wave. The articles would be compiled in a plaquette in 1910 in Monteviedo, by a local publisher, Bertani, with an Introduction by Bertotto; and would be republished incessantly throughout the twentieth century, both by political and academic editorial houses. It is included in the volume dedicated to Barrett in 1978 by Editorial Ayacucho, one of the most prestigious collections on Latin American literature, with an Introduction by Augusto Roa Bastos.

At the same time a denunciation and a call to arms, *Los yerbales* represents Barrett's main *j'accuse*. He was a great admirer of Emile Zola, an author "he quotes repeatedly, almost obsessively, as the ideal writer", according to Francisco Corral.[26] Barrett denounces the crude way yerba mate is collected in the border area between Paraguay, Brazil and Argentina; and particularly the inhuman exploitation of the workers, reduced to a form of legal slavery. The descriptions are crude, at times baroque, often interrupted by cries of indignation and apologetic comments on the harshness of his own writing, on the repugnance readers have to endure. However, Barrett also shows his ability to control his style when he chooses to describe particularly hard issues, like torture, in a more stoic way. Barrett tells the story of the companies responsible for the exploitation, the machinations to take hold of the territory and the people, the way workers are chased if they try to escape. And he gives names, the names of the politicians and corporate leaders in charge of the exploitation—one of the main reasons for the persecution he and Bertotto would soon face. These are the opening paragraphs:

---

25 Ana María VARA, *Sangre que se nos va*.

26 Francisco CORRAL, "El enigma de Rafael Barrett", 27.

> It is necessary for the world to learn for once what happens at the yerbales. It is necessary that when one wants to mention a modern example of what human greed can imagine and execute, people do not talk only about Congo, but about Paraguay.
>
> Paraguay is getting unpopulated; it is being castrated and exterminated at the 7 to 8.000 leagues given to the Paraguayan Industrial Company, the Matte Larangeira, and to the tenants and owners of Alto Paraná latifundios. The exploitation of yerba mate is based on enslavement, torment, and killing.[27]

Barrett's mention of the Congo case is revealing. As a cosmopolitan intellectual, he was familiar with the denunciation of the extreme forms of exploitation endured by local people around the African Congo river, in the hands of Belgian and British companies, in order to use them as slave labor for the extraction of ivory and rubber, as a result of which a yet undetermined number of people – several million– were killed or died of exhaustion.[28] He does not need to elaborate further on the topic, since he understands his readers were familiar with it, too. The situation Barrett denounces in Paraguay regarding yerba mate is the continuation of the colonial exploitation: the only difference is that it is in charge of the elite of the apparently free nations–nations which are not really free, but subject to neocolonialism. Barrett makes clear that what happens regarding yerba mate extraction in Paraguay is also happening in other parts of Latin America with other products, like rubber or cocoa. This reading was obvious both to contemporary critics, such as Armando Donoso,[29] as well as to critics who wrote on his work later on, such as Roa Bastos.[30]

After publishing *Los yerbales*, Bertotto is imprisoned and tortured. Barrett escapes this destiny protected by his British nationality: the British consulate in Asunción makes arrangements to help him leave Paraguay. He would head to Montevideo, Uruguay, where he would soon gain a reputation as a distinguished writer. In 1910, he briefly returned to Paraguay, soon to leave again, this time in a desperate travel to France, in search of an impossible cure for a devastating case of tuberculosis. He would die on December 10th, 1910 at a hospital in Arcachon.

27 "Es preciso que sepa el mundo de una vez lo que lo que pasa en los yerbales. Es preciso que cuando se quiera citar un ejemplo moderno de lo que puede concebir y ejecutar la codicia humana, no se hable solamente del Congo, sino del Paraguay. El Paraguay se despuebla; se le castra y se le extermina en las 7 u 8.000 leguas entregadas a la Compañía Industrial Paraguaya, a la Matte Larangeira y a los arrendatarios y propietarios de los latifundios del Alto Paraná. La explotación de la yerba mate descansa en la esclavitud, el tormento y el asesinato".

28 Cf. Adam Hochschild, *King Leopold's Ghost*.

29 Armando Donoso, *Un hombre libre*, 215.

30 Augusto Roa Bastos, "Rafael Barrett", XVIII.

But Barrett feelings for Paraguay, and particularly for its people are also present in another work, almost as much recognized and commented by critics as *Los yerbales*. *El dolor paraguayo* (*Paraguayan sorrow*) represents his most intimate work, a deeply felt depiction of this country's beauty and grief. It is a compilation of 51 short pieces published in different Paraguayan periodicals; among those, three speeches he delivered for Paraguayan workers. Although Barrett made the selection, the book was published posthumously in 1911 in Montevideo, also by Bertani. In the beginning, Barrett acknowledges the help of his wife, "whose subtle spirit enlivens some of these pages", as he puts it. She must have helped him understand much of what he saw, and certainly shared his views and supported his work.

The texts compiled in *El dolor paraguayo* are of different kinds. Fernández Vázquez has sorted them out in two groups.[31] The first one has to do with traditionalist or folkloric themes. Among them, there are impressionistic depictions of everyday life, full of color; stories related to local superstitions, which he treats humorously and sympathetically; and comments on cultural or linguistic aspects, like his sharp observations on Paraguay's bilingualism. In these articles, Barrett watches, celebrates, and rejoices in Paraguay's natural and cultural diversity.

But the second group of texts is less joyful; its somber tone certainly inspires the title of the compilation: they have to do with denunciations of the neglect and abuses suffered by vulnerable groups —such as crazy people, workers, women, and children. Among those, "Los niños tristes" ("Sad children") is one of the most beautiful and moving. Published in November 1907 in *Rojo y Azul*, it is full of Barrett's tenderness and indignation. It is a sort of allegory on Paraguay's suffering, much in the same way as "Buenos Aires" was regarding the situation of immigrants in Argentina. It talks of the infinite sadness of Paraguayan poor children, who lack the energy and even the willingness to cry, in contrast to what healthy, loved children do —that is, cry loudly because they know they will be listened to. Also as in "Buenos Aires", there is a short anecdote that introduces the climax, and a final exclamation. But this time, it is not an anarchist call to arms, but a call to help them: "Oh, unnumbered sad children! Let us dedicate ourselves to make the saint, crazy laugh appear on their red lips, and we will be saved. Let us lose any hope for us, if that makes sure hope will shine for them".[32]

---

31 José María Fernández Vázquez, "El periodista Rafael Barrett", 95 sq.
32 The complete closing paragraph reads as follows: "¡Oh, innumerables niños tristes! Consagrémonos á hacer brotar la santa, la loca risa en sus labios rojos, y nos salvaremos. Perdamos nosotros toda esperanza, con tal de que en los niños resplandezca. Evitemos que algunos se sientan en tan extremo rendidos á la pesadumbre de la fatalidad, que se duerman abandonados en medio del camino de la muerte, y no la oigan venir". Cf. Rafael Barrett, *El dolor*, 114.

Two articles complement "Los niños tristes": "Hogares heridos" ("Hurt homes"), and "El obrero" ("The worker"), published also in *Rojo y Azul* in November 1907. They offer an explanation for the sadness of children: mothers and children abandoned by careless fathers; fathers that are exploited as workers. Together, the three may be read as a kind of sociological essay on Paraguay's social crisis. There are also great political articles: on torture in Paraguay and other countries; on the dependent relationship with Argentina; on Albino Jara's revolution and its repressive aftermath; and on Paraguay's finances and indebtedness. As a whole, *El dolor paraguayo* depicts a country affected by the consequences of the horrible Paraguayan war: politically unstable, socially unequal, economically dependent; but also immensely beautiful, culturally rich, and full of life and promises. As depicted in this book, Paraguay is a country that arouses in Barrett feelings of sympathy and commiseration, of horror and love, of desperation and hope at the same time.

On communities, nationalities, and choices

Barrett's career and work challenges a traditional view on national literatures, and even on national politics. In this sense, his personal voyage could be seen as a sign and as a clue to interpret his work. Barrett is a perpetual outsider, a kind of incessant refugee, just like some of his characters. It is not surprising, then, that different authors point at different countries when discussing his intellectual nationality. This is how Jean Andreu summarizes the issue:

> Ignored by Spain as a possible member of the famous generation of 98, from which he derives, celebrated as a spirited militant among Hispanic American anarchists, recognized finally as a first class writer by Paraguayan literary historiography, Rafael Barrett stands as the problematic figure of a social, ideological and cultural transmigrant.[33]

As we have said, his books can be catalogued both as part of the Spanish or the Paraguayan literary tradition. And certainly, also as part of Argentina's and Uruguay's, because of its influence on authors such as Horacio Quiroga and Álvaro Yunque, to list just a few leading names. It is particularly telling that Yunque, one of the most prominent authors in Boedo's movement, includes Barrett in his history of Argentine social literature,[34] after dedicating him a whole book.[35]

33 Jean Andreu, "Una integración creativa", 37.
34 Álvaro Yunque, *La literatura social en la Argentina*.
35 Id., *Barrett*.

Francisco Corral is among the authors that points more clearly at Barrett's relation to the Spanish tradition, as he insists on the many characteristics he shares with the authors of the Generation of 1898. But he also acknowledges that Barrett would play a significant role in Latin America: "For Latin American literature, Rafael Barrett represents the precursor of a critical realism in which social denunciation and literary vanguard merge and enrich each other.[36] Much earlier, the Chilean critic Armando Donoso —one of the most influential critics in Buenos Aires in the 1920s— had included Barrett in his book *La otra América (The other America)* side by side with Gabriela Mistral, Arturo Cancela and Pedro Henríquez Ureña.[37] Donoso argues explicitly on why he considers Barrett a Latin American intellectual: "because he felt our pain like no one else before, and because he had the sincerity of the purest Apostle".[38] More or less the same says Yunque, who claims that "Spain completely forgot him", and that "Argentina or Paraguay, whose life inspired the ardent pages of his books, and where he lived burning, have the right to appropriate him".[39] Barrett's capacity to see under the surface of things, and to understand what happens in Latin America is the focus of Roa Bastos' comment on his work, which, according to him, represents "the revelation of an invisible reality in the virtuality of its multiple meanings".[40]

"Is Barrett an Uruguayan writer?" asks the Uruguayan critic Luis Hierro Gambardella in 1967, in the preliminary pages of an anthology of Barrett's letters. Yes, he answers, arguing that "the densest of his thought" was published in the Uruguayan newspaper *La Razón*. But he also acknowledges Barrett may be considered also a Paraguayan, since "his pages most saturated with color and love were born in Asunción, and talked about Paraguayan subjects". He concludes: "he is a writer of the Americas, thinking of the Americas as a continent that is unique and diverse".[41] In turn, Muñoz highlights the "one hundred and six days" Barrett spent in Montevideo as the most productive in his life. Montevideo is undoubtedly the city were Barrett became part of an intellectual community, and was praised as "the best chronicler of the Americas".[42] Actually, one of his most delighted admirers was Enrique Rodó, who underlined Barrett's situated internationalism, and the peripheral situation he wrote from. As he told him in a letter: "You write from a village in the Tropics, and for a Montevidean audience,

36 Francisco Corral, *El pensamiento cautivo de Rafael Barrett*, 1.
37 Quoted in Ramiro de Maetzu, "Rafael Barrett en Madrid", 10.
38 Armando Donoso,*Un hombre libre* (pp. 223–224).
39 Álvaro Yunque, *La literatura social en la Argentina*, 254.
40 Augusto Roa Bastos, "Rafael Barrett", XXXIX.
41 Luis Hierro Gambardella, *Cartas íntimas*, XX.
42 Quoted in Vladimiro Muñoz, *Barrett en Montevideo*, 18.

and giving back with personal impressions the late echoes of what happens in the world, you make things that can arouse interest wherever…".[43] Juana de Ibarbourou also talks about Barret's "internationalism", but depriving this perspective of any ideological or political edge: "Spanish… French… Uruguayan… Paraguayan…?", she wonders in 1928. "His birth certificate decides the question, but he alleges against it: 'Above motherland, lies Humankind' ". She then concludes, saying that "his soul was broadly international, universal; he was one outstanding representative of the human species".[44]

The discussion on Barrett's intellectual nationality may look like a scholarly exercise, one of merely assessing his many debts and contributions to this or that tradition. However, from the perspective of his belonging to a certain community, it is a central one. Where does he locate himself when he writes? In this sense, we find Eduardo Galeano's view on Barrett illuminating, when he grants him the Paraguayan nationality in a short text where he summarizes also his biography:

> Some of the most Latin Americans in Latin America were not born in Latin America. For example, Rafael Barrett, the most Paraguayan Paraguayan that has ever touched Paraguayan soil, was born in Spain, the son of a British father, and was raised in Paris. He arrives in Paraguay as a man, by chance or by mistake, or who knows why. As soon as he sets foot on this land, this doomed, hurt, tragic land, he finds out he belongs to it. He finds out that he is, that he has always been a Paraguayan, although he did not know. He spends in Paraguay only six years. Being —as he was— an activist, an anarchist dangerous for the system, he loves this country furiously, and denounces everything that arouses indignation at the heart; and he is expelled as a foreign agitator. He dies abroad. Barrett comes to Buenos Aires, to Montevideo, he dies expelled from the country he had chosen. Maybe a chosen nationality is more beautiful than an inherited one, because history is better than biology.[45]

Barrett declared himself Paraguayan in a deeply felt letter he wrote to his wife in 1909. He had sued the Paraguayan government for his ousting. When he com-

43 Enrique Rodó, "Las 'moralidades' de Barrett", 26.

44 Quoted in Vladimiro Muñoz, "Rafael Barrett y 'La Razón' de Montevideo", 45.

45 "Algunos de los latinoamericanos más latinoamericanos no nacieron en América Latina. Por ejemplo, Rafael Barrett, el paraguayo más paraguayo de los paraguayos que en el Paraguay han sido, nació en España, hijo de padre inglés y criado en París. Llega ya hombre hecho y derecho al Paraguay, por casualidad o por error, o quién sabe por qué. No bien pisa esa tierra, maldita, desgarrada, trágica, descubre que él es de allí. Descubre que él es paraguayo, que lo ha sido siempre, aunque no lo sabía. Vive en el Paraguay nada más que seis años. Siendo como era, un agitador, un anarquista, peligroso para el sistema, ama ese país furiosamente, denunciando todo lo que le indigna el corazón, y al cabo de seis años es expulsado por agitador extranjero. Muere afuera. Barrett viene a Buenos Aires, a Montevideo, muere desterrado del país que había elegido. Quizás es más hermosa una identidad elegida que una identidad heredada, porque la historia es mejor que la biología. Cf. Eduardo Galeano, *De Las venas abiertas de América Latina*, 4.

ments on it, he has feelings of regret, and argues as follows, mentioning his Paraguayan wife and child ("the Messiah"), but also describing how his time in Paraguay had transformed him:

> Sometimes I wonder if I was right in putting a claim on a country that is the only country *of mine*, the one I love most deeply, where I became a good man, where I met you, and where the Messiah was born. If I ever received money from my demand, I would invest it in something useful to it, for example, that school for shoeless children we have talked about.[46]

So Barrett himself proclaimed to be and feel a Paraguayan. However, I think it is David Viñas the critic who best understands the question on Barrett's intellectual nationality. And he also offers the reasons behind Galeano's ideas on the point, and a justification he is not completely able to articulate. Viñas' analysis goes one step further than Galeano's; he proposes the idea of homeland as an intellectual vindication, as a projection of Barrett's thought. He defines Barrrett's nationality not so much as a consequence of what he preferred or did, but as a consequence of what others did with his legacy: homeland is the place where Barrett's figure and texts have been taken as a symbol of resistance to hegemonic forces.

In his book on three emblematic figures of Latin American anarchism, Viñas prefers Barrett to Ricardo Flores Magón and Manuel González Prada. One way or another, Flores Magón and González Prada became national figures in their home countries, and lost their critical edge. But that was not the case with Barrett: he was not subject to "official beatifications", in Viñas' view. He considers this happened for two reasons: because he was a foreigner, and because the "managers of culture" in Argentina, Uruguay, and Brazil were merciless with anarchism, an attitude epitomized by the successive burnings of anarchist libraries in Rio de Janeiro, Montevideo, Buenos Aires and Asunción.[47] Barrett was vindicated not by the dominant culture, but by a counter culture; as we have said, the way he talked about the Paraguayan yerbales was echoed by many authors after him. In this sense, Barrett was Latin American to a degree that Flores Magón and González Prada were not able to reach. Viñas points at his role in the construction of a tradition that is counter hegemonic, that talks against the building of the nation-state, since it denounces that the new independent republics are not independent at all; that they have not put an end to slavery and exploitation; that freedom and wellbeing is still something that belongs to only a few in the new nations.

46 "A veces me pregunto si hice bien en entablar cuestiones con el único país *mío*, que amo entrañablemente, donde me volví bueno, y te conocí y nació el Mesías. Si ganara alguna suma, volvería al Paraguay y la invertiría en algo útil para él, por ejemplo, aquella escuela para niños descalzos de la que hablamos". Cf. Rafael Barrett, *Cartas íntimas*, 54. Emphasis in original.
47 David Viñas, *Anarquistas en América Latina*, 32 sq.

"*¡También América!*", he shouts, and his disappointment turns into desperation. Inequality is everywhere, so it must be resisted everywhere: there is no place to hide.

Galeano and Viñas agree on the counter hegemonic character of Barrett's position; on his identification with the poor and the excluded in Latin America: that is the reason why Barrett chooses Paraguay over Argentina —whose imperialist appetites on Paraguay he was able to perceive—, and also above Uruguay, a country that welcomed him with open arms, where he became part of the intellectuality. Barrett chooses the "*dolor paraguayo*", and by doing so, he becomes Paraguayan but also Latin American: Paraguay symbolizes Latin America better than Argentina and Uruguay, since it illustrates more clearly the region's sufferings and its dependent situation.

Barrett's life somehow epitomizes the abandonment of a homeland as a precondition for the recreation of the self, and the adoption of a new homeland. Barrett leaves his first community —Madrid's high society— in order to be free, to be able to develop his potentialities unconditioned by the prejudices and limits imposed on him. He tested the limits, and he had to pay the price of shame and public humiliation. Surprisingly to him, he found more or less the same limits on the other shore of the Atlantic, on the promising lands of the Americas. His cry "*¡También América!*" is also personal.

But how is it that this process takes place? We see it as an example of an early, tentative process of individualization, as described by Ulrich Beck; that is, a process in which there is no preset social role related to a class role. When discussing Edward Thompson, Beck agrees with him regarding the possibility of class struggle preceding class. In this sense, says Beck: "Individualization uncouples class culture from class position; as a result, there are numerous 'individualized class conflicts without classes', that is, a process in which the loss of significance of classes coincides with the categorical transformation and radicalization of social inequalities".[48]

Obviously, this is not the case in the beginning: Barrett comes of age in a social milieu that has a role for him, the role of the poor aristocrat, an ornamental face in the pretentious salons of Madrid —a role that is related to his class position. He is humiliated, a situation that destabilizes this role, as it deprives him of any dignity. This leads him to break up with society, and to leave for good. Once "in America", he tries to fit in twice; first in Buenos Aires, later in Asunción. In both occasions, he fails on a personal level. Duel as a symbol in his life seems to represent the impossibility to conform, to accept passively, silently, the role these two new societies somehow expected him to play. The two failed duels

48 Ulrich Beck, *Ulrich Beck. Pioneer*, 109.

"in America" somehow echo the first one, the original, the real one. But both seem to make a fainter mark in his life. Little by little, he begins to understand his humiliations are not merely personal but social. His personal pain recedes, almost disappears, in the face of the pain of others, the immense, the infinite "*dolor paraguayo*", which is also a Latin American pain.

In Asunción, Barrett is finally able to build the puzzle whose pieces he had been collecting since Madrid. In Paraguay he understands that an unequal social order is linked to an even more unequal international order, and that the extreme poverty endured by Paraguayan workers, the extreme pain suffered by Paraguayan women and children, has to do with a widespread situation. That is why he mentions the Congo in the opening paragraphs of *Los yerbales*, which by the time represented the ultimate example of colonial abuses, denounced by prominent figures such as Mark Twain.[49] He writes about it as a cosmopolitan in terms of access to information, showing he knows perfectly well the 'news of the day' on a world scale. But also, and most importantly, he writes as a cosmopolitan in terms of a situated ethical decision, because he sees himself as part of a community which demands his commitment to the common good. It is a kind of cosmopolitanism that somehow prefigures the "imagined communities of global risk", since there is a perception that "the global other is in our midst".[50] It is undoubtedly a heterogeneous community, since it includes those who can and who cannot talk: Barrett has to speak up because the voice of the oppressed is not being heard —one could say it is not even being articulated. As a result of his writing of *Los yerbales*, which is ultimately a political intervention, he is ousted from Paraguay, and somehow deprived of his nationality by choice. Paradoxically, the very act that induces his ousting confirms his choice and his belonging to the community.

Bibliography

Andreu, Jean, "Una integración creativa: Rafael Barrett en Paraguay". In *Hommage à Robert Jammes*, vol 1, edited by Francis Cerdan. Toulouse, Presses Universitaires du Mirail, 1994, 37–44.

Barrett, Rafael, *El dolor paraguayo*. Montevideo, O. M. Bertani, 1911.

---

49 Among other public statements, Twain published *King Leopold´s Soliloquy* in 1905, an incendiary pamphlet in the form of an imaginary monologue by the Belgian king, which went through several reprints, and whose benefits were donated to the Congo Reform Association, an advocacy association devoted to the cause. Cf. Adam Hochschild, *King Leopold's Ghost*, 242.

50 Ulrich Beck, "Cosmopolitanism as Imagined Communities", 1346.

Barrett, Rafael, *Lo que son los yerbales paraguayos*. Montevideo, Claudio García Editor, 1926.
Barrett, Rafael, *Obras completas I*, edited by Miguel Ángel Fernández and Francisco Corral. Asunción, RP Ediciones, 1988.
Barrett, Rafael, *Obras completas II*, edited by Miguel Ángel Fernández and Francisco Corral. Asunción, RP Ediciones, 1988.
Barrett, Rafael, *Obras completas III*, edited by Miguel Ángel Fernández and Francisco Corral. Asunción, RP Ediciones, 1989.
Barrett, Rafael, *Obras completas IV*, edited by Miguel Ángel Fernández y Francisco Corral. Asunción, RP Ediciones, 1990.
Barrett, Rafael, *Cartas íntimas*. Montevideo, Ministerio de Instrucción Pública y Previsión Social, 1967.
Beck, Ulrich, "Cosmopolitanism as Imagined Communities of Global Risk", *American Behavioral Scientist* 55 (2011), 1346–1361.
Beck, Ulrich (ed.), *Ulrich Beck. Pioneer in Cosmopolitan Sociology and Risk Society* (Springer Briefs on Pioneers in Science and Practice, 18). Heidelberg/New York/Dordrecht/London, Springer, 2014.
Bertotto, José G., "Rafael Barrett. Dos palabras". In Rafael Barrett, *Lo que son los yerbales paraguayos*. Montevideo, Claudio García Editor, 1926, 28–31.
Corral, Francisco, "El enigma de Rafael Barrett". In Rafael Barrett, *Obras completas I*, edited by Miguel Ángel Fernández and Francisco Corral. Asunción, RP Ediciones, 1988, 7–31.
Corral, Francisco, *El pensamiento cautivo de Rafael Barrett. Crisis de fin de siglo, juventud del 98 y anarquismo*. Madrid, Siglo XXI, 1994.
Cappelletti, Ángel J., "Anarquismo latinoamericano". In *El anarquismo en América Latina*, edited by Carlos Rama and Ángel J. Cappelletti (Biblioteca Ayacucho, 155). Caracas, Biblioteca Ayacucho, 1990, IX-CCXVI.
Donoso, Armando, *Un hombre libre. Rafael Barrett*. Buenos Aires, Ediciones Selectas América, 1920.
Fernández, Miguel Ángel, "Cuestiones preliminares". In Rafael Barret, *Germinal: Antología*, edited by Miguel Ángel Fernández. Asunción, El Lector, 1996, 9 sq.
Fernández, Miguel Ángel, "Introducción". In Rafael Barrett, *Obras completas IV*, edited by Miguel Ángel Fernández y Francisco Corral. Asunción, RP Ediciones, 1990, 7–21.
Fernández Vázquez, José María. "El periodista Rafael Barrett y *El dolor paraguayo*", *Cuadernos Hispanoamericanos* 547 (1996), 89–110.
Galeano, Eduardo, *De Las venas abiertas de América Latina a Memoria del fuego* (Colección Diálogos universitarios, 8). Montevideo, Universidad de la República, 1987.
Hierro Gambardella, Luis, "Prólogo". In Rafael Barrett, *Cartas íntimas*. Montevideo, Ministerio de Instrucción Pública y Previsión Social, 1967, VII-XXXII.
Hochschild, Adam, *King Leopold's Ghost. A Story of Greed, Terror, and Heroism in Colonial Africa*. London, Papermac, 2000.
Maetzu, Ramiro de, "Rafael Barrett en Madrid". In Rafael Barrett, *Lo que son los yerbales paraguayos*. Montevideo, Claudio García Editor, 1926, 7–13.
Muñoz, Vladimiro, "Rafael Barrett y 'La Razón' de Montevideo", *Revista de la Biblioteca Nacional* 16 (1976), 47–76.
Muñoz, Vladimiro, *Barrett en Montevideo*. Montevideo: edición del autor, 1982.
Pérez Marcevich, Francisco, *Diccionario de la literatura paraguaya. Primera parte*. Asunción, Biblioteca Colorados Contemporáneos, 1983.

Roa Bastos, Augusto, "Rafael Barrett. Descubridor de la realidad social del Paraguay". In Barrett, Rafael, *El dolor paraguayo* (Biblioteca Ayacucho, 30). Caracas, Biblioteca Ayacucho, 1978, IX-XXXII.
Rodó, Enrique, "Las 'moralidades' de Barrett". In Rafael Barrett, *Lo que son los yerbales paraguayos*. Montevideo, Claudio García Editor, 1926, 23–27.
Suriano, Juan, *Anarquistas. Cultura libertaria en Buenos Aires 1890–1910*. Buenos Aires, Manantial, 2001.
Tarcus, Horacio, *Diccionario biográfico de la izquierda argentina. De los anarquistas a la "nueva izquierda" (1870–1976)*. Buenos Aires, Emecé, 2007.
Vara, Ana María, *Sangre que se nos va. Naturaleza, literatura y protesta social en América Latina* (Colección Universos americanos, 10). Sevilla, Consejo superior de investigaciones científicas, 2013.
Viñas, David, *Anarquistas en América Latina* (Colección antología de América Latina, 1). Buenos Aires, Paradiso, 2004.
Yunque, Álvaro, *Barrett. Su vida y su obra*. Buenos Aires, Editorial Claridad, 1929.
Yunque, Álvaro, *La literatura social en la Argentina. Historia de los movimientos literarios desde la emancipación nacional hasta nuestros días* (Biblioteca de escritores argentinos, 1.) Buenos Aires, Editorial Claridad, 1941.

About the Authors

Braceras, Diana
Psychology, Medical Humanities
Faculty of Medicine/Universidad Nacional de San Martín
Buenos Aires, Argentina

Carugati, Laura S.
Philosophy
Universidad Nacional de San Martín
Buenos Aires, Argentina

De Marinis, Pablo
Sociology
Universidad Nacional de San Martín-IDAES
Buenos Aires, Argentina

Delrio, Walter
History, Ethnohistory
CONICET/Universidad Nacional de Río Negro
Río Negro, Argentina

Gianneschi, Horacio
Philosophy
Universidad Nacional de San Martín
Buenos Aires, Argentina

Gonzalez Casares, Santiago
Philosophy
Universidad Nacional de San Martín
Buenos Aires, Argentina

Grimson, Alejandro
Anthropology
Universidad Nacional de San Martín-IDAES
Buenos Aires, Argentina

Guizardi, Menara Lube
Social Science, Social Anthropology
Universidad Alberto Hurtado
Santiago de Chile, Chile

Itzigsohn, José
Sociology
Brown University
Providence, RI, USA

Kemp, Adriana
Sociology, Anthropology
Tel Aviv University
Tel Aviv-Yafo, Israel

Kleeman, Terry F.
Religious Studies, Sinology
University of Colorado
Boulder, CO, USA

Laham Cohen, Rodrigo
History
Instituto Multidisciplinario de Historia y Ciencias Humanas(IMHICIHU)
CONICET/ Universidad Nacional de Buenos Aires
Buenos Aires, Argentina

Lazzari, Axel
Sociocultural Anthropology
Universidad Nacional de San Martín-IDAES
Buenos Aires, Argentina

Melville, Gert
Medieval History
Technische Universität Dresden,
Dresden, Germany

Rehberg, Karl-Siegbert
Soziological Theory, Cultural Sociology
Technische Universität Dresden,
Dresden, Germany

Ruta, Carlos
Philosophy
Universidad Nacional de San Martín
Buenos Aires, Argentina

Scavino, Dardo
Philosophy, Latin American Studies
Université de Pau et des Pays de l’Adour
Pau, France

Schneidmüller, Bernd
Medieval History
University of Heidelberg
Heidelberg, Germany

Schwerhoff, Gerd
Modern History
Technische Universität Dresden,
Dresden, Germany

Thomaz Ribeiro, Omar
Social Anthropology
Universidade Estadual de Campinas
Campinas, Brasil

Vara, Ana María
Science and Technology Studies
Universidad Nacional de San Martín
Buenos Aires, Argentina

Wilkis, Ariel
Sociology
Universidad Nacional de San Martín-IDAES
Buenos Aires, Argentina

Zotter, Christof
Indology and Ethnology, Nepalese Studies
University of Heidelberg
Heidelberg, Germany